Building Thermal Envelope

Building Thermal Envelope

Special Issue Editors

Jorge de Brito
M. Glória Gomes

MDPI • Basel • Beijing • Wuhan • Barcelona • Belgrade • Manchester • Tokyo • Cluj • Tianjin

Special Issue Editors

Jorge de Brito
University of Lisbon
Portugal

M. Glória Gomes
University of Lisbon
Portugal

Editorial Office
MDPI
St. Alban-Anlage 66
4052 Basel, Switzerland

This is a reprint of articles from the Special Issue published online in the open access journal *Energies* (ISSN 1996-1073) (available at: https://www.mdpi.com/journal/energies/special_issues/Building_Thermal_Envelope).

For citation purposes, cite each article independently as indicated on the article page online and as indicated below:

LastName, A.A.; LastName, B.B.; LastName, C.C. Article Title. *Journal Name* **Year**, *Article Number*, Page Range.

ISBN 978-3-03928-518-1 (Hbk)
ISBN 978-3-03928-519-8 (PDF)

Contents

About the Special Issue Editors

Jorge de Brito (Full Professor) is a Full Professor of Civil Engineering in the Department of Civil Engineering, Architecture and Georesources; former head of the CERIS Research Centre (2017–2018) and director of the Eco-Construction and Rehabilitation Doctoral Programme at the Instituto Superior Técnico (IST), University of Lisbon, Portugal, from which he graduated and obtained his MSc and Ph.D. degrees. Although his research covers bridge management systems and construction technology, his main research area is sustainable construction, with emphasis on the use of recycled aggregates in concrete and mortar. He has participated in 25 competitively financed research projects and supervised 50 Ph.D. and 180 MSc theses. He is the author of 7 books, 28 book chapters, and 550 papers in peer-reviewed international journals, and has two patents. He is the Editor-in-Chief of the Journal of Building Engineering, an associate editor of the European Journal of Environmental and Civil Engineering, and a member of the editorial board of 44 international journals and of the following scientific/professional organisations: CIB, FIB, RILEM, IABMAS, and IABSE.

M. Glória Gomes (Assistant Professor) is an Assistant Professor of Civil Engineering in the Department of Civil Engineering, Architecture and Georesources at the Instituto Superior Técnico (IST), University of Lisbon, Portugal, from which she graduated and obtained her MSc and Ph.D. degrees. She performs work in the field of Building Physics, namely, in natural ventilation and the thermal and energy performance of buildings. Work in the area of the thermal and energy performance of buildings focuses on the implementation of more sustainable and energy-efficient materials and solutions such as thermal renders, nanomaterials, lightweight concrete, GFRP sandwich panels, green roofs and walls, and innovative glazing systems and shading devices. Most of those studies have a strong experimental background complemented with building energy simulations. In total, she has published 39 papers in peer-reviewed international journals and over 70 conference papers and is a reviewer for several international journals. She has participated in 13 successful research projects funded by Portuguese Fundação para a Ciência e Tecnologia (FCT), National Innovation Agency (ANI), International Energy Agency (IEA), and the European Commission. She has supervised 42 MSc students and 1 Ph.D. student to completion, and is currently supervising 6 Ph.D. students.

Preface to "Building Thermal Envelope"

This book results from a Special Issue published in Energies, entitled "Building Thermal Envelope". Its intent is to identify emerging research areas within the field of building thermal envelope solutions and contribute to the increased use of more energy-efficient solutions in new and refurbished buildings. Its contents are organized in the following sections: Building envelope materials and systems envisaging indoor comfort and energy efficiency; Building thermal and energy modelling and simulation; Lab test procedures and methods of field measurement to assess the performance of materials and building solutions; Smart materials and renewable energy in building envelope; Adaptive and intelligent building envelope; and Integrated building envelope technologies for high performance buildings and cities.

Jorge de Brito, M. Glória Gomes
Special Issue Editors

Editorial

Special Issue "Building Thermal Envelope"

Jorge de Brito * and M. Glória Gomes *

CERIS, Department of Civil Engineering, Architecture and Georresources (DECivil), Instituto Superior Técnico, Universidade de Lisboa, Av. Rovisco Pais, 1049-001 Lisbon, Portugal
* Correspondence: jb@civil.ist.utl.pt (J.d.B.); maria.gloria.gomes@tecnico.ulisboa.pt (M.G.G.)

Received: 24 February 2020; Accepted: 25 February 2020; Published: 28 February 2020

1. Introduction

The increasing requirements in building thermal and energy performance standards and the need to design nearly zero-energy buildings, while still enhancing the indoor comfort conditions, have led to a demand for more efficient thermal building envelope solutions. In fact, the effective use of building thermal envelope, as a mediator between outdoors and indoors, plays a key role in sustainable and energy efficient building design. Therefore, there is a need for continuous search of innovative materials, construction solutions and technologies that manage the energy and mass transfer between the building and the external environment taking into account not only the climatic changes but also the user preferences. Knowledge concerning the performance of building thermal envelope solutions and the existing design support tools, such as building performance simulation, is crucial for stakeholders to make informed decisions with respect to the definition and implementation of energy efficient strategies for new and refurbished buildings.

This Special Issue intends to provide an overview of the existing knowledge related with various aspects of "Building Thermal Envelope" and contributions on, but not limited to, the following subjects were encouraged: building envelope materials and systems envisaging indoor comfort and energy efficiency; building thermal and energy modelling and simulation; lab test procedures and methods of field measurement to assess the performance of materials and building solutions; smart materials and renewable energy in building envelope; adaptive and intelligent building envelope; integrated building envelope technologies for high performance buildings and cities.

So far, 13 papers have been published in the Special Issue of a total of 21 submitted. The next sections provide a brief summary of each of the papers published.

2. Building Envelope Materials and Systems Envisaging Indoor Comfort and Energy Efficiency

Kim et al. [1] analyzed the thermal performance data of the windows provided by the Korea Energy Agency and confirmed the change in the thermal performance of the windows by year and by frame material. It was found that the average U-value of the window decreased significantly from 2012 to 2015, maintaining similar values until 2017, and decreased again in 2018. This study also confirmed that the frame U-value of the PVC windows is lower than the frame U-value of the aluminum windows. The results of U-value of the windows through actual physical experiments show that, in the case of aluminum windows, the U-value corresponding to Grade 3 (1.4–2.1 W/m^2·K) was as high as about 60%, whereas in PVC windows, Grade 3 (U-value of 1.4–2.1 W/m^2·K) accounted for about 35%, and Grade 2 (U-value of 1.0–1.4 W/m^2·K) for about 29%. Moreover, a performance index of the glazing in PVC and aluminum window design was proposed.

Tywoniak et al. [2] focused on roof windows in pitched roofs. Building physics methods were used to support the design of new solutions for passive house level design, namely the avoidance of the risk of frame surface water condensation under reference conditions in the interior. The results of two-dimensional heat transfer calculations in the form of parametric studies are presented in order to express the most important factors influencing thermal transmittance and minimum surface temperatures. It was found that a combination of wood and hardened plastics in the window frame and sash is the preferred solution. The resulting thermal transmittance can be up to twice as low as usual (from 0.7 down to 0.5 W/(m^2·K) and surface temperature requirements to avoid the risk of condensation can be fulfilled. The effect of the slanting of the side lining was investigated by simulation and measurement in a daylight laboratory. It was found that the increase in thermal coupling due to slanting was negligible.

Another study was developed by Silvestre et al. [3] on the environmental, economic and energy (3E) assessment of external wall energy retrofitting with a cork-based (as recycled lightweight aggregate) thermal insulating rendering mortar (TIRM). The case study was a flat roof of an average building with the most current characteristics used in Portugal. The energy and economic costs and savings, as well as the environmental impacts, expressed with a declared unit of 1 m^2 of an external wall for a 50-year study period, were analytically modelled and compared for two main alternatives: the reference wall without any intervention and the energetically rehabilitated solution with the application of TIRM. They concluded that walls with improved energy performance (with TIRM) show lower economic and environmental impacts, with reductions from 6% to 32% in carbon emissions, non-renewable energy consumption and costs during the use stage. These results are dependent on the thickness and relative place where TIRM layers are applied.

3. Building Thermal and Energy Modelling and Simulation

In their research, Ayçam et al. [4] performed a study on the specification of traditional architectural parameters for houses in the hot-dry climatic region of Diyarbakır, Turkey. The courtyard types, settlement patterns, and street texture of traditional Diyarbakır houses were modeled through DesignBuilder energy simulation program for the case study of urban fabric of the traditional houses in Historical Diyarbakir Suriçi-Old Town settlement and the Şilbe Mass Housing Area. Annual heating, cooling, and total energy loads were calculated, and their thermal performances were compared. The aim was to create a less energy-consuming and sustainable environment with the adaptation of traditional building form-street texture to today's housing sector. The development of a settlement model, which is based on traditional houses' bioclimatic design for a hot-dry region, was intended to be applied in the modern housing sector of Turkey. The results showed that, adapting local and traditional houses forms, urban texture, and settlement patterns to modern housing sector in Turkey has significant potential for sustainable architecture and energy-efficient buildings.

A computational fluid dynamics (CFD) study was performed by Chung et al. [5] to assess the effect of a different number of awning windows and their installation locations on the airflow patterns and air contaminant distributions in restrooms in K-12 (for kindergarten to 12th grade) public schools in Taiwan, for various wind speeds and directions. A representative restroom configuration with dimensions of 10.65 m × 9.2 m × 3.2 m (height) was selected and based on the façade design feasibility, seven possible awning window configurations were considered. The numerical results were compared with the experimental results obtained in a reduced-scale model. The results indicate that an adequate number of windows and appropriate installation locations are required to ensure the natural ventilation effectiveness of awning windows. It was recommended, based on the modified odor removal efficiency (ORE) results, that K-12 school restrooms should use window configuration W2 in their north walls and that the opening angle should be set to 45° for all seasons.

4. Lab Test Procedures and Methods of Field Measurement to Assess the Performance of Materials and Building Solutions

Krause and Nowoświat [6] carried out in situ tests and laboratory studies to evaluate the impact of solar radiation on the behavior of expanded polystyrene with the addition of graphite. Temperature distributions were determined in field and laboratory conditions were determined on the surfaces of three types of panels: (A) two-layer polystyrene (gray bottom layer and white top layer); (B) polystyrene with the addition of graphite (gray polystyrene); and (C) expanded polystyrene (EPS) (white polystyrene). The distributions of temperature were recorded for different wind and solar radiation conditions. Moreover, geometric changes and deformation levels of the panels exposed to artificial sun radiation were determined in laboratory conditions. The panel entirely made of polystyrene with the addition of graphite (panel B) demonstrated high sensitivity to external factors, such as insolation and wind, whereas panels A and C are characterized by low sensitivity to solar radiation. The composite panel (A-panel) proved to avoid many adverse effects of environmental conditions compared to B-panels. The extent and nature of temperature changes on the external surface of the panels depends on the area of impact and exposure time, while the extent of deformations is also dependent on the method used to fix the panels to the ground.

An integrated measuring and control system for hot box experiments, based on a general-purpose microcontroller and on a wireless sensors network was presented by de Rubeis et al. [7]. The results of 72 h experiment on a double insulation X-lam wall revealed that the system was able to maintain stable temperature set points inside the chambers and to log the temperatures measured by the 135 probes, allowing to know both the U-value of the sample (equal to 0.216 ± 0.01 W/m^2·K) and the thermal models of all the hot box components. The U-value experimentally obtained with the hot box method was compared with the values gathered through theoretical calculation and heat flow meter measurements, showing differences lower than 20%. Furthermore, the developed data post-processing allowed creating 2D and 3D thermal models of specimen wall and chambers.

5. Smart Materials and Renewable Energy in Building Envelope

Yu et al. [8] experimentally investigated the effect of the aspect ratio (AR) of a rectangular thermosiphon loop on its natural convection performance using boundary conditions of a constant heat flux and a fixed wall temperature for the heating and cooling sections of the loop, respectively. The experimental model consisted of a loop body with an inner diameter of 11 mm, a heating section, a cooling section, located in the vertical portions of the rectangular loop, and adiabatic sections. The analyzed aspect ratios were 6, 4.5, and 3.5 (with potential differences of 41, 27, and 18, respectively, between the cold and hot ends), and the input thermal power ranged from 30 to 60 W (with a heat flux of 600 to 3800 W/m^2). The results showed that it was feasible to install a rectangular thermosiphon inside a metal curtain wall to obtain solar heated water and that increasing the height of the opaque part of the metal curtain wall could improve the heat transfer efficiency by increasing the aspect ratio of the rectangular thermosiphon installed inside the wall.

Huang et al. [9] also presented an experimental study on the natural convection heat transfer performance of a rectangular thermosiphon with an aspect ratio of 3.5 and an inner diameter of the loop of 11 mm. Different heating powers, height differences between the heating and cooling ends, and cold end temperatures were tested. The results show that the value of the dimensionless heat transfer coefficient (Nusselt number), is generally between 5 and 10 and that the heating power is the main factor affecting the natural convection intensity of the thermosiphon.

Another study carried out by Choi and Ko [10] analysed the convergence characteristics of the in situ thermal transmittance (U-value) and thermal resistance (R-value) calculation of building envelopes obtained from onsite measurements using the ISO 9869-1 average method. The criteria for determining the average method convergence, namely the test duration, are very strict, and to shorten the test duration, environmental variables should be kept constant throughout the test or an appropriate period should be selected. The results indicate that the convergence of the in situ U-value and R-value is more sensible to the length of the test duration than to the temperature difference. Moreover, no difference was found between the use of the U-value and R-value in determining the end of the test.

6. Adaptive and Intelligent Building Envelope

A hybrid heat collecting facade (HHCF) that increases the indoor air temperature and reduces the heating energy consumption was studied by Wang et al. [11]. A heat transfer model based on the heat balance method, experimentally validated, was used to analyze the thermal performance of the HHCF.

Moreover, the energy saving potential of a room with the HHCF was evaluated. The results indicated that the HHCF can reduce the heating need by 40.2% and 21.5% compared with the conventional direct solar heat gain window and the Trombe wall, respectively. Furthermore, a parametric analysis was performed and it was concluded that the thermal performance of the HHCF is mostly dependent on the window operational schedule, the width and the absorptivity of heat collecting wall, and the thermal performance of the inner double-glass window.

7. Integrated Building Envelope Technologies for High Performance Buildings and Cities

Biswas et al. [12] investigated the application of thermally anisotropic composites (TACs), composed by alternate layers of rigid foam insulation and thin and high-conductivity aluminum foil, for improving the energy efficiency of building envelopes. The TAC was coupled with copper tubes with circulating water, which acted as a heat sink and source, and the system was applied to a conventional wood-framed wall assembly. The energy benefits of the system were investigated both experimentally and numerically. Large scale test wall specimens were built with and without the TAC system and tested in an environmental chamber. Moreover, component-level and whole building numerical simulations were performed under cooling-dominated and heating-dominated climate conditions to investigate the energy benefits of applying the TAC system to the external walls of a typical, single-family residential building. It was concluded that the TAC coupled with a heat sink/source was shown to be more effective in reducing both cooling and heating loads and peak cooling loads than foam insulation of the same thickness. TAC connected to copper tubes circulating water were able to reduce cooling and heating loads by 86% and 63% compared to a baseline wall with only cavity insulation. Simulation results showed that the TAC system is able to reduce the cooling energy use by 11% under cooling-dominated climate and heating energy use by 21% in the heating-dominated climate.

An investigation on the use of Building Information Modeling (BIM) to assess the sustainability index of green buildings was performed by Liu et al. [13]. A cloud-based BIM platform was developed to digitalize the Envelope Thermal Transfer Value (ETTV) calculation, which is one of the prerequisite criteria to achieve a Green Mark score. The authors have validated the Envelope Thermal Transfer Value (ETTV) calculation, by using the developed cloud-based BIM platform in three case studies in Singapore. It was concluded that the proposed platform enhanced the productivity and accuracy of ETTV calculation and facilitated parametric capabilities that promote change management. Moreover, it allowed the relevant information to be shared and validated and project stakeholders to keep track of the GM data generated throughout the design stage and project lifecycle.

Funding: This research received no external funding.

Acknowledgments: Thanks are due to all the authors and peer reviewers for their valuable contributions to this Special Issue. The MDPI management and staff are also to be congratulated for their untiring editorial support for the success of this project.

Conflicts of Interest: The authors declare no conflict of interest.

References

1. Kim, S.-H.; Jeong, H.; Cho, S. A Study on Changes of Window Thermal Performance by Analysis of Physical Test Results in Korea. *Energies* **2019**, *12*, 3822. [CrossRef]
2. Tywoniak, J.; Calta, V.; Staněk, K.; Novák, J.; Maierová, L. The Application of Building Physics in the Design of Roof Windows. *Energies* **2019**, *12*, 2300. [CrossRef]
3. Silvestre, J.D.; Castelo, A.M.P.; Silva, J.J.B.C.; de Brito, J.M.C.L.; Pinheiro, M.D. Energy Retrofitting of a Buildings' Envelope: Assessment of the Environmental, Economic and Energy (3E) Performance of a Cork-Based Thermal Insulating Rendering Mortar. *Energies* **2020**, *13*, 143. [CrossRef]
4. Ayçam, İ.; Akalp, S.; Görgülü, L.S. The Application of Courtyard and Settlement Layouts of the Traditional Diyarbakır Houses to Contemporary Houses: A Case Study on the Analysis of Energy Performance. *Energies* **2020**, *13*, 587.

5. Chung, S.-C.; Lin, Y.-P.; Yang, C.; Lai, C.-M. Natural Ventilation Effectiveness of Awning Windows in Restrooms in K-12 Public Schools. *Energies* **2019**, *12*, 2414. [CrossRef]
6. Krause, P.; Nowoświat, A. Experimental Studies Involving the Impact of Solar Radiation on the Properties of Expanded Graphite Polystyrene. *Energies* **2020**, *13*, 75. [CrossRef]
7. de Rubeis, T.; Muttillo, M.; Nardi, I.; Pantoli, L.; Stornelli, V.; Ambrosini, D. Integrated Measuring and Control System for Thermal Analysis of Buildings Components in Hot Box Experiments. *Energies* **2019**, *12*, 2053. [CrossRef]
8. Yu, C.-W.; Huang, C.S.; Tzeng, C.T.; Lai, C.-M. Effects of the Aspect Ratio of a Rectangular Thermosyphon on Its Thermal Performance. *Energies* **2019**, *12*, 4014. [CrossRef]
9. Huang, C.S.; Yu, C.-W.; Chen, R.H.; Tzeng, C.-T.; Lai, C.-M. Experimental Observation of Natural Convection Heat Transfer Performance of a Rectangular Thermosyphon. *Energies* **2019**, *12*, 1702. [CrossRef]
10. Choi, D.S.; Ko, M.J. Analysis of Convergence Characteristics of Average Method Regulated by ISO 9869-1 for Evaluating In Situ Thermal Resistance and Thermal Transmittance of Opaque Exterior Walls. *Energies* **2019**, *12*, 1989. [CrossRef]
11. Wang, X.; Lei, B.; Bi, H.; Yu, T. Study on the Thermal Performance of a Hybrid Heat Collecting Facade Used for Passive Solar Buildings in Cold Region. *Energies* **2019**, *12*, 1038. [CrossRef]
12. Biswas, K.; Shrestha, S.; Hun, D.; Atchley, J. Thermally Anisotropic Composites for Improving the Energy Efficiency of Building Envelopes. *Energies* **2019**, *12*, 3783. [CrossRef]
13. Liu, Z.; Wang, Q.; Gan, V.J.; Peh, L. Envelope Thermal Performance Analysis Based on Building Information Model (BIM) Cloud Platform—Proposed Green Mark Collaboration Environment. *Energies* **2020**, *13*, 586. [CrossRef]

Article

A Study on Changes of Window Thermal Performance by Analysis of Physical Test Results in Korea

Seok-Hyun Kim, Hakgeun Jeong and Soo Cho *

Energy ICT·ESS Laboratory, Korea Institute of Energy Research, Daejeon 34101, Korea;
ksh7000@kier.re.kr (S.-H.K.); hgjeong@kier.re.kr (H.J.)
* Correspondence: scho@kier.re.kr; Tel.: +82-42-860-3231

Received: 10 September 2019; Accepted: 9 October 2019; Published: 10 October 2019

Abstract: The interest in zero energy buildings is increasing in South Korea. Zero energy buildings need to save energy by using passive technology. The window performance is important to the thermal insulation of the building. Also, the government regulates the window performance through regulation and standards. However, it is difficult to predict window performance because the components of the window have become complicated due to the various materials used in the glass and frame. Based on window performance standards and regulations, the quality of window performance was managed. In this research, to consider thermal performance in proper window design in South Korea, we confirmed the impact on the thermal performance of the window through various kinds of materials and shapes. The authors also propose a window shape classification and frame calculation method based on actual test results. The authors analyzed the thermal performance data of the windows provided by the Korea Energy Agency and confirmed the change in the thermal performance of the windows by year and by frame material. The average U-value of the window decreased from 2012 to 2015 and maintained similar values until 2017. In 2018, this value was decreased to comply. Also, the authors confirmed the U-value of the windows through actual physical experiments and confirmed the change in thermal performance by the construction of the windows based on the results. The results show, in the case of aluminum windows, the U-value corresponding to Grade 3 (1.4–2.1 W/m^2·K) was as high as about 60%. Regarding the analyzed results of the U-values of PVC windows, Grade 3 (U-value of 1.4–2.1 W/m^2·K) accounted for about 35%, and Grade 2 (U-value of 1.0–1.4 W/m^2·K) for about 29%. This paper also confirmed that the frame U-value of the PVC windows is lower than the frame U-value of the aluminum windows. Therefore, the authors proposed the performance index of the glazing part in PVC and aluminum window design. The results of this research can be used as basic data to identify problems in the method of determining the performance of windows in Korea.

Keywords: window thermal performance; U-value; window type; simulation performance test; actual physical performance test

---

1. Introduction

As part of a global effort to reduce greenhouse gas emissions, various methods for saving building energy have been attempted. The thermal performance of the building envelope, in particular, is a basic element. Moreover, this element has also been used in a passive energy saving method by appropriately designing the performance of windows suitable for each location. Smart window design does not imply a particular method, but this method includes energy analysis and a case study for selecting the optimal region, position, and performance of windows. However, curtain walls and façades have been used to satisfy the requirement for views, while meeting various aesthetic standards by increasing the area of the window. This has a disadvantageous thermal performance compared to a general wall with insulation. Thus, the demand for the performance improvement of the window has increased as the window area has increased in the building envelope. In the case of South Korea, interest in zero energy building is increasing and, in the case of new construction, the government regulates window performance criteria by building use and area through regulations and standards. The specified thermal performance requirement is the heat transmission coefficient (or U-value, $W/m^2 \cdot K$). This test method is an actual test of a window sample according to the Korea industrial standards. Thus, Kim proposed guidelines for the standards and regulatory requirements of South Korea and other countries in previous research [1]. In reality, however, predicting a window's performance is difficult because the components of windows have become complicated due to the various materials used in the glass and frame.

Cappelletti analyzed the influence of a window thermal bridge using a case study of clay walls that were externally insulated and had cavity insulation [2]. This study confirmed the position of the frame for reducing linear thermal transmittance. Results showed that moving the window from an internal to an external position reduced linear thermal transmittance by 70.75%. This decrease mainly depends on the installation position of the insulating layer within the window opening. This paper showed the importance of the suitable installation of a window in a wall. Adamus and Pomada [3] analyzed the effect of window installation for heat flow in a composite structure. This study experimented with simulation. Hee et al. [4] researched the importance of glazing performance on daylight and energy saving in buildings and found that the cost affects the qualities and performance of glazing proportionally. It is wise to perform techno-economic evaluations to obtain suitable glazing for a building. Tsikaloudaki et al. [5] also proposed methods to use the cooling energy saving of windows. In this study, cooling energy performance was assessed about ISO 18292 through the calculation of the cooling energy. This paper concluded that the results of the statistical analysis provided mathematical expressions, which were used in practice, with moderate errors, for predicting the cooling energy performance of windows concerning their thermal and optical properties. Kim et al. [6] verified the improved thermal performance of the double window system. In this paper, the patterns of airflow were also examined according to the operating mode change. The thermal performance analysis of conventional window systems showed that heat loss was reduced by 49% compared to the double window system. Carlosa and Corvacho [7] showed that the provided solar heat gain coefficient (SHGC) values depend on the airflow rate passing through the system and the portion of the glazed area about the glass portion of the window. These authors confirmed the effect of the SHGC, but these results also show the importance of the glass ratio of a window. According to these results, the designer needs to consider the ratio of glazing and frame in a window area. Wen et al. [8] proposed a new method of the window-to-wall ratio by case study. The authors also recommended that the window-to-wall ratio (WWR) maps for open office buildings in Japan be created, and noted that the impact of window selection is very important in building energy saving. Ihm et al. [9] studied the energy performance of residential buildings to determine the impact of window selection. In particular, the window-to-wall ratios were varied to determine the effect of window properties on heating, cooling, and total energy consumption.

Weather differences between Ulsan and Inchon, in South Korea, are also important; these cities are representative of the two major climate types in South Korea. Based on a life-cycle cost analysis, the cost-effectiveness of double Low-E clear glass filled with argon gas is greater for residential buildings in South Korea. Furthermore, this glass should be required by the building energy efficiency code. Thus, the designer must consider the design regulations and the energy performance of the window. For window design optimization, window size in the wall is very important [10]. Generally, optimizing the size of the window for energy-saving does not meet any of the predetermined visual acceptance criteria. Mangkuto et al. [11] also undertook a case study about design optimization considering window size, orientation, and wall reflectance. The analysis evaluated the effect of daylight dynamic metrics by the geometry and position of the opening, as well as the inside surface reflectance of the room. The study reached several conclusions applicable to window design [12]. In previous studies, researchers confirmed the effect of window installation in buildings. However, the optimization of window design for energy-saving and indoor comfort is difficult due to variations in climates and environments. Thus, the testing method of windows needs to be standardized through the use of official methods.

Prediction of window performance by various types of the window is important for energy consumption calculations in buildings. Thus, many researchers have reviewed the various types of window design and shapes, such as a complex window [13], design parameter for energy performance [14], window firm [15], transmittance [16], and current thermos chromic window [17]. These results highlight some of the preconditions. An important prerequisite is that the performance of all windows needs to be tested with the test organization's official testing method. In South Korea, government officials responsible for construction regulations require the test paper of the actual window performance for permission of a new building. The actual test method of the window confirms the U-value of a window sample of the same size and configuration as the actual window to be installed. Furthermore, the Korean government manages a label system according to the level of performance. This system allows customers to easily select window products. Based on these standards and regulations, the quality of window performance is managed. However, studies on the performance of changing windows are not sufficient, relating only to changes in the components and appearance of the window. For efficient use of windows, smart window design must be used [18,19]. This method calculates the ratio of the window and frame in a window for energy-saving effects, such as passive control. In addition, this result should be applied to the appropriate position considering the performance of actual window products. Designers need to see how to calculate the thermal performance of windows through traditional testing methods. Window performance classification should separate the performance of each type of glazing and frame. In this research, to use thermal performance for smart window design in South Korea, we confirmed the tendency of the thermal performance of the window according to various kinds of materials and shapes. In this paper, we propose a window shape classification and frame calculation method based on actual test results. We confirm the trend of window thermal performance using the government database of window performance to provide a practical smart window design. Through window certification and window labeling system analysis in Korea, the authors show the minimum performance requirements of windows. Also, the authors confirmed the standard of the test environment and method and determined the variation of window thermal performance by year using the window database. This database was managed under the certification of the Korea Energy Agency. We also performed a complex performance analysis of the shape and type of glazing and frame according to the actual test results of the window. Finally, using these research results, we have a better understanding of the current level of window thermal performance. Figure 1 shows a flow chart of our method of research.

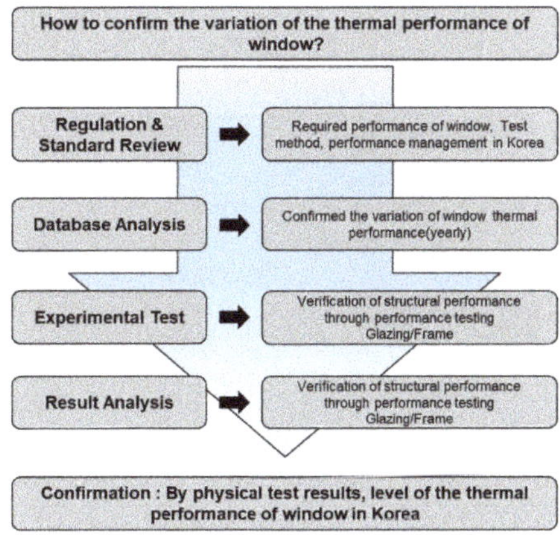

Figure 1. Flow chart of method of research.

2. Window Performance-related Laws and Regulations in Korea

2.1. Building Energy Conservation Design Standard

In Korea, the designer or owner of a new build submits their plans to the energy conservation plan to meet the thermal performance requirements of the outer wall of the building. The Building Energy Conservation Design Standards were published by the Ministry of Land, Transport, and Maritime Affairs and enacted on 11 January 2008 [20]. Because Korea emphasizes heating energy consumption relative to cooling energy consumption, it is working to reduce heat loss through windows. Window insulation performance standards (U-values) were restricted by laws and regulations. Furthermore, to reduce cooling energy, it is recommended that a low solar heat gain coefficient (SHGC) be applied only when the window area of a certain level is large. There is no performance limited by any laws and regulations. Therefore, there are no data stored in the database of Korean Energy Agency on the U-value of the window.

For energy saving in buildings, this standard (Building Energy Conservation Design Standard) was proposed as part of the building standard. The purpose is energy-saving and the management of the building's saving plan, and the energy savings that will be achieved in terms of the efficient management of energy. This regulation was applied to buildings such as apartment blocks containing more than 50 dwellings, as well as welfare facilities, research institutes, hospitals, bathhouses, swimming pools, and large stores. Furthermore, wall thickness should comply with the technical regulations and technical standards.

For energy saving, technical standards need to change. The plans classify the demand level of residential and nonresidential buildings. In addition, for the purpose of efficient energy management of buildings by "The Support System for Composition of Green Architecture", the Building Energy Conservation Design Standards have been revised for establish the criteria for energy-saving designs concerning the decrease of heat loss and the standards in order to writing energy-saving plans and design review reports, and to determine the things related to the comport of building standards for the promotion of the construction of green architecture. They also proposed the demand level of window thermal performance by each area. These areas were divided into Central 1, Central 2, South, and Jeju. The standards specified the proposed U-values (heat transmittance coefficient, $W/m^2 \cdot K$). For energy saving, these standards have been strengthened due to advances in technology. Characteristics of the Building Energy Conservation Design Standards are the quantitative value that meets the required performance level and the purpose of confirmation. These standards specify the minimum required level for window thermal performance. This is separated into outdoor and indoor surfaces. Table 1 shows the performance levels proposed by the Building Energy Conservation Design Standards.

Table 1. U-value requirement of window by Building Energy Conservation Design Standard.

Positon / Region		Central 1	Central 2	South	Je-ju
Window	Surface at outdoor	Residential	0.90 or Less	1.00 or Less	1.20 or Less	1.60 or Less
		Non-Residential	1.30 or Less	1.50 or Less	1.80 or Less	2.20 or Less
	Surface at indoor	Residential	1.30 or Less	1.50 or Less	1.70 or Less	2.00 or Less
		Non-Residential	1.60 or Less	1.90 or Less	2.20 or Less	2.80 or Less

* All value is U-value($W/m^2 \cdot K$).

2.2. Energy Standard and Labeling Program

The Korean government has undertaken three energy efficiency management programs to increase the energy efficiency of appliances: Energy standards and labeling, high-efficiency equipment certification, and e-Standby [21]. Implemented in 1992, the energy standards and labeling program mandate all manufacturers to attach an energy efficiency label with rank from first to fifth class to their energy-intensive and highly disseminated appliances. Appliances failing to meet minimum energy performance standards (MEPS) shall be prohibited from production and sale. The program targets 37 appliances, including home appliances, lighting products, vehicles, and tires. Windows are also subject to this program. A window is defined as having "the product size of upper 1 m^2 by KS F 3117 and this installation at the outdoor surface", and "the target is that the case of combined frame and glazing, supplying to a product by domestic company" [22]. The actual test method is KS F 2278 (thermal performance) and KS F 2292 (airtightness) [23,24]. Simulation (via numerical analysis program) is also approved. The simulation uses software, such as WINDOW and THERM, to calculate window performance. These have a thermal transmittance equation based on ISO 15099. In the case of the window performance test, it is necessary to perform actual tests when changing the frame material, opening, and type. Simulation tests may also be approved for changes in glazing type, including changes in glass and air gap. In this program, the performance of a window is indicated by R, where R means the U-value (W/m^2·K). The grades of the energy standard and labeling program are shown in Table 2.

Table 2. Window grades of the energy and labeling program.

R*	Air Tightness	Grade
R ≤ 1.0	Under 1.0 m^3/h·m^2	1
R ≤ 1.4	Under 1.0 m^3/h·m^2	2
1.4 < R ≤ 2.1	Under 2.0 m^3/h·m^2	3
2.1 < R ≤ 2.8	-	4
2.8 < R ≤ 3.4	-	5

* R: U-value (W/m^2·K).

2.3. Korean Industrial Standards

The Korean Industrial Standards are the government standards by law for industrial standardization in Korea. These standards were announced by the chief of the Korean Agency for Technology and Standards and, thus, were marked KS. KS consists of 21 areas ranging from the basic section (A) to the information section (X). These are further divided into three parts. First are the standards of the product (shape, size, quality), second is the standards of a method (test, analysis, examination, operation), and last are the transmission standards (term, technique, unit) [25].

The type, symbol, performance, quality, size, material, and parts were determined for the window based on KS F 3117. A general environment and test method of the actual test were proposed. The test method of window thermal performance was proposed by KS F 2278: Standard test method for thermal resistance for windows and doors. This standard contains the details of the test, equipment, cartridge, and specimen. It also proposes the settings of surface thermal resistance, heat flux through hot box and cartridge, and the equation for results.

Under KS F 2278, the environmental conditions of the laboratory are maintained. A cold chamber, warm chamber, and hot box are contained in the equipment. The air temperature of the cold chamber must be maintained at 0 ± 1 °C, and the air temperature of the warm chamber and hot box need to be maintained at 20 ± 1 °C. This steady state is confirmed regarding the air temperature and the heater input (W). The result of this test was logged data to three times at 30 min each during the measuring period. The size of the specimen is 2000 mm (W) × 2000 mm (H). Figure 2 shows the scheme of equipment in KS F 2278.

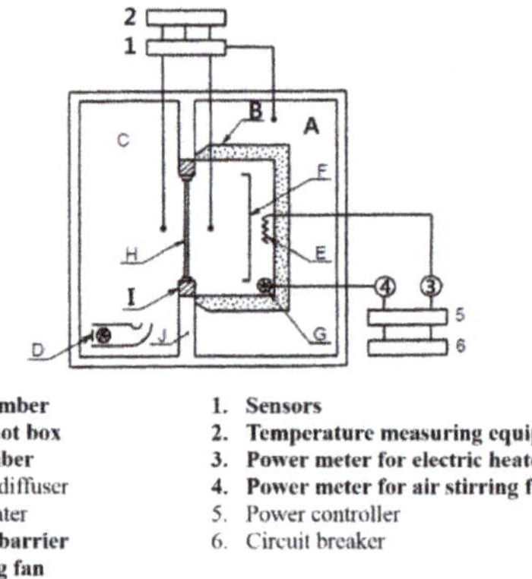

A. Warm chamber	1. Sensors
B. Guarded hot box	2. Temperature measuring equipment
C. Cold chamber	3. Power meter for electric heater
D. Chilled air diffuser	4. Power meter for air stirring fan
E. Electric heater	5. Power controller
F. Radiation barrier	6. Circuit breaker
G. Air stirring fan	
H. Specimen	
I. Surrounding panel	
J. Partition panel	

Figure 2. Scheme of equipment.

In Korea, the KS Standard confirms the physical test performance of windows. The results are used in all laws and systems, but direct comparison is difficult because ISO test methods and measurement methods are different. It is difficult to confirm the U-value through calculation because there is no heat transmissivity of the simple thermal insulation member. However, for the glass part whose heat transmission coefficient is relatively easy to calculate, some simulation results are adopted. The U-value of the window is calculated by checking the calorie movement due to the temperature difference after installing a certain specimen size (2 m × 2 m) according to the method of KS standard. Therefore, it is difficult to compare the ISO method with the KS standard, the U-value of the window proposed in this study is calculated based on the KS standard measurement method. Equation (1) shows the theoretical calculation way of the U-value of windows where U_W is the U-value of window($W/(m^2 \cdot K)$), and Q_H, Q_F, and Q_I are the heat input (W) to the heater, the air circulating fan, and calibrated heat flow, respectively; T_{Ha} and T_{Ca} are the average air temperature (K) in the hot box and cold chamber. Also, A_W is the window area (m^2).

$$U_W = \frac{(Q_H + Q_F - Q_I)}{(T_{Ha} - T_{Ca}) \cdot A_W}$$
(1)

2.4. Numerical Analysis Simulation Tools

In the case of the testing of a similar model of window set, the test laboratory uses a simulation tool like WINDOW (LBNL) through the energy standard and labeling program [26].

WINDOW, from Berkeley National Lab, is a publicly available computer window analysis program. For calculating window thermal performance like U-values, solar heat gain coefficients (SHGC), shading coefficients (SC), and visible transmittances (VT), WINDOW provides an adaptable heat transfer analysis method so the updated rating procedure. This was developed by the National Fenestration Rating Council (NFRC). This program is based on the ISO 15099 standard. Using the WINDOW program, designers can draw window shapes and develop new products. This also assists educators in teaching about heat loss through windows and helps public officials in developing building energy codes. By reference to the International Glazing Database (IGDB), this simulation is used in the calculation of the glazing performance of a window. Frame design can also be entered into THERM. Figure 3 shows the interface of WINDOW 7.4. In this UI, a glazing combination was shown.

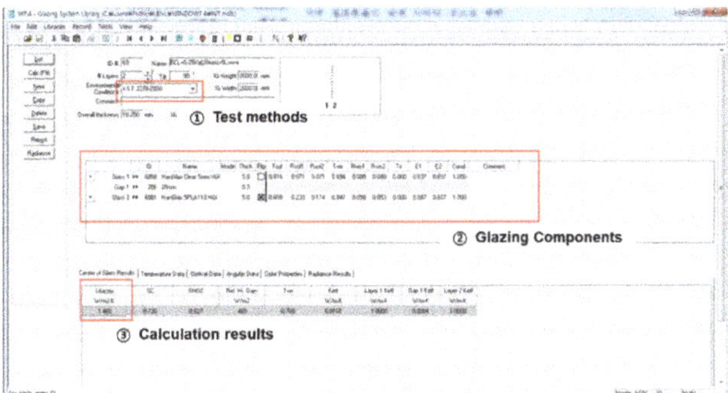

Figure 3. Interface of WINDOW 7.4.

3. Changes in Window Thermal Performance through Database Analysis Results

In this section, the authors confirm the change in window thermal performance by year based on the database of the Energy standard and labeling program of the Korea Energy Agency. We also assess the effect of the variation of the frame on the thermal performance of a window.

3.1. The Variation of the Number of Registered Window by Year

Window data used in this chapter are managed by the Korean Energy Agency, and data on manufacturing and performance information are provided by the window set product category of manufacturer web pages. The performance was divided according to the grades of the 'Energy standard and labeling program', which provides U-values. Thus, all information on window products were registered from 2012, and the database presented to the web page by product registration of each manufacturer [27].

As a result of confirming the number of registered products by year, we obtained the number of windows registered as 490 EA, 624 EA, 914 EA, 1037 EA, 848 EA, 1065 EA, and 975 EA in each year from 2012 to 2018, respectively. The number of registrations was highest in 2015 and decreased in 2016, but was nonetheless maintained at around 1000 per year. As a result of confirming the differences in frame material, the percentage of PVC windows was 56% in 2012. The ratio was 53% and 66% in 2013 and 2014, respectively. Percentages of 62%, 46%, 54%, 36% from 2015 to 2018, respectively, indicate that it occupies a considerably high proportion as compared with window frames made of other materials. Aluminum window percentages were 31%, 29%, 23%, 29%, 44%, 37%, and 53% from 2012 to 2018, respectively. In 2018, the aluminum window percentage of registered windows was greater than that of PVC windows. Figure 4 shows the results of the registration quantity of window frame materials.

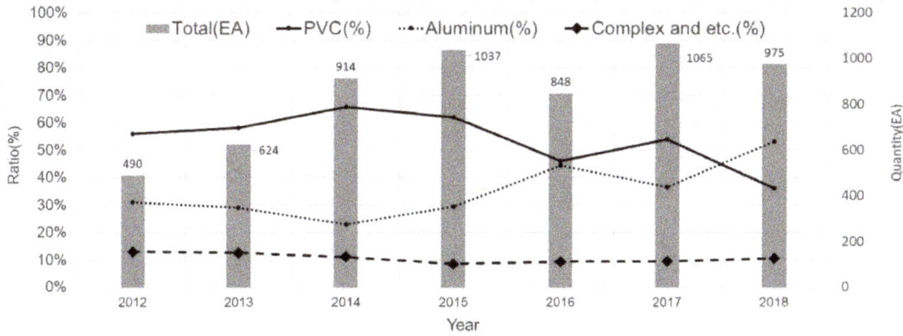

Figure 4. Registration quantity of window by year.

3.2. Change in Window Thermal Performance by Year

From 2012 to 2018, the average U-value of windows was found to be 1.710 W/m^2·K, 1.518 W/m^2·K, 1.384 W/m^2·K, 1.331 W/m^2·K, 1.367 W/m^2·K, 1.328 W/m^2·K, and 1.190 W/m^2·K, respectively. According to these results, we confirmed the U-value decreased by 0.192 W/m^2·K and 0.134 W/m^2·K in 2013 and 2014, respectively. From 2014 to 2017, the U-values of windows were nearly constant at 1.3 W/m^2·K. Moreover, the U-value in 2018 was 1.190 W/m^2·K, which is a very low value. Figure 5 shows the results of U-value variation by frame type.

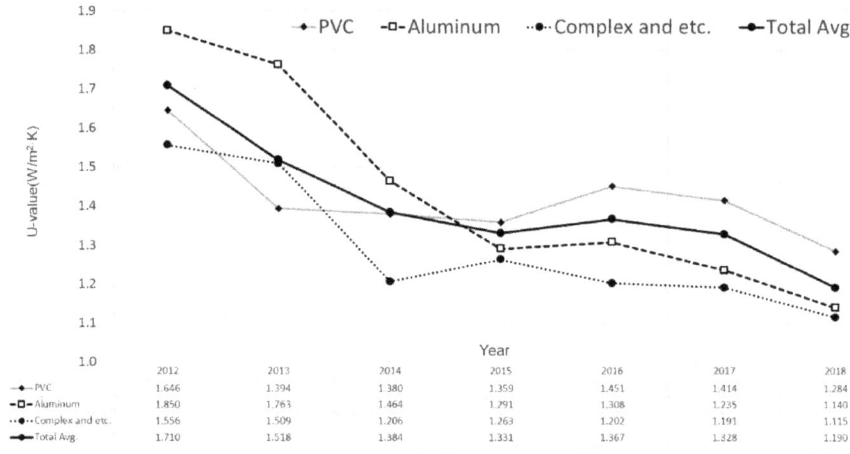

	2012	2013	2014	2015	2016	2017	2018
PVC	1.646	1.394	1.380	1.359	1.451	1.414	1.284
Aluminum	1.850	1.763	1.464	1.291	1.308	1.235	1.140
Complex and etc.	1.556	1.509	1.206	1.263	1.202	1.191	1.115
Total Avg.	1.710	1.518	1.384	1.331	1.367	1.328	1.190

Figure 5. Variation of U-value by frame type in each year.

By confirming the differences in window frames we confirmed that, in the case of PVC windows, from 2012 to 2018 the average U-values were 1.646 W/m^2·K, 1.394 W/m^2·K, 1.380 W/m^2·K, 1.359 W/m^2·K, 1.451 W/m^2·K, 1.414 W/m^2·K, and 1.284 W/m^2·K, respectively. By the results of aluminum windows, from 2012 to 2018 we confirmed that the average U-values were 1.850 W/m^2·K, 1.763 W/m^2·K, 1.464 W/m^2·K, 1.291 W/m^2·K, 1.308 W/m^2·K, 1.235 W/m^2·K, and 1.140 W/m^2·K, respectively. Also, in the case of complex, etc. window frames, from 2012 to 2018 the average U-values were 1.556 W/m^2·K, 1.509 W/m^2·K, 1.206 W/m^2·K, 1.263 W/m^2·K, 1.202 W/m^2·K, 1.191 W/m^2·K, and 1.115 W/m^2·K, respectively.

3.3. Analysis of Window Thermal Performance Variation

According to the analysis of the performance data, in this paper, we confirmed that the average U-value of the window decreased from 2012 to 2015 and maintained a similar value until 2017. The window test results in 2016 do not tend for the U-value to be constantly compared to the 2015 window test results. The reason is that the tendency of the data decreases as the amount of performance testing of the windows decreases. Besides, the amount of performance testing of aluminum windows has increased, to be similar to the number of PVC windows, and it is considered that the latter has been exceeded. In 2018, this value decreased. In consideration of the decreasing tendency from 2016, as the registration number steadily increased from the start of registration in 2012 to 2015, it can be judged that the required performance value of the windows is generally satisfied based on 2015. Following the strengthening of the method of the heat transmission coefficient of Korea in 2018, it can also be judged that the value of the heat transmission coefficient is low because the thermal characteristics of the newly registered windows are improved. In terms of the frame material classification of windows, the aluminum frame was used because the heat transmission coefficient of aluminum frame windows steadily decreased from 1.4050 W/m^2·K in 2012 to 1.140 W/m^2·K in 2018. It is considered that the thermal performance of the window steadily increased. However, the performance of the window cannot be analyzed in detail based on the registered materials, such as the material of the frame, and the performance of the glass used for the configuration of the windows and the glass depends on the classification of the materials used. Therefore, the performance of the glass used for the actual composition and the ratio of the glass to the frame were confirmed [28], and that it was necessary to check the thermal performance according to the configuration of the windows [29].

4. Analysis of Changing Thermal Performance According to Window Composition

4.1. Constituent Classification of Windows and Physical Experiment Results

In general, the window can be divided into a transmission glass part (glazing) and a frame part. In order to confirm the thermal performance of the windows, significant effort is made to enhance the thermal performance of each part; however, in the case of physical testing of the thermal performance of the window, due to size restrictions of the windows, test windows may be composed of a ratio different from that of the windows of a building [30]. In particular, to register in the "Energy Standards and Labeling Program", specimens of 2 m width and 2 m length are required for testing laboratories. Therefore, there may be differences in thermal performance due to the difference between the glass area and the frame area. Also, the configuration of the frame and the glass portion may be different due to the form of the window, and the fragility of heat (this mean is more conductive than other parts) can be displayed differently [31]. In this study, the influence of the construction of the windows on the heat transmission coefficient derived from the physical examination was also analyzed for the glass applied to the fitting, and the frame material, the glass portion and the area ratio of the frame, the form of the windows, etc., were examined. The windows were classified as sliding, double cut-up sliding, quadruple cut-up sliding, fixed window, fixed window with a project (etc.), and project window. For the opening type and composition of glazing, we divided windows into single windows and double windows. Figure 6 shows the equipment of the window thermal performance and the classification of window type.

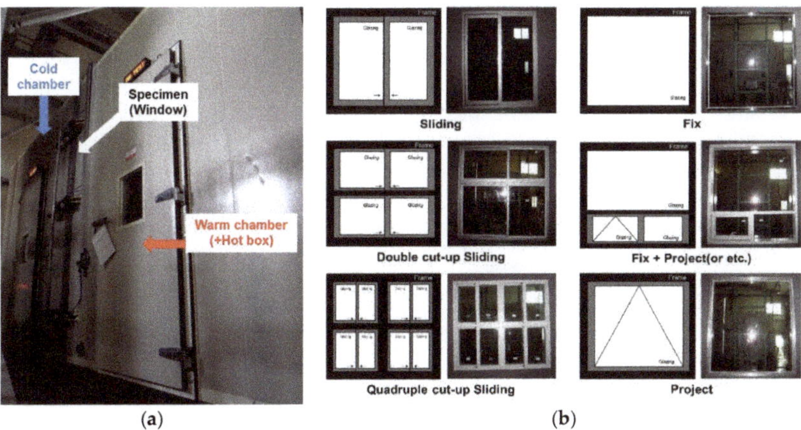

Figure 6. Equipment of the window thermal performance test (**a**) and shapes of various window types (**b**).

The total number of windows used for the research was 134, and windows were utilized based on the actual physical examination data. Single windows numbered 109 and double windows numbered 25. As a result of examining the heat transmission coefficient from the data, the average heat transmission coefficient of the total number of windows tested was 1.597 W/m^2·K; for single windows, the average heat transmission coefficient was 1.673 W/m^2·K, and for double windows, it was 1.263 W/m^2·K. Thus, the heat transmission coefficient of the double windows was about 24% lower than that of the single windows, and the lower thermal performance of the single window compared to the double window was confirmed. From the checking of the heat transmission coefficient of the windows by frame material, the U-values were found to be in a range of 0.675–3.560 W/m^2·K, with an average U-value of 1.519 W/m^2·K. The U-values of aluminum windows ranged from 0.811 to 3.246 W/m^2·K, with an average U-value of 1.616 W/m^2·K. As a result, the performance distribution of the material-specific windows showed that the U-value distribution of PVC windows appeared to be greater and that the average U-value of the PVC window was about 0.097 W/m^2·K lower. Figure 7 shows the results of U-value distribution by frame material type.

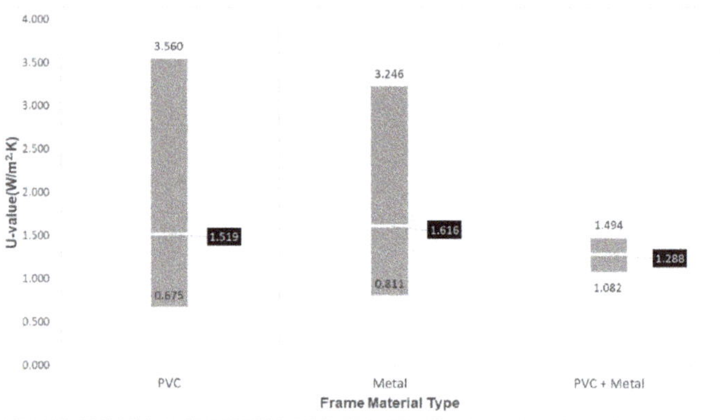

Figure 7. U-value distribution by frame material type.

4.2. Thermal Performance Analysis of Glazing by Simulation

In the case of glazing applied to windows recently, heat insulation performance is enhanced using double glazing in which a gas layer is formed between two single glass plates or triple glazing in which gas layers are formed between three single glass plates [32]. In Korea, it is difficult to separately assess changes in the U-value of windows according to the type of glass since the U-value of the entire window is confirmed without separating the glass and the frame. In particular, the necessary thermal performance is derived through various constitutional differences, such as using low-E glass, which is a functional glass, and vacuum glass, and filling using an inert gas such as argon or others. Figure 8 shows an example scheme for multiple glazing components.

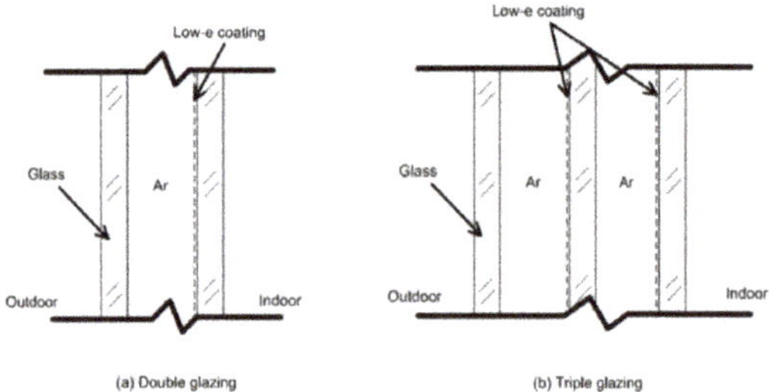

Figure 8. Examples of glazing components.

In the case of glass, the IGDB provides a certified performance from each international manufacturer so that it can be used for simulations and other purposes. The thermal performance of glass can be confirmed according to the product model of each manufacturer using the simulation program. In this study, we confirmed the detailed product information of the glass used for the actual physical test windows. Thus, with reference to information on the applied glasses, double glazing was divided into six types, and each product model was applied. Triple glazing was classified into three types, and each product model was also applied. A U-value was derived for each glass configuration through simulation. The simulation tool used the WINDOW program provided by LBNL, and the environmental conditions, such as applied indoor and outdoor temperature and heat transfer resistance, were provided by KS F 2278.

As a result of confirming the U-value of the glass part by simulation, the average U-value was 1.401 W/m^2·K. In the case of the aluminum window, the average U-value was 1.240 W/m^2·K, and in the case of the PVC window, it was 2.136 W/m^2·K. Thus, in the case of PVC windows, since the ratio of double windows is higher than that of aluminum windows, it is considered that the glazing performance of the latter is relatively low.

4.3. Analysis of Thermal Performance Due to Differences in Construction of Windows

The physical test results of the windows and the simulation results of the used glass part were integrated, and the difference in thermal performance according to the form and configuration of the windows was classified by frame material and analyzed. In the case of aluminum windows, the double window was formed as a quadruple cut-up sliding window, and the depth of the window was set to 235 to 250 mm. The proportion of the glass portion of the frame ratio averaged 65%, the average U-value of windows was 1.315 W/m^2·K, and the average U-value of the glazing was 1.268 W/m^2·K. Since it was a double window, it was considered that the ratio of the glazing part was low relative to the whole area and that the influence of the frame was large, although the performance of the glazing part could be set higher. In the case of an aluminum single window, the fixed and other ratio appears as its highest at about 48%, and in the fixed window the ratio is about 30%. When the ratio of the glazing portion exceeds 80%, it is confirmed that the area of the glazing portion of the window is relatively large, along with occupying a ratio of about 60% or more as a whole. The ratio of the U-value of windows corresponding to Grade 3 (1.4–2.1 W/m^2·K) of the "Energy Standards and Labeling Program" was as high as about 60%. Glazing with U-values of 1.1–1.3 W/m^2·K represented approximately 57% of the total, and it was confirmed that the U-value of the windows was higher than the performance of the glazing section. Figure 9 shows the analysis results of the aluminum window.

In the case of PVC windows, the depth of the double window was set to 170 to 260 mm. In the case of the quadruple cut-up sliding windows, the proportion of glazing was generally about 52%–54%. In the case of sliding windows, the proportion of glazing was about 68%–76%, and the depth of the window was set to 235 to 250 mm. Also, we confirmed that the proportion of glazing was about 52%–54%, and the proportion of glazing in sliding windows was about 68%–76% in the case of quadruple cut-up sliding windows. By these results, the authors confirmed the U-values of the window and glazing as 1.460–1.613 W/m^2·K and 3.353–3.952 W/m^2·K, respectively. The performance of the glass was set to low, but it was constructed with double windows to improve the performance of the frame. Moreover, thereby, the performance of the whole window was set to the required level. When the U-value was 0.675 to 0.735 W/m^2·K, the U-value of glazing was 1.217 to 1.300 W/m^2·K. Therefore, the double window requires high-performance glazing for a window less than 1.0 W/m^2·K. The proportion of the single PVC window was the highest at 71%. Compared to the frame, the proportion of glazing corresponding to a 70%–80% window was highest at 71%, and to an 80%–90% window was 18%. In the PVC window, we confirmed that the proportion of glazing was very high. From the results of the U-value of the windows, Grade 3 (U-value of 1.4–2.1 W/m^2·K) represented about 35%, and Grade 2 (U-value of 1.0–1.4 W/m^2·K) represented about 29%. From this result, it was confirmed that the U-value of 1.1 to 1.3 W/m^2·K in the glazing range has the highest share at 33%. Figure 10 shows the analysis results of PVC windows.

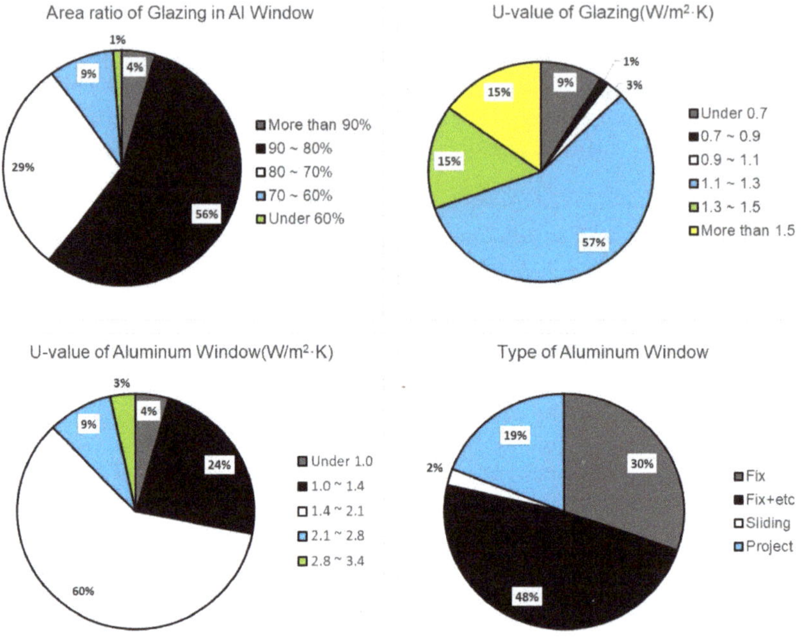

Figure 9. Analysis results of the aluminum window.

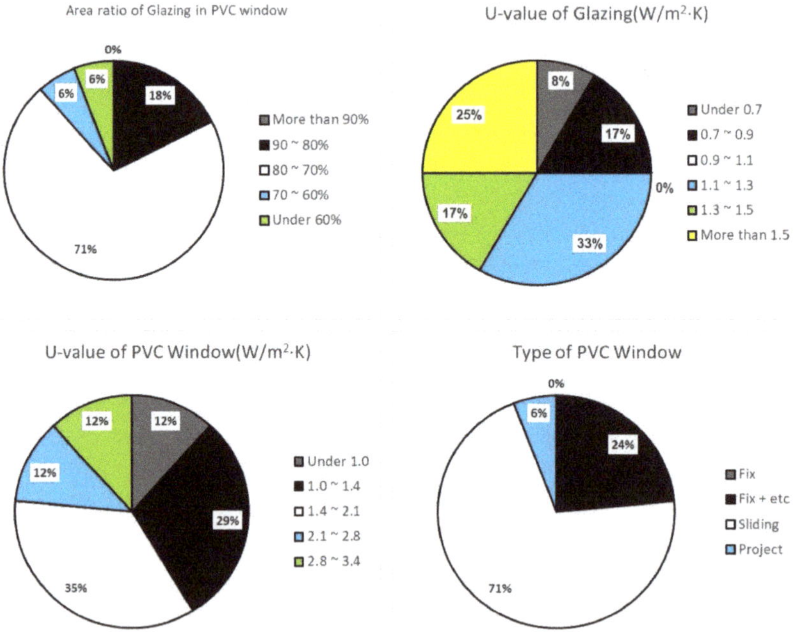

Figure 10. Analysis results of PVC windows.

Equation (2) shows the frame U-value formula for each part's performance, where U_F is the U-value of the frame, and U_W and U_G are the U-values of the window and glazing, respectively. A_W, A_G, A_F are the areas of the window, glazing, and frame, respectively. Since the value of the test result utilized in this study is the U-value of the entire window for which the type of spacer and the theoretical performance of the glass and frame cannot be confirmed, the spacer is installed and only the area is excluded from the glass U_G calculation. In the test method conducted in this study, the frame performance of the window includes the performance of the spacer because the effect of the spacer cannot be measured separately. At this time, the area of the glass excludes the area where the spacer was used. The part of frame and the glass overlap is divided into the performance of the frame.

$$U_F = \frac{(U_W \times A_W) - (U_G - A_G)}{A_F} \tag{2}$$

By the results of the aluminum window, the frame performance was confirmed considering the ratio of the window glass. If the glazing ratio was 90% or more, the average U-value of the frame was 3.872 W/m^2·K; when the glazing ratio of 80% to 89%, it was 2.934 W/m^2·K; also, the case of 70% to 79%, U_F was 3.308 W/m^2·K; and, the glazing ratio of 60% to 69% was 4.276 W/m^2·K. As a result, it was confirmed that the frame U-values of the glazing ratio of 80% to 89%, which accounted for the highest proportion of frames among the total test results of the aluminum windows, was displayed at the lowest level. When PVC windows were analyzed, the frame performance was confirmed considering the ratio of the window glass. In the case of the glazing ratio of 80% to 89%, the U-value of the frame was 3.838 W/m^2·K; and the portion of glazing of 70% to 79% has 2.658 W/m^2·K. Also, in the case of the glazing ratio of 60% to 69%, U-value was 2.332 W/m^2·K. It was confirmed that the frame U-value of the PVC windows was shown to be lower than the frame U-value of the aluminum windows.

4.4. Propose to the Performance Index of Glazing Parts in Window Design

According to the performance level of the energy and labeling program to distinguish the thermal performance of window in Korea, to confirm the performance of glazing that can achieve the evaluation of window, by using the grade of the Korean window obtained through this study, an index for selecting the performance of the glazing part was proposed. The performance index of the glazing part provided was based on the performance of the frame proposed in Equation (2), using the thermal performance of the surveyed Korean windows as a guide. The performance index of the glazing part is divided according to the ratio of the area of the glazing part in the window; also, this index proposed the minimum U-value of the glazing part for the suitable performance level. By experimental test results, the average U-value of the frame was applied to the frame U-value of each window. Moreover, the illustration of glazing components was described together in the figure.

The authors proposed the performance index of the glazing part to achieve the graded performance of the PVC window, as shown in the Figure 11. This index showed glazing area (each axis and example picture), glazing U-value (color spot on axis and value), frame U-value (U_f), and Grade level (color rhombus). By this index, a designer could consider the glazing U-value level by material type of window and glazing area level. In the case of the glazing part being 85%, a designer uses a glazing with a U-value of 0.5 W/m^2·K or less in order to construct a window of Grade 1. In addition, in order to construct Grade 2 and 3 windows, the glazing sections must satisfy 0.97 W/m^2·K and 1.79 W/m^2·K, respectively. Cases of 75%, 65%, 55% area of glazing could use a glazing part corresponding to 0.45 W/m^2·K, 0.28 W/m^2·K, 0.28 W/m^2·K to make a Grade 1 window. Therefore, it is not suitable for constructing a single sliding window using a general double-layer glazing or triple-layer glazing. In order to solve this, the construction of the window must be chosen for a double sliding window or the structure of the frame must be improved.

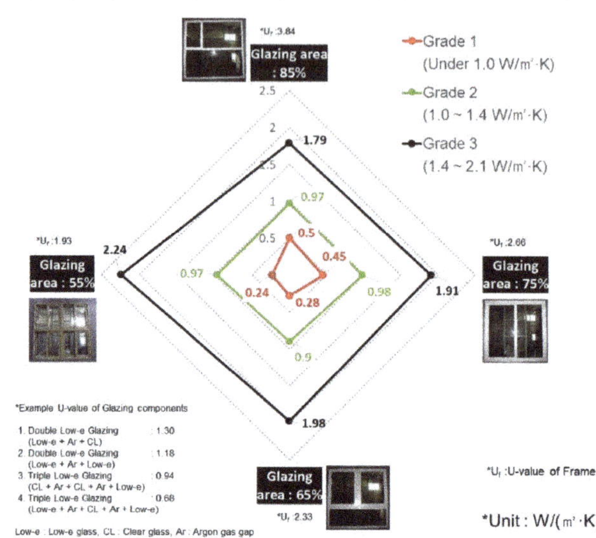

Figure 11. Performance index of glazing part in PVC windows design.

As shown in Figure 12, the performance index of the aluminum window glazing part is different from the performance index of the glazing part of the PVC window. To achieve the Grade 1 window, it is necessary to set the glazing part to 0.85 W/m^2·K when it corresponds to 95% of the glazing part. It is necessary to set the Grade 2 and Grade 3 window with glazing parts 1.27 W/m^2·K and 2.01 W/m^2·K, respectively. When the area of the glazing part corresponds to 65%, it was confirmed that the evaluation of each window cannot be achieved through the improvement of the performance of the glazing part. This is thought to be due to the low frame thermal performance of aluminum windows used in Korea.

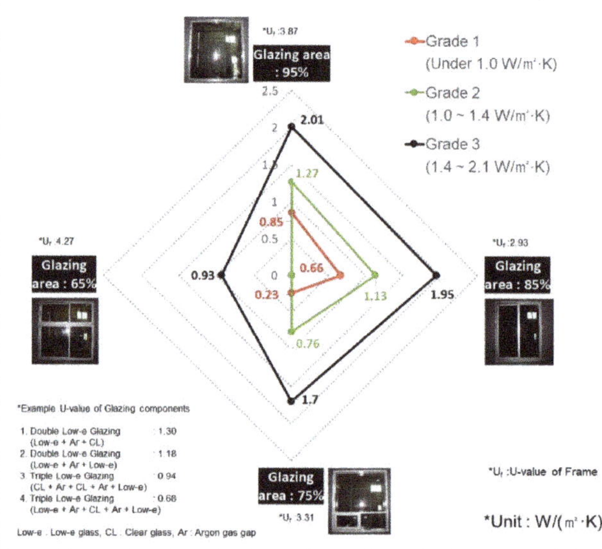

Figure 12. Performance index of glazing part in aluminum windows design.

Based on these results, when designing a window, the designer can roughly grasp the performance of the glazing part according to the frame type, configuration, and form of the window. When the designer does not find suitable glazing performance about window design, they could apply the median value of similar glazing area or glazing components. According to the improvement of the frame performance, the area ratio can be predicted.

5. Conclusions

In this paper, we confirmed the regulation and certification of window performance in Korea and analyzed the method and management system for testing this performance. Based on this, the performance data of the managed windows were analyzed, and the form and performance of the windows were classified. Also, we confirmed the change in the heat insulation performance by window construction through actual physical experiment. The results of this study are as follows.

(1) Based on the results of the research on the performance of existing windows, we confirmed the method of performance assessment of the windows, and the performance required to meet both the Building Energy Conservation Design Standards related to windows and the energy standard and labeling program. To assess the performance of the windows required previously, we confirmed the method of measuring the performance of the windows used in Korea and confirmed that it is necessary to classify the factors affecting the performance of the windows. So, by these results, we could identify the limits of the measurement methods (specimen size, the ambiguity of performance division about glazing, and frame).

(2) In this study, we analyzed the thermal performance data of the windows provided by the Korea Energy Agency and confirmed the change in the thermal performance of the windows by year and the change in thermal performance by frame material. The average U-value of the windows decreased from 2012 to 2015 and maintained a similar value until 2017. In 2018, this value decreased. The aluminum frame was used because the U-value of the aluminum frame windows steadily decreased from 1.4050 W/m^2·K in 2012 to 1.140 W/m^2·K in 2018. It is considered that the thermal performance of the window steadily increased. By these results, we confirmed that the thermal performance of the window was affected by the grade of the energy labeling program. This means that the thermal performance of the window was determined by government requirements.

(3) The authors confirmed the U-value of the windows through actual physical experiments and confirmed the change in thermal performance by the construction of the windows based on the results. Besides, based on the results of the thermal performance analysis of glazing through simulation, the thermal performance was analyzed by the difference in the construction of the glazing and the frame. The proportion of aluminum windows with a U-value corresponding to Grade 3 (1.4–2.1 W/m^2·K) was as high as about 60%. Glazing with a U-value of 1.1–1.3 W/m^2·K accounted for approximately 57%, and it was confirmed that the U-value of the windows was higher than the performance of the glazing section. Regarding the analyzed results of the U-values of PVC windows, Grade 3 (U-value of 1.4–2.1 W/m^2·K) accounted for about 35%, and Grade 2 (U-value of 1.0–1.4 W/m^2·K) for about 29%. From this result, it was confirmed that glazing with a U-value range of 1.1 to 1.3 W/m^2·K accounted for the highest share, of 33%. This paper also confirmed that the frame U-value of the PVC windows is lower than the frame U-value of the aluminum windows.

(4) By these results, the authors proposed the performance index of the glazing part in PVC and aluminum window design. In the case of PVC window design, the designer can roughly predict the performance of the glazing part according to the frame type, configuration, and form of window. Also, authors confirmed that the performance index of the glazing part in aluminum window design could not propose the performance of glazing for Grade 1 and Grade 2 window configuration.

Through this study, we confirmed the change in the performance of windows in Korea and the changes in thermal performance due to the composition and material. Especially the material of the frame is important for window shape and glazing area in the window area, because the cases occur where it is impossible to determine the combination of glazing parts to achieve the performance requirement value of the window.

Based on this result, it is expected to be useful for future Korean smart window design solutions. The results of this research can be used as basic data to identify problems in the method of determining the performance of windows in Korea. Since Korea specifies the required performance of the entire window, it is difficult to grasp the detailed thermal performances of changes in the glass and frame materials that make up the window. Also, although it is possible to compare the performance of the same specimen with the window performance test, it is difficult to predict the performance change of the window if the size of the actual applied window differs. Therefore, the authors would like to propose consideration of the prediction of the change of performance with the change of the size of the window in the future, using the analysis in this research for predicting the form of the window and the performance of the frame. So, the performance index of glazing part in PVC and aluminum window design was proposed for suitable chose of glazing thermal performance. Also, to consider the impact of cost and energy consumption, as with the results of this study, we will further understand the impact of cost and energy consumption associated with performance differences in window; further research is underway to confirm in future research.

Author Contributions: Conceptualization, methodology, writing—original draft preparation, writing—review and editing, S.-H.K. and H.J.; project administration, funding acquisition, S.C.

Funding: This research was funded by the Korea Institute of Energy Technology Evaluation and Planning (KETEP) and the Ministry of Trade, Industry & Energy (MOTIE) of the Republic of Korea, grant number 20172010105690 and The APC was funded by the Korea Institute of Energy Technology Evaluation and Planning (KETEP) and the Ministry of Trade, Industry & Energy (MOTIE) of the Republic of Korea (No. 20172010105690).

Acknowledgments: This work was supported by the Korea Institute of Energy Technology Evaluation and Planning (KETEP) and the Ministry of Trade, Industry & Energy (MOTIE) of the Republic of Korea (No. 20172010105690).

Conflicts of Interest: The authors declare no conflict of interest.

References

1. Kim, S.; Kim, S.; Kim, K.; Cho, Y. A study on the proposes of energy analysis indicator by the window elements of office buildings in Korea. *Energy Build.* **2014**, *73*, 153–165. [CrossRef]
2. Cappelletti, F.; Gasparella, A.; Romagnoni, P.; Baggio, P. Analysis of the influence of installation thermal bridges on windows performance: The case of clay block walls. *Energy Build.* **2011**, *43*, 1435–1442. [CrossRef]
3. Adamus, J.; Pomada, M. Analysis of heat flow in composite structures used in window installation. *Compos. Struct.* **2018**, *202*, 127–135. [CrossRef]
4. Hee, W.J.; Alghoul, M.A.; Bakhtyar, B.; Elaye, O.; Shameri, M.A.; Alrubaih, M.S.; Sopian, K. The role of window glazing on daylighting and energy saving in buildings. *Renew. Sustain. Energy Rev.* **2015**, *42*, 323–343. [CrossRef]
5. Tsikaloudaki, K.; Theodosiou, T.; Laskos, K.; Bikas, D. Assessing cooling energy performance of windows for residential buildings in the Mediterranean zone. *Energy Convers. Manag.* **2012**, *64*, 335–343. [CrossRef]
6. Kim, J.; Kim, T.; Leigh, S. Double window system with ventilation slits to prevent window surface condensation in residential buildings. *Energy Build.* **2011**, *43*, 3120–3130. [CrossRef]
7. Carlosa, J.S.; Corvacho, H. Evaluation of the performance indices of a ventilated double window through experimental and analytical procedures: SHGC-values. *Energy Build.* **2015**, *86*, 886–897. [CrossRef]
8. Wen, L.; Hiyama, K.; Kogane, M. A method for creating maps of recommended window-to-wall ratios to assign appropriate default values in design performance modeling: A case study of a typical office building in Japan. *Energy Build.* **2017**, *145*, 304–317. [CrossRef]
9. Ihm, P.; Park, L.; Krarti, M.; Seo, D. Impact of window selection on the energy performance of residential buildings in South Korea. *Energy Policy* **2012**, *44*, 1–9. [CrossRef]
10. Ochoa, C.E.; Aries, M.B.C.; van Loenen, E.J.; Hensen, J.L.M. Considerations on design optimization criteria for windows providing low energy consumption and high visual comfort. *Appl. Energy* **2012**, *95*, 238–245. [CrossRef]

11. Mangkuto, R.A.; Rohmah, M.; Asri, A.D. Design optimisation for window size, orientation, and wall reflectance with regard to various daylight metrics and lighting energy demand: A case study of buildings in the tropics. *Appl. Energy* **2016**, *164*, 211–219. [CrossRef]

12. Acosta, I.; Campano, M.; Molina, J.F. Window design in architecture: Analysis of energy savings for lighting and visual comfort in residential spaces. *Appl. Energy* **2016**, *168*, 493–506. [CrossRef]

13. Sun, Y.; Wu, Y.; Wilson, R. A review of thermal and optical characterisation of complex window systems and their building performance prediction. *Appl. Energy* **2018**, *222*, 729–747. [CrossRef]

14. Dussault, J.; Gosselin, L. Office buildings with electrochromic windows: A sensitivity analysis of design parameters on energy performance, and thermal and visual comfort. *Energy Build.* **2017**, *153*, 50–62. [CrossRef]

15. Li, C.; Tan, J.; Chow, T.; Qiu, Z. Experimental and theoretical study on the effect of window films onbuilding energy consumption. *Energy Build.* **2015**, *102*, 129–138. [CrossRef]

16. Fasi, M.A.; Budaiwi, I.M. Energy performance of windows in office buildings considering daylight integration and visual comfort in hot climates. *Energy Build.* **2015**, *108*, 307–316. [CrossRef]

17. Ye, H.; Meng, X.; Xu, B. Theoretical discussions of perfect window, ideal near infrared solar spectrum regulating window and current thermos chromic window. *Energy Build.* **2012**, *49*, 164–172. [CrossRef]

18. Kiran Kumar, G.; Saboor, S.; Kumar, V.; Kim, K.; Ashok Babu, T.P. Experimental and theoretical studies of various solar control window glasses for the reduction of cooling and heating loads in buildings across different climatic regions. *Energy Build.* **2018**, *173*, 326–336.

19. Egan, J.; Finn, D.; Soares, P.H.D.; Baumann, V.A.R.; Aghamolaei, R.; Beagon, P.; Neu, O.; Pallonetto, F.; O'Donnell, J. Definition of a useful minimal-set of accurately-specified input data for Building Energy Performance Simulation. *Energy Build.* **2018**, *165*, 172–183. [CrossRef]

20. Ministry of Land, Transport and Maritime Affairs. *Building Energy Conservation Design Standard*; Ministry of Land, Transport and Maritime Affairs: Sejong, Korea, 2008.

21. Ministry of Knowledge Economy. *Energy Efficiency Equipment Operating Regulations*; Ministry of Knowledge Economy: Sejong, Korea, 2012.

22. Korea Agency for Technology and Standards. KS F 3117: Windows Set, 2015. Available online: http://www.kssn.net/en/search/stddetail.do?itemNo=K001010105704 (accessed on 10 October 2019).

23. Korea Agency for Technology and Standards. KS F 2278: Standard Test Method for Thermal Resistance for Windows and Doors. 2017. Available online: http://www.kssn.net/en/search/stddetail.do?itemNo=K001010113352 (accessed on 10 October 2019).

24. Korea Agency for Technology and Standards. KS F 2292: The Method of Air Tightness for Windows and Doors. 2018. Available online: http://www.kssn.net/en/search/stddetail.do?itemNo=K001010122343 (accessed on 10 October 2019).

25. Korea Agency for Technology and Standards. Available online: http://www.kats.go.kr/content.do?cmsid=27 (accessed on 10 October 2019).

26. Lawrence Berkeley National Laboratory, COMFEN. Available online: http://www.lbl.gov/ (accessed on 10 October 2019).

27. Korea Agency for Energy. Available online: http://eep.energy.or.kr/ (accessed on 10 October 2019).

28. Alves, T.; Machado, L.; de Souza, R.G.; de Wilde, P. Assessing the energy saving potential of an existing high-rise office building stock. *Energy Build.* **2018**, *173*, 547–561. [CrossRef]

29. Buenoa, B.; Cejudo-Lopez, J.M.; Kuhn, T.E. A general method to evaluate the thermal impact of complex fenestration systems in building zones. *Energy Build.* **2017**, *155*, 43–53. [CrossRef]

30. Gou, S.; Nik, V.M.; Scartezzini, J.; Zhao, Q.; Li, Z. Passive design optimization of newly-built residential buildings in Shanghai for improving indoor thermal comfort while reducing building energy demand. *Energy Build.* **2018**, *169*, 484–506. [CrossRef]

31. Cho, S.; Kim, S. Analysis of the Performance of Vacuum Glazing in Office Buildings in Korea: Simulation and Experimental Studies. *Sustainability* **2017**, *9*, 936. [CrossRef]

32. Gustavsena, A.; Grynning, S.; Arasteh, D.; Jelle, B.P.; Goudey, H. Key elements of and material performance targets for highly insulating window frames. *Energy Build.* **2011**, *43*, 2583–2594. [CrossRef]

Article

The Application of Building Physics in the Design of Roof Windows

Jan Tywoniak *, Vítězslav Calta, Kamil Staněk, Jiří Novák and Lenka Maierová

Department of Architecture and Environment, University Center of Energy Efficient Buildings, Czech Technical University in Prague, Třinecká 1024, 273 43 Buštěhrad, Czech Republic; vitezslav.calta@fsv.cvut.cz (V.C.); kamil.stanek@fsv.cvut.cz (K.S.); Jiri.Novak.4@fsv.cvut.cz (J.N.); lenka.maierova@uceeb.cz (L.M.)
* Correspondence: tywoniak@fsv.cvut.cz

Received: 22 April 2019; Accepted: 12 June 2019; Published: 16 June 2019

Abstract: This paper deals with a small but important component in a building envelope, namely roof windows in pitched roofs. Building physics methods were used to support the search for new solutions which correspond to the maximum extent for requirements for passive house level design. The first part of the paper summarizes the key phenomena of heat transfer, mainly based on a comparison of vertical windows in walls. The results of repeated two-dimensional heat transfer calculations in the form of parametric studies are presented in order to express the most important factors influencing thermal transmittance and minimum surface temperatures. Several configuration variants suitable for technical design are discussed. It was found that a combination of wood and hardened plastics in the window frame and sash is the preferred solution. The resulting thermal transmittance can be up to twice as low as usual (from 0.7 down to 0.5 W/(m^2·K), with further development ongoing. Surface temperature requirements to avoid the risk of condensation can be safely fulfilled. Concurrently, it is shown that the relative influence of thermal coupling between the window and roof construction increases with the improvement of window quality. Specific attention was given to the effect of the slanting of the side lining, which was analyzed by simulation and measurement in a daylight laboratory. The increase in thermal coupling due to slanting was found to be negligible. Motivations for specific building physics research are mentioned, such as the need to study the surface heat transfer in the case of inclined windows placed in a deep lining.

Keywords: roof window; thermal performance; passive building; building component development

---

1. Introduction

Building physics, e.g., [1], including heat and moisture transfer, building-energy performance, energy assessments of elements and buildings, building acoustics, daylighting, the distribution of contaminants, etc., can be perceived as a set of rules for the assessment of constructions and buildings [2]. It can also be perceived as a set of sub-tools that is used actively during preparation and development work. Individual requirements need to be seen as interconnected, such as the requirement for daylight versus the requirement to limit the risk of a room overheating. Such a process accompanying development work should be perceived as multilevel. When applying building physics knowledge and practical tools, it is necessary to choose the key phenomena that will be prioritized.

The paper discusses the use of building physics methods to develop a roof window suitable for passive buildings. This was the subject of a joint project between a research organization and an innovative company specializing in the production of windows [3]. The contribution of this paper is to highlight selected phenomena and show possible applications of the technical solution of the new component. At the same time, it should be remembered that, in addition to the obvious technically oriented approaches, the user's customs (window opening, cleaning) and aesthetic requirements must be considered in the design of the final product.

Passive buildings [4] have become a clearly defined category of energy-efficient buildings [5], which is in line with the trend of sustainable solutions [6] over more than 25 years of development. Design methods and voluntary certification procedures are steady, and in some countries and regions they may be linked to subsidy policies or building regulations. At the same time, building solutions together with technical systems (minimizing heat penetration through the building envelope, minimizing heat losses by ventilation with heat recovery from the ventilating air) are the basis for the downstream categories of energy-optimized buildings obtaining a significant share of their energy from renewable sources (passive house premium [7], nearly zero-energy buildings (NZEBs) [8], Effizienzhaus Plus [9], etc.). Even a seemingly small element, such as a window, can be significant in terms of the reduction of energy use, and can thus be part of the solution.

In order to meet the requirements for energy-optimized buildings in cold to moderate climates, it is generally necessary to use windows with a thermal transmittance of around 0.8 W/(m^2·K) [10]. However, roof windows of this quality are not yet normally available. At the same time, the effect of installation into the roof construction should be taken into account, since there is a considerably larger additional thermal flow than for windows in perimeter walls.

For roof windows, the condensation of water vapor on their inner surfaces, especially at the edges of the glass, can be observed at low external air temperatures. Together with an effort to reduce the heat transfer of the windows, the spacing frames of the glazing units, the shape of the window frames, and the materials used in their construction have also been gradually improved.

It is known that window frames typically have a higher thermal transmittance than modern glazing units [11]. This is a particularly significant phenomenon due to the smaller dimensions of the roof windows (in the case of openable windows, the window frame means an assembly of parts, the fixed frame, and the movable frame of the sash).

One of the possible ways to reduce heat leakage at the edge of a window is to fit additional thermal insulation elements during the installation of the window from the outside of the frames [12].

Roof windows are the weakest elements of the building envelope in terms of heat losses. For this reason, they are not popular with designers involved with passive buildings. However, they must be used in some cases, and the consequence of increased heat transmission must be compensated for in order to achieve the passive house criteria. This critical situation can be illustrated by the fact that, in the database of components certified for passive houses [13], of 268 positions for windows (frames and connections), only two deal with roof windows (in one case equipped with quadruple glazing, in the other case equipped with triple and double glazing combined in one assembly).

The aim of the project in [3] was to find a solution for a roof window so that the thermal transmittance of the entire window corresponds to the requirements for passive buildings and so that the temperature on the frame surface meets the requirement to eliminate the risk of surface water condensation under reference conditions in the interior. The possible increased risk of the overheating of rooms during the summer due to solar gain was not explicitly addressed in this project.

This paper discusses methods used during research and development studies in Section 3. It then introduces and comments on the main findings in Section 4. Appendix A shows, through a simple case study, the effects of roof windows of different quality on the overall thermal transmittance of the building envelope of a typical family house. Preliminary studies about heat transfer phenomena close to window surfaces are introduced in Appendix B.

2. Problem Analysis

To analyze the problem, a comparison of a roof window with a vertical window in a perimeter wall was performed (see Figures 1 and 2).

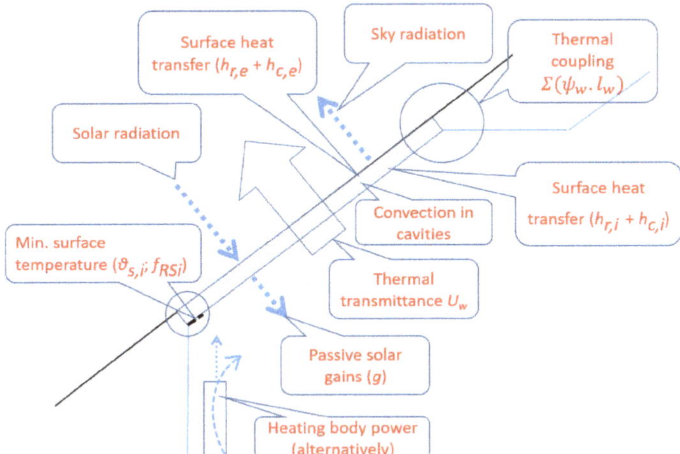

Figure 1. Significant thermal phenomena related to roof windows.

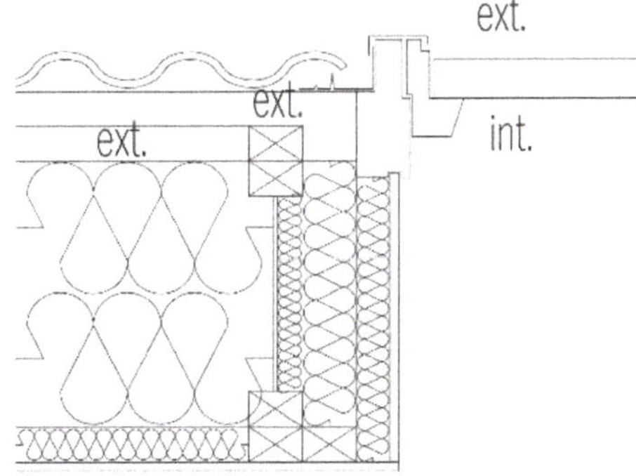

Figure 2. Schematic horizontal cross-section of a typical position of a roof window in a pitched roof. ext.: exterior air temperature; int: interior air temperature.

Heat transfer in the air cavities between glazing panes is generally larger due to increased heat convection caused by air movement (the more inclined the more significant; nonlinear). As a result, thermal transmittance is increased in the range of 0.1 to 0.2 W/(m²·K) relative to the whole window if calculated according to [14].

Considering the fact that frames are weaker parts of a window in terms of the total thermal transmittance, attention is given to a more detailed description of the thermal properties of the frames [15], and to their improvement. Such an approach cannot be clearly conducted by the study of roof windows.

Heat is transferred due to a connection of the window to the opaque part of the building envelope: In the design of vertical windows, it is attempted to minimize such thermal coupling by searching for the optimal position of a window in the wall [16] and the optimal geometry and materials for the construction detail. On the contrary, in roof windows, a very significant problem arises due to the windows' geometrical situation (Figure 2). The external perimeter of the window is situated in the cold area of the roof. Therefore, it is not possible to achieve the so-called thermal-bridge free solution (see Figure 3). The result is expressed as linear thermal transmittance [17], ψ_w (W/(m·K)). Although the values recommended in the Czech national standard [18] are larger than those for vertical windows in perimeter walls, they may not be easily reached (Table 1).

Table 1. Standard values of linear thermal transmittance, ψ_w (W/(m·K)), as a result of thermal couplings. Adapted from the Czech national standard [18].

	Required	Recommended	Recommended for Passive Buildings
Window in wall	0.1	0.03	0.01
Window in pitched roof	0.3	0.1	0.02

Simple calculations performed in [19] (Figure 3) have shown that unavoidable heat transfer due to the thermal coupling between the window and the roof plays an important role. For this reason, it is recommended to integrate the additional heat transfer due to thermal coupling in the (extended) thermal transmittance, $U_{w,inst}$, in order to obtain a "full picture" in one value [10]:

$$U_{w,inst} = \frac{A_g.U_g + A_f.U_f + \Sigma(\psi_g.l_g) + \Sigma(\psi_w.l_w)}{A_g + A_f},$$

(1)

where $\Sigma(\psi_w.l_w)$ describes the influence of the installation, U_g is the thermal transmittance of the glazing unit (W/(m²·K)), U_f is the thermal transmittance of the frame (W/(m²·K)), the thermal bridges of the glazing edge are expressed by the linear thermal transmittance, ψ_g (W/(m·K)), and the thermal bridges due to the installation in the roof are expressed by the linear thermal transmittance of the window, ψ_w (W/(m·K)). Figure 3 shows the results of a preliminary calculation for a hypothetical window of excellent quality: U_g 0.60 W/(m²·K), U_f 0.60 W/(m²·K), ψ_g 0.03 W/(m·K), ψ_w 0.05 W/(m·K), reference window size. It can be seen that, for improvements to roof windows, all parts are of high importance, i.e., the glazing, frame, installation method, and overall geometry.

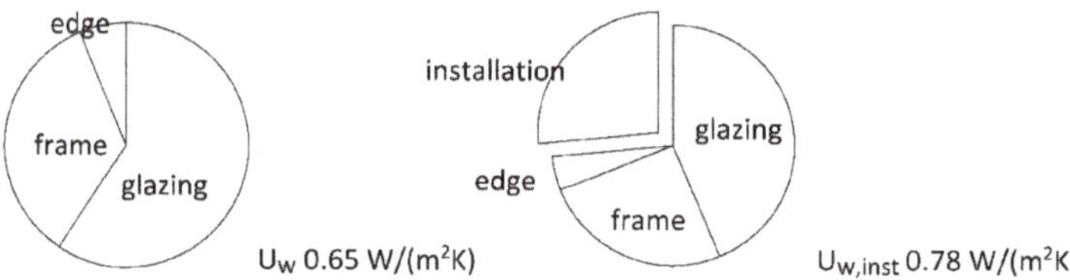

Figure 3. The result of a preliminary calculation for a (hypothetical) roof window of excellent quality. Heat transfer (**left**) and heat transfer including the effect of installation on roof thermal coupling (**right**) based on 2D calculations for all relevant cross-sections [19].

As a rule, heat transfer is calculated and measured vertically according to standardized test procedures [20], even for later use in an inclined position. Thus, if the goal is to achieve identical heat transfer to that in vertical windows, the requirement should be more stringent, being reduced at least by 0.1 W/(m²·K). The standard [18] is partly based on this approach, as shown in Table 2, for the required and recommended values of thermal transmittance. During preparation of the standard [18] this approach was not used for passive house recommendations because of the fact that windows of corresponding low thermal transmittance were not available.

Table 2. Standard values of thermal transmittance, U_w (W/(m²·K)), adapted from the Czech national standard [18].

	Required	Recommended	Recommended for Passive Buildings
Window in wall	1.5	1.2	Range: 0.8–0.6
Window in pitched roof	1.4	1.1	0.9

Another important fact is that roof windows are usually smaller than windows in walls, and therefore the proportion of the frame, as the weaker element in terms of heat transfer, will be significant.

The reference size of the roof window [21] is 1.14 m × 1.40 m, and the conversion to different window dimensions is only performed in practice in exceptional cases.

Excessively low interior surface temperatures can create conditions for the surface condensation of water vapor and mold growth (expected above a critical air humidity of 80%). This applies in particular to window frames and glazing edges. The surface temperature factor, $f_{Rsi} = (\theta_i - \theta_e)/(\theta_{ai} - \theta_e)$, is used for the assessment. Its required minimum values [18] are shown in Figure 4, and depend on the outdoor air temperature, indoor air temperature, and relative humidity.

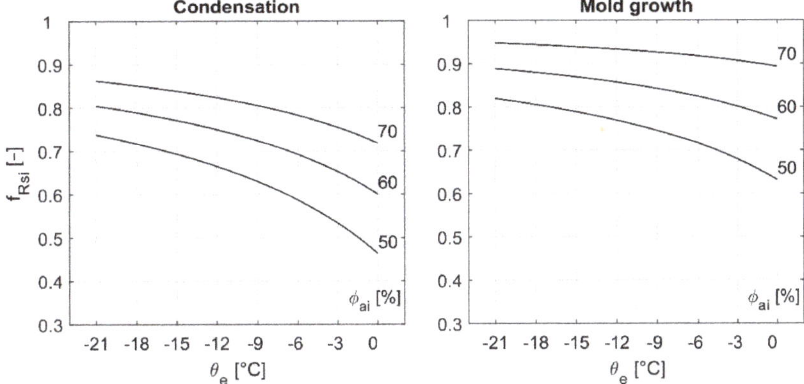

Figure 4. Minimum values of the surface temperature factor to avoid water vapor condensation (**left**) and mold growth (**right**) assuming an interior air temperature of 20 °C for different air humidity.

The surface heat transfer coefficient, h_{si} (W/(m²·K)), describing the heat transfer between the internal surface of the window and surroundings can be different for inclined roof windows and for vertical windows. Smaller inclined roof windows are additionally often influenced by heating bodies close by, and the situation can therefore be very different.

Requirements on the thermal transmittance of the roof construction result in an overall thickness of approximately 400 mm or more. This can negatively influence the daylight quality due to the very deep side lining. Therefore, the distribution of daylight in rooms as a primary function of each window should also be studied very carefully.

Additionally, in the case of roof windows, the situation of radiative heat exchange between the glazing and the two perpendicular side linings is fundamentally different from that of a wall window. Heat transfer can also be significantly affected by the presence of a heating body below the window (differently depending on the power, temperature, proportion of the radiant and convective heat-to-room transfer, and overall space situation). Moreover, a deep lining contributes to a different air flow in the room and in the space in front of the window.

The radiative heat exchange between the external window surface and the (clear) sky [1] is higher for roof windows (e.g., multiplied by a factor of 1.5 for 45° sloped windows). This leads to an increase in the total external surface heat transfer coefficient, h_{se} (W/(m²·K)).

In summer, specific phenomena should be considered (although they are not the subject of this paper). These include passive solar gains in the rooms and the resulting risk of room overheating, which are primarily influenced by the orientation of the façade/roof, window size, and the coefficient of the permeability of total solar radiation (solar factor), g (-), of the glazing unit, by the shading from external obstacles and shading devices. The overall effect depends on several other parameters of the (occupied) room, including the thermal inertia, ventilation strategy, and actual climatic data. The external air can be significantly warmer close to the roof surface (heated by the roof covering).

Generally, a higher passive solar gain can be expected due to the inclination of roof windows. Moreover, efficient external shading, such as venetian blinds, are not applicable to roof windows.

3. Methods

3.1. Parametric Studies

Introductory parametric studies [3] involving 2D-heat transfer calculations using the COMSOL Multiphysics [22] modeling software were carried out to map the key dependencies of the overall layout and geometry of window and frame material on the window's thermal performance. Initially, two different configurations were studied: (i) Usual assembly with high-performance glazing near the exterior (configuration A); and (ii) alternative assembly with a window casing equipped with high-performance glazing near the interior and an additional single-glazed pane at the exterior side (configuration B) (Figure 5).

The calculation model followed the rules given in [23]. Connection to the roof is considered here as adiabatic; this means that the effects of thermal couplings were not included. For that reason, the estimation of the minimum interior surface temperature is only indicative.

A glazing unit consisting of more panes and cavities filled by Argon was simplified for the calculation as a homogeneous solid material having equivalent thermal conductivity back-calculated from the known thermal transmittance of the glazing unit, U_g. The connection between the frame and the glazing unit was simplified by using a two-box-method [24] for spacers of known characteristics. Air gaps between the fixed and movable part of the frame were simulated according to [23].

In this case, high-performance homogeneous materials for the frame were considered. The thermal conductivities of the fixed and movable part were 0.039 W/(m·K) and 0.065 W/(m·K), respectively. This corresponds to the use of hardened polystyrene [25] with a density of 100 and 400 kg/m^3, respectively. The thermal transmittance of the glazing unit was 0.6 W/(m^2·K) for triple glazing and 1.0 W/(m^2·K) for double glazing. Calculations were performed for indoor and outdoor air temperatures of 20 and 0 °C, respectively.

During the parametric studies, several geometrical parameters and material parameters were changed over a relatively wide range in order to determine their importance within the whole configuration. The thermal transmittance of the glazing was fixed.

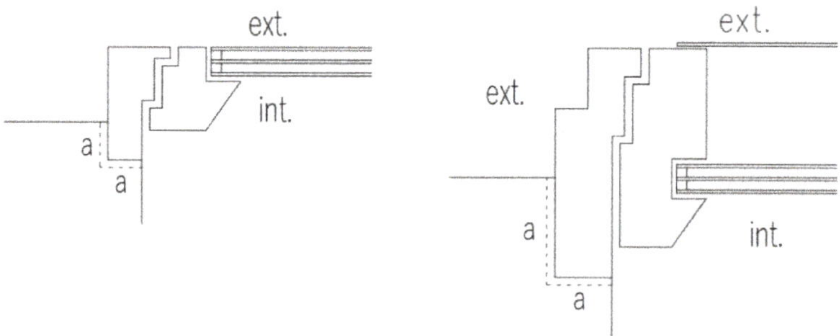

Figure 5. Configuration A (**left**) and alternative configuration B (**right**) that were tested in parametric studies. a: adiabatic condition [23].

3.2. Real Design Solution

Repeated 2D-heat transfer calculations in steady state conditions were performed for selected configurations according to consultations with the development team. The HT-Flux software [26] was used to analyze representative cross-sections (for head and sill, over and under the hinge). The thermal transmittance of the frame, the linear thermal transmittance due to the connection between the glazing and frame, and the resulting thermal transmittance for the reference window size (1.14 m × 1.40 m) were estimated for each variant. Additionally, a preliminary estimate was made of the minimum surface temperature. The simplifications for modeling were identical to those described in Section 3.1. Material parameters and boundary conditions are summarized in Section 4.2.

The effects of thermal coupling between the window and the roof construction were analyzed for selected window variants. Again, the HT-Flux software was used to analyze representative cross-sections (e.g., Figure 2) for a typical configuration of a pitched roof for passive house quality.

The thermal transmittance of the installed window, $U_{w,inst}$, was estimated for the reference window size. Moreover, the minimum window surface temperature and corresponding factors, f_{Rsi}, were evaluated [27].

3.3. Daylight Simulation

In this study, the problem of daylighting was preliminarily investigated by means of numerical simulations. The influence of the possible slanting of the roof window lining was investigated. The quantity of daylight and its spatial distribution may change with the degree of lining slanting (that is, the angle, α, between the lining plane and the plane perpendicular to the roof, see Figure 6). In order to quantify this effect, the daylight factor, D (%), was calculated by means of a validated software tool [28] using the CIE standard overcast sky model at a horizontal reference plane in an attic room lit by a roof window. It was calculated repeatedly with the angle, α, varying stepwise from 15 to 45°.

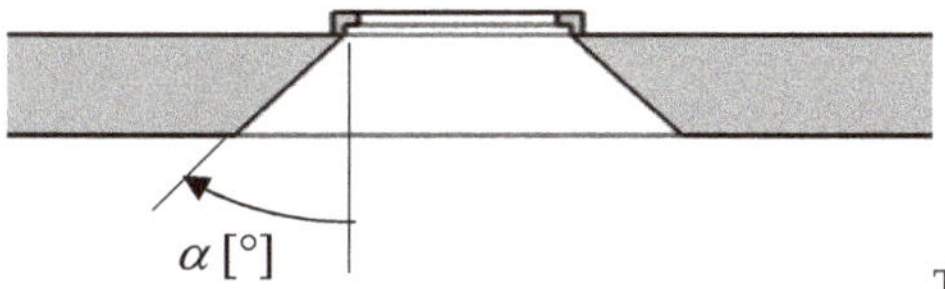

Figure 6. A cross-section through the roof window showing the inclination (slanting; α) of the window lining.

The geometry and dimensions of the studied room represent a typical inhabited attic room (Figure 7). The dimensions of the roof window are 860 × 1180 mm, and its triple glazing has a light transmittance, τ, of 0.74. Except for the floor, all of the internal surfaces are supposed to be painted a bright white with a corresponding light reflectance, ρ, of 0.84. The floor reflectance, ρ, is 0.66 (gray carpet). The reflectance should correspond to the physical model prepared for measurement in Section 3.5. The internal surface coatings of the physical model prepared for daylight measurement in Section 3.4 were selected in order to reproduce such high values of reflectance.

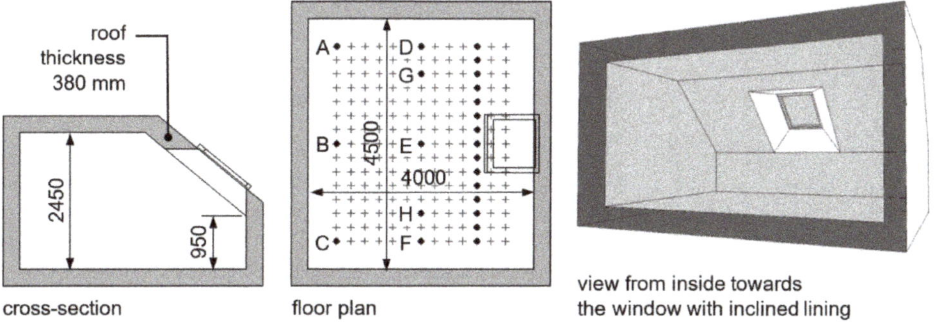

Figure 7. The geometry of the room studied for daylighting distribution (letters represent the control points for investigation). (Unit: mm).

3.4. Daylight Measurements

The daylight distribution was measured using a model with a 1:4 scale (Figure 8) in the daylight laboratory in Danube University Krems, Austria. The simulator had a diameter of 6 m (Figure 9). The horizontal daylight level in the middle of the simulator, where the model was placed, was set to 8000 lx. The model of the room had a changeable roof. The roof window was either placed centrally or near to the side partition wall. Additionally, alternatives for different geometries of side lining were analyzed, namely perpendicular and slanted with $\alpha = 45°$. Surfaces were finished using high-reflectance coatings corresponding to the reflectance values used in the simulation in Section 3.3.

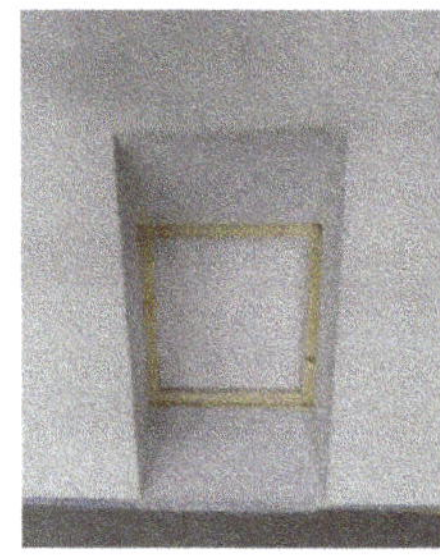

Figure 8. Model of a room for daylight measurement with changeable pitched roof.

Figure 9. Experimental setup for daylight measurement in a laboratory.

3.5. Formulation of General and Detailed Recommendations

Additionally, as a result of all the analyses, a set of recommendations for the technical design of the roof window was formulated. Moreover, further steps for the implementation of new products were followed, namely (i) the creation of a catalogue of the overall solution and (ii) the support of an independent assessment. Furthermore, a certification body (Centrum stavebního inženýrství a.s., Zlín, Czech Republic) performed measurements of the following mandatory declared parameters [20]: Thermal transmittance, airtightness, and water tightness to wind-driven rain.

4. Results

4.1. Parametric Studies

The results of the heat transfer calculations for two different window configurations (see Figure 5) are presented in Figure 10. Configuration A has a higher thermal transmittance of the frame, U_F, for the relevant geometries, however, the use of a high-performance glazing at the exterior side leads to an overall lower thermal transmittance of the window compared to configuration B.

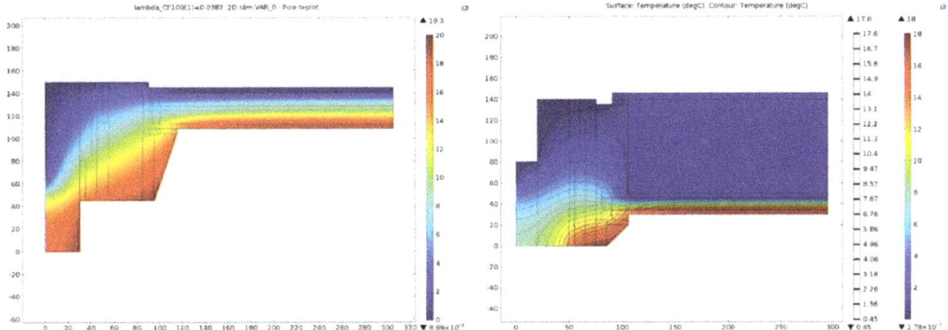

Figure 10. Temperature distribution for window configuration A (**left**) and B (**right**).

More detailed results of the parametric studies for configuration A are presented in Figures 11 and 12. Each parameter was tested for a wide range of geometries, separately assuming homogeneous material for the frame with very low thermal conductivity. It can be concluded that the most significant parameters are (i) the setting depth of the window frame into the thermally insulated roof layer, (ii) the overall width of the frame and sash, and (iii) the material of the frame and sash. Other material and geometrical parameters are less important. These results were considered in the next developmental steps.

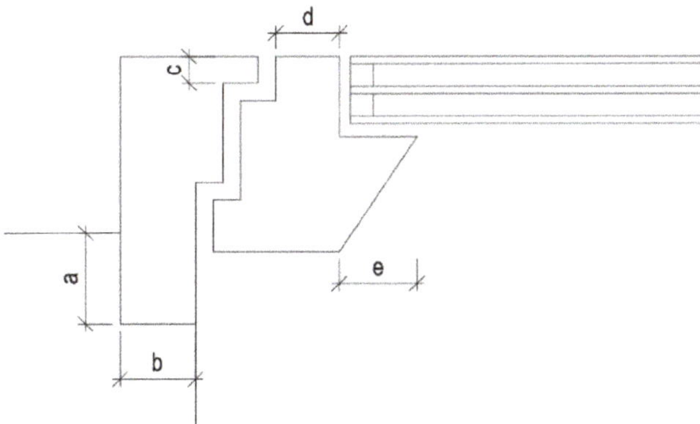

Figure 11. Schematic of a window cross-section showing the most important parameters analyzed in parametric studies.

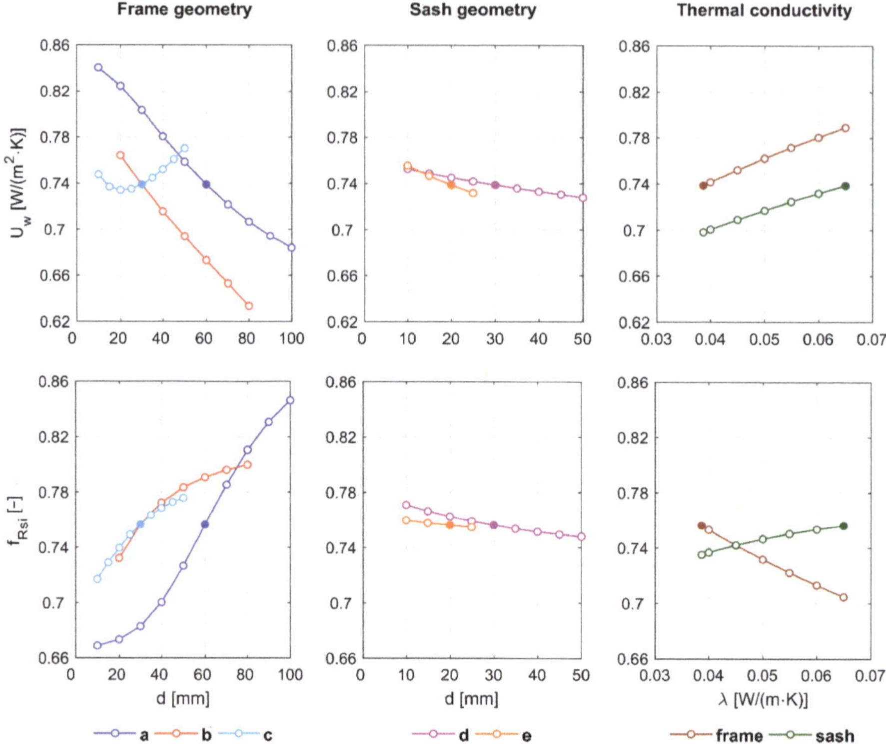

Figure 12. Key tendencies discovered in the parametric studies, expressed as the thermal transmittance of the whole window and as the surface temperature factor. Filled points represent basis values. For legend, see Figure 11.

4.2. Real Design Solution

Figures 13 and 14 show the configuration, dimension, and material properties as suggested by the development team for two selected final design variants, considering the results in Section 4.1 and technical limitations. The figures represent the cross-sections over the hinge. The frame and sash were made of a combination of soft wood and hardened plastics. The thermal conductivities used in the 2D calculation are summarized in Table 3. The thermal conductivity of the hardened polystyrene, Compacfoam [25], was measured in accordance with EN 12667 [29] using the HFM 300 apparatus (Linseis GmbH, Selb, Germany) using a sample with a size of 300 mm × 300 mm. The sample, declared as CF100, had a bulk density of 122 kg/m^3. The dry thermal conductivity was determined as 0.037 W/(m·K), and the thermal conductivity in the wet state (the moisture content of the water-saturated sample over four weeks was 1.6 m%), λ_{char}, was 0.039 W/(m·K). Other values in Table 3 are taken from the literature. The boundary conditions used in the 2D calculations are summarized in Table 4.

Several types of glazing unit can be used in each design variant. The window in Figure 13 is equipped with triple glazing with a U_g of 0.5 W/(m^2·K). The window in Figure 14 has a U_g of 0.3 W/(m^2·K), and is equipped with a special glazing unit in which two transparent polyester foils divide the space between two glazing panes into three cavities to reduce the overall heat transfer. For construction reasons, their spacers have a higher thermal conductivity than those used in the best available triple glazing. The thermal performances of the presented window variants are summarized in Table 5.

Table 3. Material properties used in the 2D heat transfer calculation.

Material	Thermal Conductivity, λ (W/(m·K))	Pattern According to Figures 13 and 14
Glazing pane	1.00	
Seals (EPDM profiles)	0.25	
Wood (soft)	0.12	
Compacfoam	0.039	
Aerogel	0.014	
Extruded polystyrene	0.032	

Table 4. Boundary conditions and heat flux scale used in the 2D heat transfer calculations.

Boundary Condition	Air Temperature θ (°C)	Surface Heat Transfer Resistance R_{si} (m²K/W)
Interior—Building construction	20	0.10
Interior—Window	20	0.13
Interior—Edges of window	20	0.20
Interior—Edges of window for the evaluation of minimum surface temperature	20	0.25
Exterior	−17	0.04
Ventilated cavity connected to the exterior (under roof covering)	−17	0.10

Table 5. Thermal performance for window variants *I* and *II* according to Figures 13 and 14.

Variant	Heat Transfer Coefficient (W/K)			Mean Linear Thermal Transmittance for Glazing Edge (W/m·K)	Thermal Transmittance for Reference Window Size (W/(m²·K))	
	Glazing $H_{T,g}$	Frame $H_{T,f}$	Edge $H_{T,\psi,g}$	ψ_g	U_F	U_W
I	0.530	0.503	0.103	0.025	0.94	0.71
II	0.303	0.344	0.148	0.037	0.59	0.50

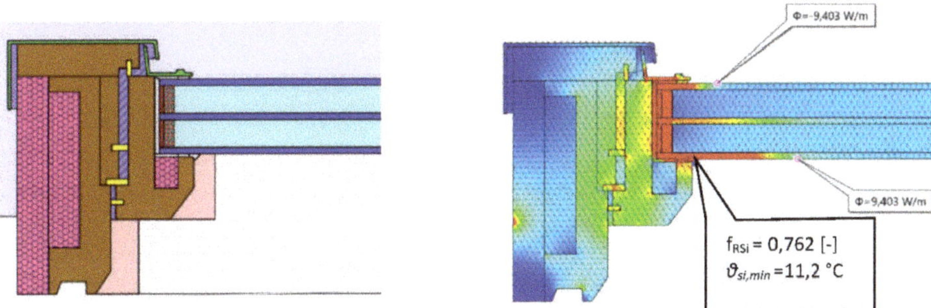

Figure 13. Schematic cross-section of the window variant, *I*, with triple glazing (**left**) and calculated heat flux distribution (**right**).

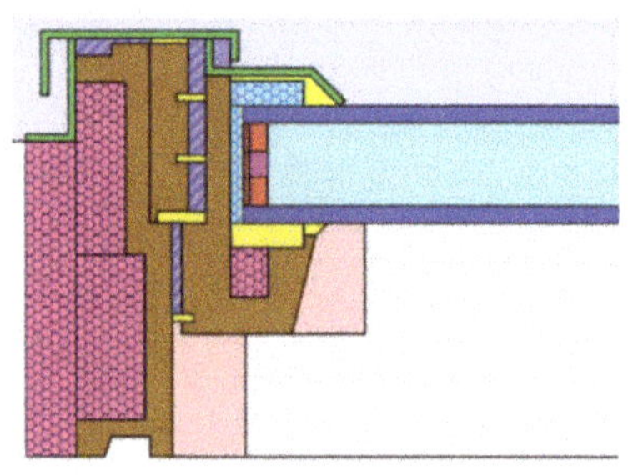
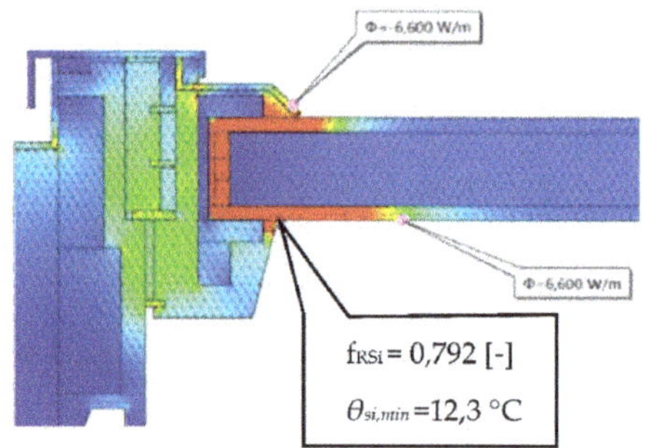

Figure 14. Schematic cross-section of the window variant, *II*, with special glazing using two foils to divide the cavity between two glazing panes (**left**) and calculated heat flux distribution (**right**).

Table 6 shows the effects of thermal coupling for perpendicular and slanted window linings for variant *II*. The values of $U_{w,inst}$ were derived from the calculated thermal transmittance of the window and the linear thermal transmittance at both sides, the sill and the head. The values of ΔU represent the difference between $U_{w,inst}$ and U_w. A significant influence of thermal coupling is evident. The slanting itself plays a minor role in the heat transfer.

Table 6. Effects of thermal coupling for the perpendicular and slanted linings.

Perpendicular Lining			Slanted Lining ($\alpha = 45°$)		
ψ_w (W/(m·K))	$U_{w,inst}$ (W/(m²·K))	ΔU (W/(m²·K))	ψ_w (W/(m·K))	$U_{w,inst}$ (W/(m²·K))	ΔU (W/(m²·K))
0.071	0.72	0.22	0.093	0.79	0.29

The minimum surface temperature is presented for variant *II* (Table 7) in the most critical combination: (i) For the slanted lining and (ii) for the glazing unit with a metallic spacer. The lowest surface temperature is still at the edge of the glazing, similar to in the previous calculation without thermal coupling. The values fulfill the requirements (see Figure 4), and therefore the risk of water vapor condensation is minimized.

Table 7. Minimum surface temperature and surface temperature factor for variant II, coupled to the roof structure. Slanted lining, indoor temperature of 20 °C, exterior temperature of −17 °C.

Area	Minimum Surface Temperature $\theta_{si,min}$ (°C)	Surface Temperature Factor f_{Rsi} (-)
Sill	10.8	0.752
Head	11.0	0.756
Side lining	10.9	0.755

4.3. Daylight Simulations

Figure 15 shows the calculated daylight factors, D, at the selected control points on the horizontal reference plane 850 mm above the floor level, and the mean daylight factor, D_m, as a function of the window lining inclination, α. By slanting the lining, it is possible to achieve an improvement in the overall level of daylight (expressed in terms of D_m) as well as a significant local improvement in the level of daylight in the area close to the window. Due to the local increase in the daylight factor, a significantly larger portion of the floor area will be available for activities which require higher daylight levels, i.e., $D > 1.5\%$.

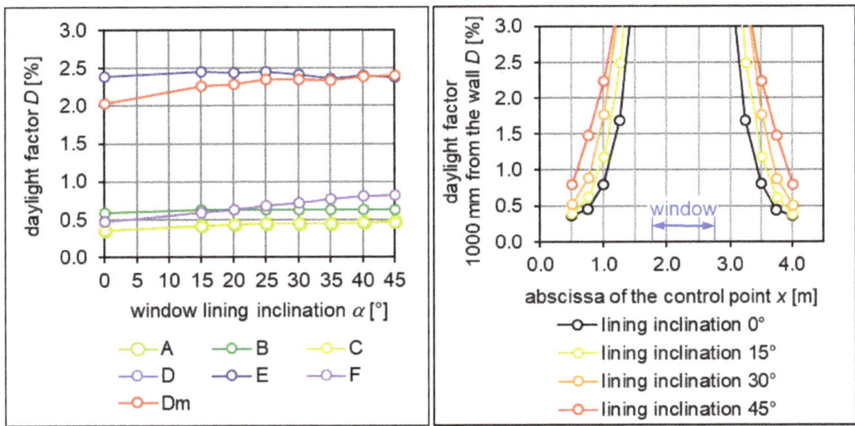

Figure 15. The results of the daylight calculation as a function of the lining inclination. The daylight factor at selected control points (see Figure 7) and the mean value (D_m) (**left**) and daylight factors in the line of control points at a distance of 1 m from the external wall (**right**).

4.4. Daylighting Measurements

In the model, the values of external and internal illumination at points on the horizontal plane were measured at a height corresponding to measurements on real buildings. Figure 16 shows the results expressed as a daylight factor for a series of points closest to the window (row A) for a perpendicular and slanted side lining. Significant differences in the daylight factor were found for the slanted side lining. Only negligible differences were found in rows B and C.

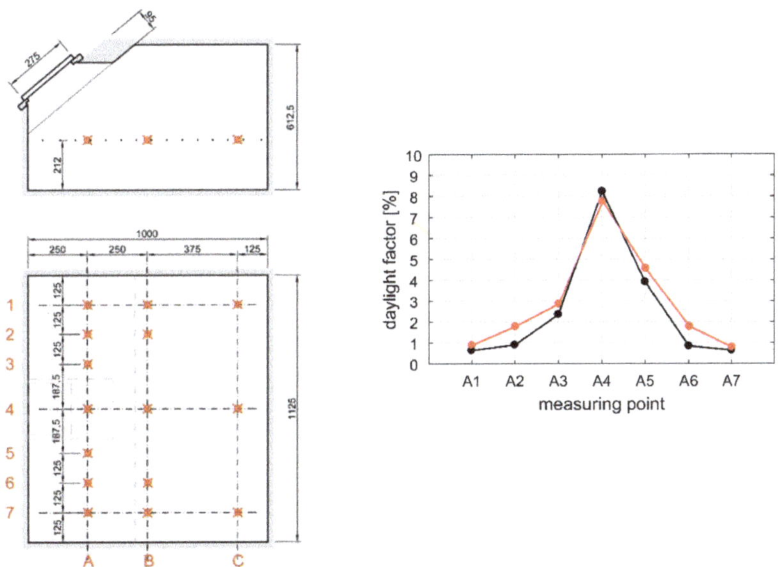

Figure 16. Laboratory-measured values of the daylight factor. Measured points (**left**) and values for points A1 to A7 (**right**). The black line represents the perpendicular side-lining and the red line represents the slanted side-lining with a slant angle of 45°. (Unit: mm).

5. Discussion

This paper illustrates the practical use of building physics tools for studying heat transfer and the daylight situation. The aim was to find a technical solution for roof windows suitable for passive buildings and other energy-optimized buildings in cooperation with an industrial partner. An iterative multilevel process using two-dimensional heat conduction calculations, from initial parametric studies to specific detailed calculations, was applied in collaboration with the development team. The usability of the results for practical window design was discussed at each step of development.

Additionally, a limited degree of verification was performed through measurements in a climatic chamber. Final (formal) verification was achieved through certification by an independent testing laboratory. For practical use, an interactive tool was developed which calculates the resulting thermal transmittance according to the basic selection of the frame and glazing and refers to the elaborated window details to ensure that the surface temperature requirements are met.

The simulated values of thermal transmittance of 0.7 to 0.5 $W/(m^2{\cdot}K)$ can be considered very good results. In any case, the further reduction of heat transmission would be accompanied by an increase in the relative importance of the thermal coupling between the window and roof construction.

Slanted lining has a very positive effect on the quality of daylighting. Measurements made using a physical model in the laboratory show similar tendencies to those observed in simulations, although the simulated results show slightly higher daylight values. It is possible to accept an increase in thermal coupling to some extent. A slant of 30° seems to be a good compromise in the majority of cases.

A higher level of daylight will probably also bring a greater independence from artificial lighting, and thus allow greater energy savings or (hypothetically) possibly the use of smaller roof windows.

In further research, the authors aim to focus on several related tasks:

(a) The risk of overheating. Overheating and/or cooling energy demand of attic rooms and, in this context, the possibility of advanced controlled shading of roof windows [30].

(b) 3D heat transfer models. Due to the fact that windows contain parts for which it is not possible to perform a 2D calculation, especially in corners, it might be useful to use the 3D calculation of a whole window including coupling, taking advantage of the symmetry. However, to achieve this, some simplifications would have to be made, for example, for the opening mechanism, handle, and eventually ventilation flaps.

(c) Values of the heat transfer between the local surface and surroundings. Aside from the complexity of the calculations, the results were further burdened by uncertainties regarding the boundary conditions. Very detailed studies of heat transfer near to the surfaces of roof windows are needed.

Author Contributions: J.T., as a head of the project, was responsible for the general methodology and writing the paper. V.C. performed the 2D parametric studies and heat transfer evaluation. J.N. performed the daylighting simulations. K.S. supported the development of the project by making thermal performance measurements. L.M. was responsible for daylight measurements.

Funding: This research was funded by the Ministry of Education, Youth and Sports, Czech Republic, grant number LO1605.

Acknowledgments: This work was supported by the Ministry of Education, Youth and Sports, Czech Republic, within National Sustainability Programme I (NPU I), project LO1605—University Centre for Energy Efficient Buildings—Sustainability Phase. Our special thanks go to Slavona Company for its inspiring cooperation and support. We also thank for the cooperation of the daylighting laboratory at Danube University in Krems, Austria.

Conflicts of Interest: The authors declare no conflict of interest.

Nomenclature

A	Area, m^2
D	Daylight factor, %
U	Thermal transmittance, $W/(m^2 \cdot K)$
f	Surface temperature factor, dimensionless
g	Solar factor, dimensionless
h	Surface heat transfer coefficient, $W/(m^2 \cdot K)$
l	Length, m
ψ	Linear thermal transmittance, $W/(m \cdot K)$
ρ	Light reflectance, dimensionless
τ	Light transmittance, dimensionless
Indices	
a	Air
c	Convection
g	Glazing
e	Exterior
f	Frame
i	Interior
inst	Installed
r	Radiation
s, S	Surface
R	Required
w	Window

Appendix A

Table A1. Examples of the distribution of transmittance heat loss for a pitched roof (total area 140 m^2) of a family house with six roof windows (6 × 1.0 m^2). Alternative **A** corresponds to a typical solution around the year 2000, alternative **B** corresponds to a passive house quality roof with traditional roof windows, and alternative **C** corresponds to a passive house quality with high-performance windows.

	Thermal Transmittance	Heat Transfer Coefficient		Increased Heat Transfer (%) (100% = No Windows)
		(W/K)	(%)	
A				
Roof	0.3 $W/(m^2 \cdot K)$	40.2	69	69	
Roof windows	1.8 $W/(m^2 \cdot K)$	10.8	19	} 31	
Window–roof thermal coupling	0.3 $W/(m \cdot K)$	7.2	12	
Total		58.2	100	139
B				
Roof	0.1 $W/(m^2 \cdot K)$	13.4	49	49	
Roof windows	1.5 $W/(m^2 \cdot K)$	9.0	33	} 51	
Window–roof thermal coupling	0.2 $W/(m \cdot K)$	4.8	18	
Total		27.2	100	194
C				
Roof	0.1 $W/(m^2 \cdot K)$	13.4	71	71	
Roof windows	0.6 $W/(m^2 \cdot K)$	3.6	19	} 29	
Window–roof thermal coupling	0.08 $W/(m \cdot K)$	1.9	10	
Total		18.9	100	135

Appendix B Results of Observations of Heat Transfer at an Internal Surface under Real Conditions—Preliminary Comparison of a Large Vertical Window and a Small Roof Window (Case Study)

Two windows in an occupied family house were monitored for heat transfer between the interior glazing and the surroundings. There were two principally different situations: (a) A roof window with standard double glazing and with dimensions of 0.86 m × 0.61 m fitted in an insulated sloping roof; and (b) a balcony door containing a glazed window with dimensions of 2.4 m × 0.8 m, with visible glazing dimensions of 2.15 m × 0.66 m. Under the roof window is a heating body (a hot water radiator) reaching up to half the width of the window. The room with a balcony door has a high-quality envelope at the passive house level [5]; the windows here are equipped with triple glazing. There is no heating body at the balcony door.

In both cases, the values of the heat flux density, the temperature at the surface of the glass, and the temperatures at a small distance from the surface were measured at selected locations (Figure A1).

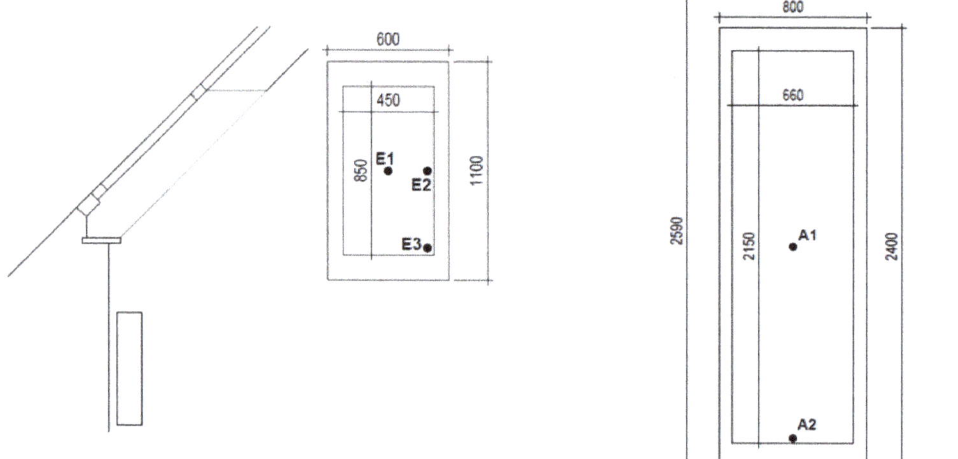

Figure A1. Schematic of a roof window (**left**) and a glazed balcony door (**right**). Points A1, A2, E1, E2, and E3 indicate where the surface heat transfer coefficient was preliminarily estimated. (Unit: mm).

Figure A2 is an infrared image of a balcony door, with a relatively uniform drop in surface temperatures from the top downwards, more pronounced at the bottom, over a height of approximately 200 mm. The air in this room has very little thermal stratification (up to 1.6 K at a height corresponding to the height of the window). Figure A3 is an infrared image of a roof window showing unevenness in the surface temperature distribution, including an apparent direct exposure of the heating body (surface temperature at heating body top by 50 °C).

The complicated situation in the immediate vicinity of the roof window is also reflected by a number of other air temperature measurements made at different distances from the window (Figure A4). This is demonstrated by the presence of inhomogeneous turbulent air streams with temperatures that are very different from the air temperature measured in the middle of the room. Air temperatures near the surface are approximately 2 to 5 K higher when the heating body is in normal operation than in the case when the upwards heat exchange is blocked. Figure A5 clearly shows the effect of the heating body on the surface and air temperature in the layer close to the window surface. In cases with an unblocked heating body function, the temperature at the window surface and its surroundings is higher than the air temperature in the center of the room.

Figure A2. Illustrative infrared picture of a glazed balcony door. **a**: a thin textile ribbon installed 15 mm in front of the glazing surface; **b**: a thin textile ribbon at a distance of 1.4 m from the glazing. Both ribbons were used for indirect informative measurement of the temperature of the surrounding air.

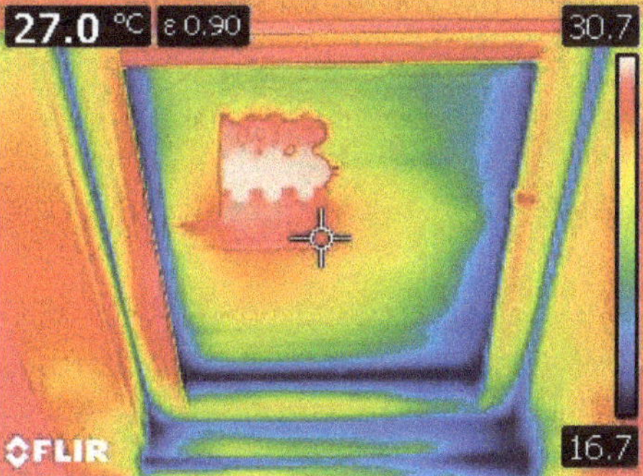

Figure A3. Illustrative infrared picture of a roof window. A multiple reflection of a heating body is clearly visible, which affects the window, in particular the radiative heat transfer component.

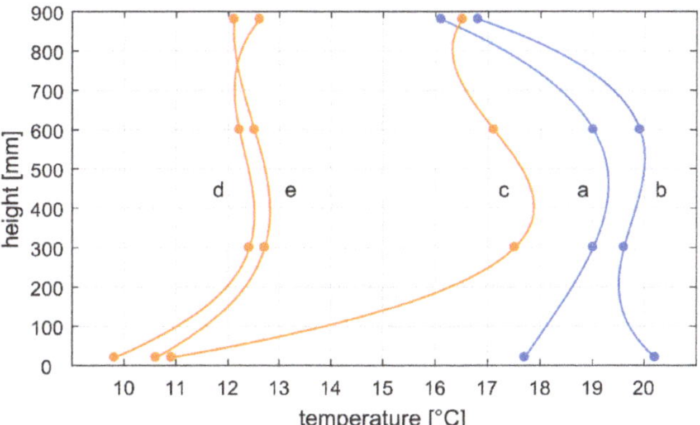

Figure A4. Illustrative single temperature measurement for the roof window—vertical temperature profile along the window. a: a distance of 15 mm from the glazing. b: a distance of 80 mm from the glazing. c, d, and e: glass temperature in the center of the window, at the left edge of the window, and at the right edge of the window, respectively. The indoor air temperature was 21.5 °C, the surface temperature of the heating body was 27 °C, and the exterior air temperature was 0 °C.

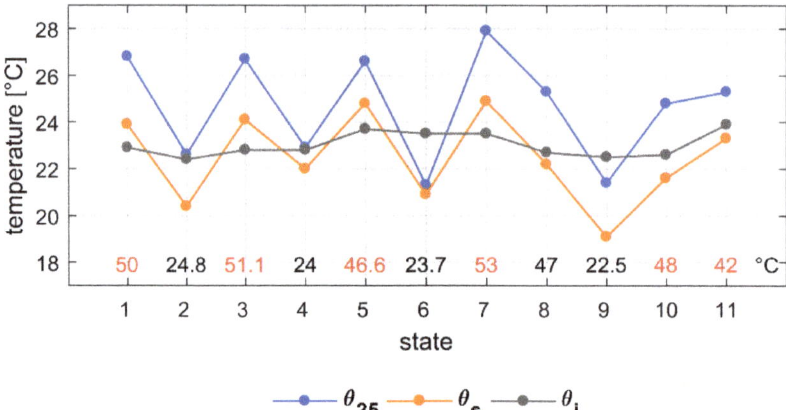

Figure A5. Measured temperatures in the center of the roof window for different states of the heating body. θ_{25} is the temperature at a distance of 25 mm from the surface, θ_{s} is the temperature at the surface, and θ_{i} is the air temperature at the center of the room. The temperature of the heating body is shown in the bottom row.

An evaluation of the surface heat transfer coefficient from the heat flow density measurements and the temperature differences between the air and the window surface is presented in Figure A6 and Table A2. The surface heat transfer coefficient of the balcony door generally corresponds well to the expected value (standard value h_{si} = 1/0.13 W/(m²·K)). For the roof window, the corresponding value of this method, proven by the measurement of the balcony door, seems to be quite different, being up to twice as high as the value in the center of the window.

The studied case is not representative of other possible configurations of roof windows (due to the presence of a heating body, different window size, different geometry of the side lining, etc.). Further measurements in laboratory conditions and other analyses should be used to study this effect more precisely.

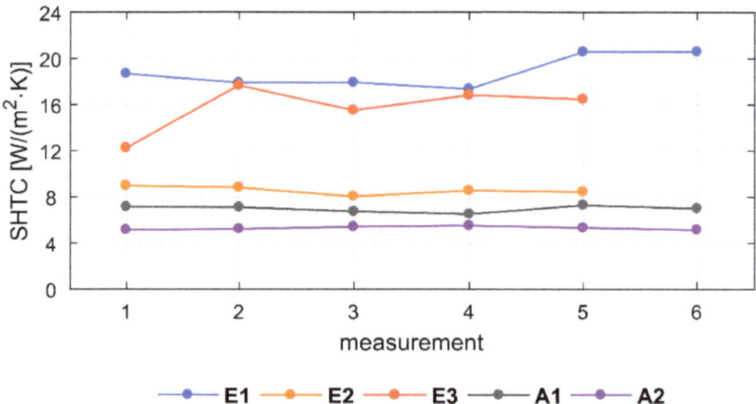

Figure A6. Surface heat transfer coefficient (SHTC; h_{si}) estimated from preliminary measurements at the roof window and the balcony door window (see Figure A1).

Table A2. Surface heat transfer coefficient—overview of measured values.

Window	Position		Mean Value h_{si} W/(m^2·K)
Glazed balcony door	Center	A1	7.0
	Bottom	A2	5.3
Roof window	Center	E1	18.8
	Center, right	E2	8.6
	Bottom, right edge	E3	15.7

References

1. Hens, H. *Building Physics. Heat, Air and Moisture. Fundamentals and Engineering Methods with Examples and Exercises*; Wilhelm Ernst & Sohn: Berlin, Germany, 2012.
2. European Committee for Standardization. *Set of European Standards in Responsibility of CEN TC89 Thermal Performance of Buildings and Building Components*; European Committee for Standardization (CEN): Brussels, Belgium, 2019.
3. Technological Agency of the Czech Republic. Project TH01021120 SONG—Střešní Okna Nové Generace (Skylights of New Generation). Technological Agency of the Czech Republic: Prague, Czech Republic. Available online: https://www.uceeb.cz/projekty/stresni-okna-nove-generace (accessed on 16 June 2019).
4. Schnieders, J.; Feist, W.; Rongen, L. Passive Houses for different climate zones. *Energy Build.* **2015**, *105*, 71–87. [CrossRef]
5. Kriterien für den Passivhaus-, EnerPHit- und PHI-Energiesparhaus-Standard. Passivhaus Institut. Available online: https://passiv.de/downloads/03_zertifizierungskriterien_gebaeude_de.pdf (accessed on 1 March 2019).
6. Grove-Smith, J.; Aydin, V.; Feist, W.; Schnieders, J.; Thomas, S. Standards and policies for very high energy efficiency in the urban building sector towards reaching the 1.5 degrees C target. *Curr. Opin. Environ. Sustain.* **2018**, *30*, 103–114. [CrossRef]
7. The New Passive House Classes. Available online: https://passipedia.org/certification/passive_house_categories (accessed on 1 March 2019).
8. D'Agostino, D.; Mazzarella, L. What is a nearly zero energy building? Overview, implementation and comparison of definitions. *J. Build. Eng.* **2019**, *21*, 200–212. [CrossRef]
9. Erhorn-Kluttig, H.; Erhorn, H.; Reiss, J. Plus energy—A new energy performance standard in Germany for both residential and non-residential buildings. *Adv. Build. Energy Res.* **2015**, *9*, 73–88. [CrossRef]
10. Informationen, Kriterien und Algorithmen für Zertifizierte Passivhaus Komponenten: Transparente Bauteile und Öffnungselemente in der Gebäudehülle. Available online: http://www.passiv.de/downloads/03_zertifizierungskriterien_transparente_bauteile.pdf (accessed on 1 March 2019).
11. Gustavsen, A.; Arasteh, D.; Jelle, B.P.; Curcija, C.; Kohler, C. Developing low conductance window frames: Capabilities and limitations of current window heat transfer design tools—State-of-the-art review. *J. Build. Phys.* **2008**, *32*, 131–153. [CrossRef]
12. Okna stresni, zateplovací sada thermos. Available online: https://www.oknastresni.cz/zatepleni-oken/zateplovaci-sada-thermos (accessed on 20 October 2018).
13. Component Database. Roof Windows. *Passivhaus Institut*. Available online: https://database.passivehouse.com/de/components/list/roof_window (accessed on 1 March 2019).
14. European Committee for Standardization. *EN 673 Glass in Buildings*; European Committee for Standardization (CEN): Brussels, Belgium, 2011.
15. Van de Bossche, N.; Buffe, L.; Janssens, A. Thermal optimization of window frames. In Proceedings of the 6th International Building Physics Conference, IBPC 2015, Torino, Italy, 14–17 June 2015; Volume 78.
16. Misiopecki, C.; Bouquin, M.; Gustavsen, A.; Jelle, B.P. Thermal modeling and investigation of the most energy-efficient window position. *Energy Build.* **2018**, *158*, 1079–1086. [CrossRef]
17. European Committee for Standardization. *EN ISO 13789 Thermal Performance of Buildings—Transmission and Ventilation Heat Transfer Coefficients—Calculation Method*; European Committee for Standardization (CEN): Brussels, Belgium, 2018.

18. UNMZ. *CSN 73 0540-2 2011 Thermal Protection of Buildings. Part 2—Requirements*; UNMZ: Prague, Czech Republic, 2011.

19. Tywoniak, J.; Staněk, K.; Calta, V. Roof windows for passive houses what can be improved? In Proceedings of the 7th International Building Physics Conference IBPC 2018, Syracuse, NY, USA, 23–26 September 2018; Syracuse University: Syracuse, NY, USA, 2018; pp. 1449–1454.

20. European Committee for Standardization. *EN 14351-1 Windows and EXTERNAL Pedestrian Doorsets without Resistance to Fire and/or Smoke Leakage*; European Committee for Standardization (CEN): Brussels, Belgium, 2018.

21. European Committee for Standardization. *EN ISO 12567-2. Thermal Performance of Windows and Doors—Determination of Thermal Transmittance by Hot Box Method—Part 2: Roof Windows and other Projecting Windows*; European Committee for Standardization (CEN): Brussels, Belgium, 2006.

22. COMSOL Multiphysics 4.4. Available online: https://www.comsol.com (accessed on 5 March 2016).

23. European Committee for Standardization. *EN ISO 10077-2 Thermal Performance of Windows, Doors and Shutters—Calculation of Thermal Transmittance—Part 2: Numerical Method for Frames*; European Committee for Standardization (CEN): Brussels, Belgium, 2017.

24. Svend, S.; Lautsen, J.B.; Kragh, J. Linear thermal transmittance of the assembly of the glazing and the frame in windows. In Proceedings of the 7th Symposium on Building Physics in the Nordic, Reykjavik, Iceland, 13–15 June 2005; IBRI: Reykjavik, Iceland, 2005; pp. 995–1002.

25. Compacfoam. Available online: http://www.compacfoam.at (accessed on 20 April 2019).

26. Software HTflux. Available online: https://www.htflux.com (accessed on 20 January 2019).

27. European Committee for Standardization. *EN ISO 10211 Thermal Bridges in Building Construction—Heat Flows and Surface Temperatures—Detailed Calculations*; European Committee for Standardization (CEN): Brussels, Belgium, 2018.

28. Velux Daylight Visualiser. Available online: https://www.velux.com/article/2016/daylight-visualizer (accessed on 20 January 2019).

29. European Committee for Standardization. *EN ISO 12667 Thermal Performance of Building Materials and Products. Determination of Thermal Resistance by Means of Guarded Hot Plate and Heat Flow Meter Methods. Products of High and Medium Thermal Resistance*; European Committee for Standardization (CEN): Brussels, Belgium, 2001.

30. Intellectual Property Office. *Exterior Shading System with Integrated Functions for Roof Windows*; Utility Model CZ 32088; Intellectual Property Office: Prague, Czech Republic, 2018.

Article

Energy Retrofitting of a Buildings' Envelope: Assessment of the Environmental, Economic and Energy (3E) Performance of a Cork-Based Thermal Insulating Rendering Mortar

José D. Silvestre [1,*], André M. P. Castelo [1], José J. B. C. Silva [2], Jorge M. C. L. de Brito [1] and Manuel D. Pinheiro [1]

[1] CERIS, Department of Civil Engineering, Architecture and Georresources, Instituto Superior Técnico, Universidade de Lisboa, Avenida Rovisco Pais 1, 1049-001 Lisboa, Portugal; andre.castelo@tecnico.ulisboa.pt (A.M.P.C.); jb@civil.ist.utl.pt (J.M.C.L.d.B.); manuel.pinheiro@tecnico.ulisboa.pt (M.D.P.)

[2] Architecture Department, University of Évora, Apartado 94, 7002-554 Évora, Portugal; jcs@uevora.pt

[*] Correspondence: jose.silvestre@tecnico.ulisboa.pt; Tel.: +351-21-841-9709

Received: 25 November 2019; Accepted: 22 December 2019; Published: 27 December 2019

Abstract: This paper presents an environmental, economic and energy (3E) assessment of an energy retrofitting of the external walls of a flat of an average building with the most current characteristics used in Portugal. For this intervention, a cork-based (as recycled lightweight aggregate) TIRM (Thermal Insulating Rendering Mortar) was considered. The declared unit was 1 m^2 of an external wall for a 50-year study period and the energy and economic costs and savings, as well as the environmental impacts, were analytically modelled and compared for two main alternatives: the reference wall without any intervention and the energetically rehabilitated solution with the application of TIRM. Walls with improved energy performance (with TIRM) show lower economic and environmental impacts: reductions from 6% to 32% in carbon emissions, non-renewable energy consumption and costs during the use stage, which depends on the thickness and relative place where TIRM layers are applied. A worse energy performance is shown by reference walls (without TIRM) during the use stage (corresponding to energy used for heating and cooling), while the improved walls present economic and environmental impacts due to the application of TIRM (including the production, transport and application into the building) that do not exist in the reference walls. The comparison between reference walls and energy-retrofitted ones revealed that reference wall become be more expensive when more demanding operational energy requirements are analysed over a 50-year period, even if renewable materials are more expensive.

Keywords: cork; energy retrofitting; life cycle assessment; life cycle costs; thermal insulating rendering mortar

---

1. Introduction

Thermal Insulating Rendering Mortars (TIRM) are an important solution that has been used to improve the energy efficiency of the envelope of refurbished and new buildings. It can be applied as an external or internal rendering and insulation of external walls (Figure 1). The main quality of TIRM is their low thermal conductivity, which is provided by the incorporation of lightweight aggregates instead of sand in the mix.

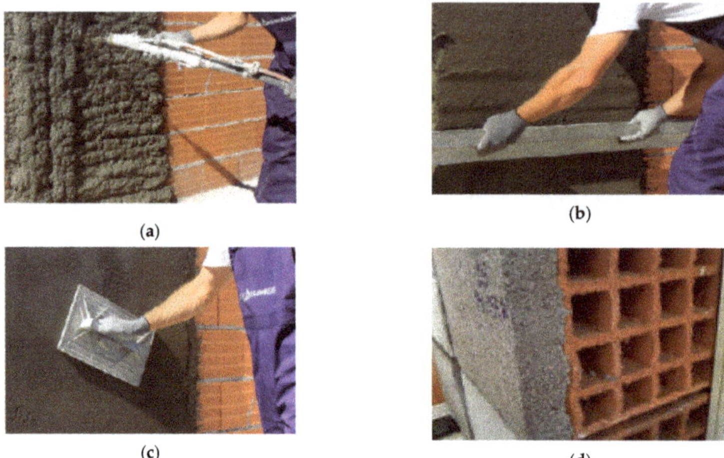

Figure 1. (**a**) Spraying application of thermal insulating rendering mortar (TIRM) with cork as lightweight aggregate; (**b**,**c**): Smoothing of each coat of mortar; (**d**) TIRM with cork as lightweight aggregate applied as external rendering of the external wall of a building [1,2].

This research study includes the assessment of the 3E (environmental, economic and energy) performance during the life cycle of a TIRM applied in the thermal renovation of the external walls of a flat of an average building with the most current characteristics used in Portugal. The characterisation of the 3E performance of this construction material when used in the energy renovation of buildings' envelope was based on data from companies, previous research works, reference literature, and software databases. This approach is of the outmost importance to provide reliable results to support energy retrofitting interventions. The energy benefits of these measures are well known, but it is important to discuss the environmental and economic balance between the expense of resources in the materials used and the savings during the building's use after the intervention.

The environmental life cycle assessment (LCA) is made from cradle-to-cradle (C2C), being focused on the consumption of non-renewable primary energy and carbon footprint of the TIRM and of its components. The economic assessment C2C is based on market prices (e.g., the cost of manufacture, transportation to site and installation, corresponding to the market acquisition cost) and on the savings that result from using this material in the retrofit of envelopes [3].

In the TIRM considered for this work, cork is used as a recycled lightweight aggregate. This TIRM is available in the Portuguese market only in recent years and is produced by two Portuguese and one Italian company. The use of TIRM with cork as a lightweight aggregate has, on average, a thermal conductivity of approximately 0.093 W/m °C, which corresponds to 20% of the thermal conductivity of common rendering mortars. For this study, an average value of thermal conductivity of 0.095 W/m °C was considered in order to take into account the various cork-based (as lightweight aggregate) TIRM found in the market of the Mediterranean region [1,2].

2. State-of-the-Art on the LCA of Building Envelope

The application of LCA to study alternatives for the buildings' envelope is gaining an increasing importance all around the world, since it could help to determine the solution that most improves the overall performance of the buildings' envelope [4,5]. The building' envelope is one of the parts that significantly influences the 3E performance of a building, and the external walls influence directly the 3E performance of the buildings' envelope due to their large weight within the envelope's initial whole-life cost, life cycle energy consumption, embodied energy and users' comfort.

The 3E impacts of each alternative of the external wall solution results directly from the characteristics of the materials selected (e.g., initial embodied energy, thermal properties, design choices and construction procedures). Thus, it is very important to make available a method that provides the comparison of alternatives and helps to determine the best one to implement in every design of new or refurbishment of a building envelope [6].

There are ongoing developments and studies with this aim all over the world. From the studies identified, the methods being applied to find the alternatives for the buildings' envelope with the best performance are:

- In Portugal, seven exterior wall alternatives with almost the same thermal performance, and seven different heating systems, were considered in the calculation of the LCA of a house. The study period was defined as 50 years, including the production stage and the heating energy and maintenance actions [7]. Two external claddings' alternatives (stone claddings and renders) were compared in terms of environmental LCA and service life prediction [8], in another study completed in the same country;
- In China, five façade solutions were considered for an office building, considering their environmental load (and corresponding environmental cost), including operational energy, and economic cost, in order to provide their general payback time and green payback [9,10];
- In Indonesia, a life cycle energy assessment study regarding middle- to high-class residential high-rise apartments in Jakarta identified the best alternatives among common wall types regarding the energy consumption from cradle to the use phase over a 40-year study period [5];
- In Lebanon, a multiscale life cycle energy analysis framework was used to determine the profile of energy use of new residential buildings by considering embodied, operational and user transportation energy requirements, over 50 years. It also identified the most effective ways to reduce energy use along the various life cycle stages, and at different scales, of the built environment [11];
- In Spain, an LCA study compared five constructive systems for the building envelope of a modular house: hollow brick; hollow brick with Phase Changing Materials (PCM); conventional brick; conventional brick with polyurethane insulation and PCM; conventional brick with polyurethane insulation [12]. The environmental impacts over an 80-year study period were considered for each alternative;
- In Italy, alternative envelope solutions, including type and width of masonry and insulating materials, were considered in the LCA study of a conventional house and an office building [13].

Other methodologies and case studies have already been developed worldwide to optimize the buildings performance. In this study, the Environmental, Economic and Energy from Cradle to Cradle (3E-C2C) methodology was used, developed in the University of Lisbon's Instituto Superior Técnico [14]. This method estimates the 3E performance of the alternatives for the buildings envelope during all life cycle stages in order to select the best one.

In the research for this study, it was found that most studies of the building's envelope only comprise part of the LCA, mainly Life Cycle Energy Assessment (LCEA) studies and there are very few studies that comprise the 3E LCA, including the economic assessment, and all its different stages [15–17]. Therefore, using this method along with this case study is an innovative approach, and an improvement compared to previous studies with analogous building assemblies and objectives.

3. Materials and Methods

The object of the study was an intermediate flat of a model building denominated "Hexa". The building has six residential floors and a ground floor for commerce [18] and it has the most current building and architectural characteristics used in Portugal [19]. The apartment located at the right of Figure 2 was selected as case study, without any adjacent building to the façade on the east. Évora was the location chosen for the Hexa building in this study, which is located in the south of Portugal, with extreme temperature in summer (up to 40 °C) and cold in winter (0 °C).

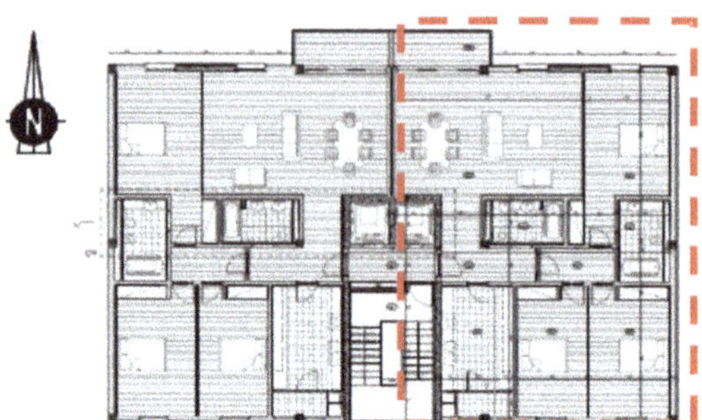

Figure 2. Residential model flat considered in this study (on the right).

The external walls used in this study were the south and north façades of the flat, using as declared unit/reference flow one square metre of external wall. Fifty years was set as the reference study period [18], since it is the service life of a building usually considered in structural design and was also the one considered in most of the studies identified in the state-of-the-art (see Section 2).

To use TIRM for the thermal retrofit of the façades of the "Hexa" building [19], two reference solutions were considered, without insulation: a single-leaf wall of hollow fired-clay bricks with 0.22 m of thickness (W1); a cavity wall with two leaves with 0.15 m and 0.11 m of thickness (W11), of the same material. Then, twelve enhanced solutions with TIRM were considered using the same thickness, but applied on the interior, exterior or on both sides of the wall, as shown in Table 1.

Table 1. Label, thickness and U-Value of the wall solutions analysed.

Wall Type	Insulation Material	Designation	Thickness (m)			U-Value [W/(m^2 °C)]
			Interior	Exterior	Total	
Single Leaf	–	W1	–	–	0.26	1.36
	TIRM	W3	0.10	–	0.34	0.57
	TIRM	W4	0.15	–	0.39	0.44
	TIRM	W6	0.10	0.10	0.42	0.36
	TIRM	W7	0.15	0.15	0.52	0.26
	TIRM	W8	–	0.04	0.88	0.28
	TIRM	W9	–	0.10	0.34	0.57
	TIRM	W10	–	0.15	0.39	0.44
Cavity Wall	–	W11	–	–	0.35	0.95
	TIRM	W13	0.10	–	0.43	0.48
	TIRM	W14	0.15	–	0.45	0.36
	TIRM	W16	0.10	0.10	0.51	0.32
	TIRM	W17	0.15	0.15	0.58	0.23
	TIRM	W18	–	0.04	0.62	0.34
	TIRM	W19	–	0.10	0.43	0.48
	TIRM	W20	–	0.15	0.45	0.36

The energy renovation of reference walls (W1 and W11) is important. Nevertheless, the cooling and heating needs of the flat in each year of the study period depend either on the reduced thermal performance (U-value) achieved after this intervention and on the surface (inner or outer) of the external wall in which the TIRM is applied. Actually, a lower U-value maximizes its effect on decreasing the needs of energy for heating and cooling when the insulation material is applied on the outer surface of the external wall. This occurs because, when compared with internal insulation, higher efficiency to avoid the heating losses during winter along linear thermal bridges on the external wall is achieved by external insulation.

Table 2 describes the replacement, repair and maintenance operations of each internal coating and external cladding along the life cycle (after the retrofit operation).

Table 2. Maintenance, repair and replacement operations of each external cladding and internal coating of external wall evaluated.

Cladding or Coating	Maintenance, Repair and Replacement Operations
ECS1—Adherent (0.02-m-thick render; water-based paint)	Total cleaning and repainting every 5 years and, after 25 years, repair of 35% of the area.
ECS2—Thermal insulating rendering mortar with cork and water-based paint	
ICS1—Adherent (0.02-m-thick render; water-based paint)	Total cleaning and repainting every 5 years and, after 10 years, repair of 5% of the area.
ICS2—Thermal insulating rendering mortar with cork and water-based paint	

3.1. 3E-C2C Method

Regarding the goal and scope, this research study applied an approach for the integrated 3E's (Environmental, Energy and Economy) assessment from cradle to cradle (3E-C2C) of the life cycle performance of construction materials or assemblies related with the thermal performance of buildings, which was already used in the use of External Thermal Insulation Composite Systems (ETICS) in the energy retrofit of buildings [20].

The 3E-C2C method assesses the 3E's impacts of each construction material or assembly for the whole life cycle (C2C) by analytically modelling and considering all the factors that can affect them (e.g., the operational performance of the assembly, and its service life and recycling potential, as shown in Table 3).

Table 3. Detailed life cycle stages of building materials classification [21].

LCA Boundaries			Life Cycle Stages/LCA Information Modules		Life Cycle Stage Designation and Description
Cradle to Cradle	Cradle to Grave	Cradle to Gate	Product Stage (A1–A3)	A1	Raw material extraction and processing, processing of secondary material input
				A2	Transportation to the manufacturer
				A3	Manufacturing
		Gate to Grave	Construction process stage (A4, A5)	A4	Transportation to the building site
				A5	Installation in the building
			Use stage - information modules related to the building fabric (B1–B5)	B1	Use or application of the installed product
				B2	Maintenance
				B3	Repair
				B4	Replacement
				B5	Refurbishment
			Use stage—information modules related to the operation of the building (B6, B7)	B6	Operational energy use
				B7	Operational water use
			End-of-life stage (C1–C4)	C1	D-construction, demolition
				C2	Transport to waste processing
				C3	Waste processing for reuse, recover and/or recycling (3R)
				C4	Disposal
			Benefits and loads beyond the system boundary (D)	D	Reuse, recovery and/or recycling (3R) potentials

This study considered "one square metre of external wall for 50 years from thermal retrofit (TIRM application)" as the declared unit, not considering a functional unit, and considering the use stage, reference service life and end-of-life stages of each alternative. This approach can compare external wall alternatives with different U-values because the corresponding LCA study considers the environmental and costs of their relative thermal performance over 50 years and of the production of the corresponding thermal insulation thickness.

3.1.1. Environmental Performance

The LCA standardised method [22,23], its four main steps (goal and scope definition, inventory analysis, impact assessment, and interpretation) and most of the principles from European standards [24,25], are complied with by the 3E-C2C method for the quantification of the environmental performance from cradle to cradle. Regarding the system boundaries of this study, at each life cycle stage the environmental performance is defined by:

- Product Stage (A1–A3): For each product or construction material, the inventory of the LCA data of the production resulted either from the studies completed in Portuguese plants [14] or from the application of Native LCA in the selection of coherent LCA data sets on TIRM to be used [26]. The composition considered for this product was based on a Portuguese producer [2]. The LCA of the production of each construction material (cradle-to-gate approach) was calculated with SimaPro and environmental impact results were achieved by using an environmental impact assessment method with a mid-point approach—CML 2001 baseline method;
- Construction process stage (A4–A5): The thermal retrofit includes the installation of the product in the building: removal, and transportation to waste processing and disposal, of the old render and paint; external and/or internal rendering of the external wall with TIRM, and application of the corresponding coating;
- Use stage—maintenance, repair and replacement (B2–B4): During the study period, the environmental impacts of the materials applied in replacement, repair and maintenance operations, and of the corresponding waste flows, were considered;
- Use stage—energy cost (B6): The energy performance is based on the estimation of the energy needs for heating and cooling during the buildings' operation. These needs are then divided by the total area of external wall to result in a value related to the declared unit considered. This value, and the corresponding environmental impacts, are based on the residential consumption for heating and cooling considering an updated Portuguese electricity mix [27];
- End-of-life stage (C): The transportation of the discarded product as part of the waste processing and transportation of waste (C2), the waste processing (C3) and the waste disposal, including physical pre-treatment and management of the disposal site (C4) are considered; since the environmental impacts of demolition (C1) are similar for all alternatives, they are not considered.

3.1.2. Economic Performance

The Whole-Life Cost (WLC) method [28] and most of the principles included in the European Standards [21] support the economic module of the 3E-C2C approach.

The NPV (Net Present Value), considering the study period, the heating and cooling needs and the operation costs of each substage, is the comparison unit between the alternatives. The formulas presented in Table 4 support the estimation of the NPV.

The economic performance is defined, for each life cycle stage, by:

- Product and construction process stages (A1–A5): For the installation of the TIRM in the building, the cost of the renovation described in the construction process as to be considered, except for the costs of workmanship to remove the old render and the paint and for the costs of installation of any scaffolding, on the external surface of the external wall, to complete this operation. These costs were collected from: a Portuguese producer of TIRM with cork as a lightweight aggregate [1]; previous research studies [29]; construction firms, market surveys and building materials suppliers [18]; reference national documents [30];

- Use stage—maintenance, repair and replacement (B2–B4): The cost of replacement, repair and maintenance operations completed in each year defines the economic cost of this stage in year "n" per square metre of external wall;
- Use stage—energy cost (B6): The energy used for heating and cooling [31], calculated by the method used in Portuguese codes [32], permits the calculation of the energy cost in year "n" per square metre of external wall;
- End-of-life stage (C): Only the costs for transportation and disposal of the materials or building assemblies and the expenses and/or revenues from recycle, reuse, and energy recovery are considered for the economic cost in year 50 per square metre of external wall [33,34].

Table 4. Equations used to determine the NPV of each solution.

Equation	Unit	List of Abbreviations
$NPV = \sum\limits_{n=0}^{50} \frac{C_n}{(1+d)^n}$	(€/m^2)	- C_n, cost in year n (€/m^2); - d, real discount rate applied (3%), without considering risk.
$C_n = Cev_n + Cec_n + Ceg_n$	(€/declared unit)	–
$Ceg_n = 0.1 \times T \times \left(\frac{N_{ic}}{\eta_i} + \frac{N_{vc}}{\eta_v}\right) \times \frac{A_{ap}}{A_{ew}}$	(€/year *m^2 of external wall)	- T, cost of 1 kWh of electricity in Portugal, for household consumers, without VAT or standing charges (€/kWh) (0.139 €/kWh, for an installation of more than 2.3 kVA); - N_{ic}, nominal annual heating needs per square metre of net floor area of the flat (kWh/m^2 *year); - η_i, nominal efficiency of the heating equipment (which is 1, considering the reference value [32]); - N_{vc}, nominal annual cooling needs per square metre of net floor area of the flat (kWh/m^2 *year); - η_v, nominal efficiency of the cooling equipment (which is 3, considering the reference value [32]); - A_{ap}, net floor area of the flat assessed (129.96 m^2); - A_{ew}, total area of the external wall assessed (40.27 m^2).

4. 3E Assessment of Energy Retrofitting Alternatives

To evaluate and compare the 3E performance of the energy renovation alternatives considered in this case study for two reference external walls without insulation (W1 and W11), the 3E-C2C method was used. The envelope renovation results from the application of a TIRM with cork as a lightweight aggregate and considering various relative locations and thicknesses.

The 3E-C2C method is in accordance with European and international standards and performance labels. C2C LCA studies were considered to assess the environmental performance, which is focused on the consumption of nonrenewable primary energy (PE-NRe) and on the carbon footprint (reflected on the environmental impact category "Global Warming Potential"—GWP). The economic performance from C2C considered market prices and also the "economic savings" (lower cooling and heating energy demand) resulting from the application of TIRM in the envelope renovation of buildings.

The main thermal insulating characteristics of the TIRM using cork as a lightweight aggregate, including the improvements in the thermal performance of the building envelope after its installation and the corresponding reduction of energy demand, were considered in the assessment of the energy performance. Ongoing changes in comfort demands and in building occupancy have led to a higher consumption of operational energy. Therefore, higher values (from 30 and 50%, while 10% of the energy needs is the default value) were considered to simulate future representative scenarios for apartments [35] or multifamiliar residential buildings [36].

4.1. Carbon and Energy Consumption Balances

Figure 3 shows the results from the C2C environmental and energy carbon footprint of the external wall alternatives, expressed by the GWP. This figure expresses an environmental impact between 74% and 93% at stages A1–A5 and between 1% and 2% at stages C2–C4 and D (of the total C2C GWP, without considering the energy for heating or cooling), both stages having an impact directly proportional to the thickness of TIRM applied in the renovation. The GWP in the B2–B4 stages does not differ much between the alternatives due to their common maintenance strategy, and represent 4% to 24% of their C2C GWP.

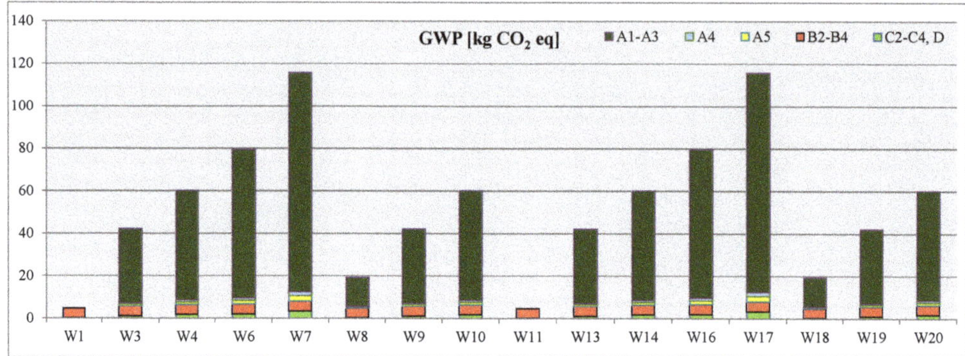

Figure 3. C2C Global Warming Potential (GWP, in kg CO$_2$ eq, with the energy needed for heating and cooling not being considered) of each alternative of external wall.

The C2C consumption of PE-NRe (Figure 4) has a trend similar to GWP. Between 52% and 86% of the impact comes from stages A1–A5 and between −5% and 3% corresponds to the end-of-life, being directly proportional to the thickness of TIRM applied. The entire "positive" C2C PE-NRe for the reference solution is within the B2–B4 stages, varying between 11% and 53% for the remaining solutions, with the energy needed for heating and cooling not being considered.

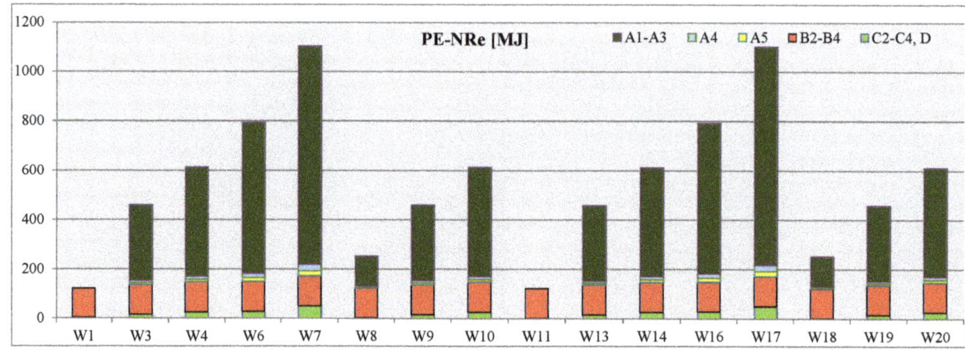

Figure 4. C2C consumption of nonrenewable primary energy (PE-NRe, in MJ, with the energy needed for heating and cooling not being considered) of each alternative of external wall.

4.2. Economic Costs and Benefits

Figure 5 shows the results of the economic balance regarding the use of TIRM as external and/or internal rendering of the walls defined for this study. It was found that the NPV of the C2C cost of each external wall alternative: is directly proportional to the thickness of TIRM applied, and varies from 19% to 60%, at stages A1–A3, A4 and A5; is between 1% and 3% at the end-of-life; represents around 54% for the reference wall, and varies from 23% to 46% for the remaining solutions, for the replacement, repair and maintenance operations (stages B2–B4), even if being similar in absolute terms for all alternatives. The remaining contribution of the NPV results from the cost of energy used for heating and cooling (B6 substage), about 44% for the reference solutions and between 14% and 35% of the NPV of the remaining ones.

Figure 5 allows concluding that there is no wall alternative where TIRM was applied that provides "economic savings" in comparison to the reference solutions. Nevertheless, these results were obtained for a consumption of energy during the B6 substage necessary to fulfil only 10% of the heating and cooling needs. However, these "economic savings" can become more significant, as shown in Figure 6, if higher values are used to simulate future realistic scenarios for dwellings or multifamiliar residential buildings.

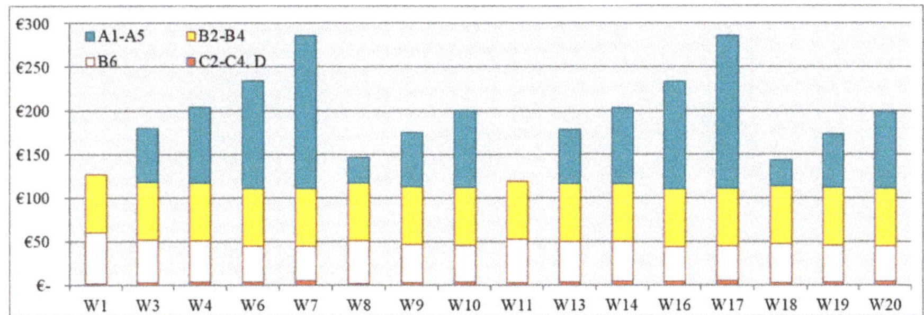

Figure 5. NPV of the economic (including A1–A5, B2–B4 and C2–C4 and D stages) and energy (for B6 substage) costs of each external wall alternative.

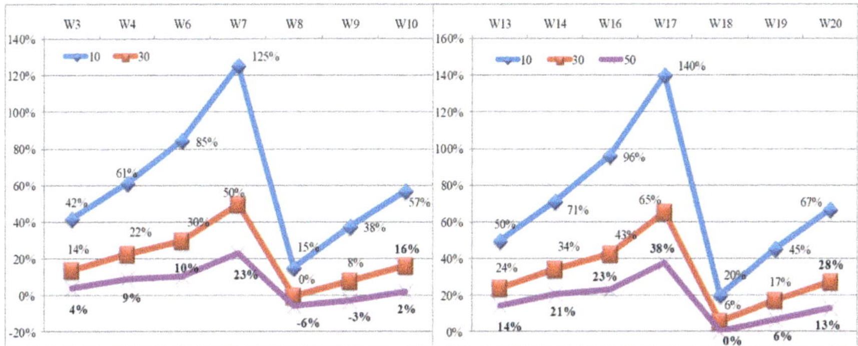

Figure 6. Difference between the NPV of the economic (including A1–A5, B2–B4 and C2–C4 and D stages) and energy (B6 substage) costs of each external wall alternative and the NPV of reference solution (W1, for single-leaf walls, at left; W11, for cavity walls, on the right), considering different consumption patterns for the use stage (providing 10%, 30% or 50% of the energy needs).

4.3. Energy Savings in Heating and Cooling

The "environmental impact savings" results show that the application of TIRM on the external surface of these walls could result on carbon savings from 6% to 32%. For the "environmental impact savings" of consumption of PE-NRe for the operational energy consumption during the study period, similar results were achieved.

In this study, these "environmental impact and economic savings" during the B6 substage are expressed per m^2 of the external wall of the flat chosen. However, the corresponding savings at a national scale provided by the implementation of these energy retrofit operations in Portugal or in other countries can be extrapolated but the thermal performance characteristics of most of existing buildings have to be considered and adapted to each specific case.

5. Discussion

For the single-leaf wall group, the C2C environmental LCA showed that W9 has the lowest C2C PE-NRe, considering the consumption of energy necessary to satisfy 10% of the heating and cooling needs, as shown in Table 5. If this value is increased to 30% or 50%, then W10 becomes the best alternative. The best economic alternative considering the consumption of energy needed to fulfil 10% of the heating and cooling needs is W1 (no renovation) but, when the energy consumption increases to 30% or 50%, the best alternative becomes W8 (Table 5 and Figure 6).

Within the cavity wall group, the C2C environmental LCA showed that W18 has the lowest C2C PE-NRe, considering the consumption of energy necessary to fulfil 10% of the heating and cooling needs, as shown in Table 6. If this value is increased to 30% or 50%, then W20 becomes the best alternative. The best economic alternative considering the consumption of energy needed to fulfil 10% or 30% of the heating and cooling needs is W11 (no renovation) but, when the energy consumption increases to more than 50% (53%), the best alternative becomes W18 (Table 6 and Figure 6).

Table 5. Single-leaf external wall solutions that offer the best performance, depending on the dimension analysed.

Approach	Life Cycle Stages Considered	Performance Aspects	Heating and Cooling Needs Fulfilled (%)	Best Performance	Difference to the Second and to the Reference Alternatives
LCA: GWP	C2C (A1–A3; A4; A5; B2–B4; C2–C4 and D), without energy use for heating and cooling	Environmental	–	W1 (no renovation)	310% (W2/W8)
LCA: PE-NRe					110% (W2/W8)
LCA: GWP	C2C (A1–A3; A4; A5; B2–B4; B6; C2–C4 and D)	Environmental	10	W8	1% (W9); 11% (W1)
			30	W10	1% (W9); 21% (W1)
			50		1% (W9); 24% (W1)
LCA: PE-NRe			10	W9	1% (W10); 17% (W1)
			30	W10	2% (W9); 24% (W1)
			50		3% (W7/W9); 26% (W1)
WLC	A1–A3; A4; A5; B2–B4; B6; C2–C4 and D	Economic	10	W1 (no renovation)	15% (W8)

Table 6. External cavity wall solution that offers the best performance, depending on the dimension analysed.

Approach	Life Cycle Stages Considered	Performance Aspects	Heating and Cooling Needs Fulfilled (%)	Best Performance	Difference to the Second and to the Reference Alternatives
LCA: GWP	C2C (A1–A3; A4; A5; B2–B4; C2–C4 and D), without energy use for heating and cooling	Environmental	–	W11 (no renovation)	310% (W12/W18)
LCA: PE-NRe					111% (W12)
LCA: GWP	C2C (A1–A3; A4; A5; B2–B4; B6; C2–C4 and D)	Environmental	10	W18	4% (W11)
			30	W20	0% (W19); 9% (W11)
			50		1% (W19); 12% (W1)
LCA: PE-NRe			10	W18	0% (W19); 6% (W11)
			30	W20	1% (W19); 12% (W11)
			50		2% (W19); 14% (W1)
WLC	A1–A3; A4; A5; B2–B4; B6; C2–C4 and D	Economic	10	W11 (no renovation)	19% (W18)
			30		5% (W18)
			53	W18	0% (W11)

6. Conclusions

This paper presents the life cycle performance (3E) assessment of the application of a TIRM (Thermal Insulating Rendering Mortar) with cork in the energy rehabilitation of a model building. The three vectors of sustainability were considered in order to provide a true Life Cycle Sustainability Assessment: environmental, economic and social (represented here by the thermal comfort expressed by the energy needs, even if the energy costs and environmental loads are being considered in the other two pillars).

A worse energy performance on the operational stage (energy used for heating and cooling) of reference walls (without TIRM) was found, while walls with improved energy performance (with TIRM) show lower environmental and economic impacts: 6% to 32% reduction in carbon emissions, costs and in nonrenewable energy consumption during the use stage, depending on the relative position of the layers and on the thickness of TIRM applied. However, environmental and economic impacts due to the removal of the ancient coating and application of TIRM during production, transport and execution on site, that do not exist in the reference walls, have to be considered for improved walls.

At the end, it was found that although renewable materials are more costly, the reference wall tends to be more expensive when a 50-year period is considered with an higher energy demand for heating and cooling, namely for single walls without insulation. If the reference wall is a cavity wall without insulation, only an external economic incentive can encourage an energy retrofitting intervention.

Author Contributions: Data curation, J.D.S.; Methodology, J.D.S.; Validation, A.M.P.C. and J.J.B.C.S.; Writing-original draft, J.D.S.; Writing-review & editing, A.M.P.C., J.J.B.C.S., J.M.C.L.d.B. and M.D.P. All authors have read and agreed to the published version of the manuscript.

Funding: This research work was completed in the scope of the project MARIE—Mediterranean Building Rethinking for Energy Efficiency Improvement (2011–2014), cofinanced by the European Regional Development Fund (ERDF) and the MED Programme.

Acknowledgments: The authors are also grateful for the support from FCT (Foundation for Science and Technology) and from CERIS Research Institute from Instituto Superior Técnico, Universidade de Lisboa. Special thanks are due to the Portuguese manufacturers of construction materials, for providing the necessary data to complete this research work.

Conflicts of Interest: The authors declare no conflict of interest.

References

1. Secil. Secil Mortars. Available online: http://www.secilargamassas.pt/pt (accessed on 22 October 2013). (In Portuguese).
2. THW. Thermowall. Available online: http://thermowall.pt/ (accessed on 22 October 2013). (In Portuguese).
3. Garrido, R.; Silvestre, J.; Flores-Colen, I. Economic and energy life cycle assessment of aerogel-based thermal renders. *J. Clean. Prod.* **2017**, *151*, 537–545. [CrossRef]
4. Utama, A.; Gheewala, S. Life cycle energy of single landed houses in Indonesia. *Energy Build.* **2008**, *40*, 1911–1916. [CrossRef]
5. Utama, A.; Gheewala, S. Indonesian residential high rise buildings: A life cycle energy assessment. *Energy Build.* **2009**, *41*, 1263–1268. [CrossRef]
6. Ferreira, J.; Pinheiro, M.D.; Brito, J. Refurbishment decision support tools review—Energy and life cycle as key aspects to sustainable refurbishment projects. *Energy Policy* **2013**, *62*, 1453–1460. [CrossRef]
7. Monteiro, H. Life cycle assessment of a Portuguese house with different exterior wall solutions and alternative heating systems. In *Master Thesis in Energy for Sustainability*; University of Coimbra: Coimbra, Portugal, 2010.
8. Silvestre, J.D.; Silva, A.; Brito, J. Uncertainty modelling of service life and environmental performance to reduce risk in building design decisions. *J. Civ. Eng. Manag.* **2015**, *21*, 308–322. [CrossRef]
9. Gu, L.J.; Gu, D.J.; Lin, B.R.; Huang, M.X.; Gai, J.Z.; Zhu, Y.X. Life cycle green cost assessment method for green building design. In *Building Simulation 2007*; Tsinghua Press: Beijing, China, 2007; pp. 1962–1967.
10. Gu, L.J.; Lin, B.R.; Zhu, Y.X.; Gu, D.J.; Huang, M.X.; Gai, J.Z.L. Integrated assessment method for building life cycle environmental and economic performance. In *Building Simulation 1*; Tsinghua Press: Beijing, China, 2008; Volume 2, pp. 169–177.

11. Stephan, A.; Stephan, L. Reducing the total life cycle energy demand of recent residential buildings in Lebanon. *Energy* **2014**, *74*, 618–637. [CrossRef]
12. Rincón, L.; Castell, A.; Pérez, G.; Solé, C.; Boer, D.; Cabeza, L.F. Evaluation of the environmental impact of experimental buildings with different constructive systems using material flow analysis and life cycle assessment. *Appl. Energy* **2013**, *109*, 544–552. [CrossRef]
13. Asdrubali, F.; Baldassarri, C.; Fthenakis, V. Life cycle analysis in the construction sector: Guiding the optimization of conventional Italian buildings. *Energy Build.* **2013**, *64*, 73–89. [CrossRef]
14. Silvestre, J.D.; Brito, J.; Pinheiro, M.D. From the new European Standards to an environmental, energy and economic assessment of building assemblies from cradle-to-cradle (3E-C2C). *Energy Build.* **2013**, *64*, 199–208. [CrossRef]
15. Chau, C.; Leung, T.; Ng, W. A review on life cycle assessment, life cycle energy assessment and life cycle carbon emissions assessment on buildings. *Appl. Energy* **2015**, *143*, 395–413. [CrossRef]
16. Zimek, M.; Schober, A.; Mair, C.; Baumgartner, R.; Stern, T.; Füllsack, M. The third wave of LCA as the "decade of consolidation". *Sustainability* **2019**, *11*, 3283. [CrossRef]
17. Fauzi, R.; Lavoie, P.; Sorelli, L.; Heidari, M.; Amor, B. Exploring the current challenges and opportunities of life cycle sustainability assessment. *Sustainability* **2019**, *11*, 636. [CrossRef]
18. Real, S.F. Contribution of the life cycle cost analysis to design sustainability in construction (in Portuguese). In *Masters Dissertation in Civil Engineering, Instituto Superior Técnico*; Technical University of Lisbon: Lisbon, Portugal, 2010; p. 153.
19. Ferreira, J.; Pinheiro, M.D. In search of a better energy performance in the Portuguese buildings—The case of the Portuguese regulation. *Energy Policy* **2011**, *39*, 7666–7683. [CrossRef]
20. Silvestre, J.D.; Castelo, A.; Correia da Silva, J.; de Brito, J.; Pinheiro, M.D. Retrofitting a building's envelope: Sustainability performance of ETICS with ICB or EPS. *J. Appl. Sci.* **2019**, *9*, 1285. [CrossRef]
21. CEN. *Sustainability of Construction Works—Environmental Product Declarations—Core Rules for the Product Category of Construction Products*; Comité Européen de Normalisation: Brussels, Belgium, 2012.
22. ISO. *Environmental Management—Life Cycle Assessment—Principles and Framework*; International Organization for Standardization: Genève, Switzerland, 2006.
23. ISO. *Environmental Management—Life Cycle Assessment—Requirements and Guidelines*; International Organization for Standardization: Genève, Switzerland, 2006.
24. CEN. *Sustainability of Construction Works—Assessment of Buildings—Part 2: Framework for the Assessment of Environmental Performance*; Comité Européen de Normalisation: Brussels, Belgium, 2011.
25. CEN. *Sustainability of Construction Works—Assessment of Environmental Performance of Buildings—Calculation Method*; Comité Européen de Normalisation: Brussels, Belgium, 2011.
26. Silvestre, J.D.; Lasvaux, S.; Hodkova, J.; de Brito, J.; Pinheiro, M.D. NativeLCA—A systematic approach for the selection of environmental datasets as generic data: Application to construction products in a national context. *Int. J. Life Cycle Assess.* **2015**, *20*, 731–750. [CrossRef]
27. ERSE. Electric Energy Labelling. ERSE—The Energy Services Regulatory Authority. Available online: http://www.erse.pt/pt/desempenhoambiental/rotulagemenergetica/comparacaoentrecomercializadores/Paginas/default.aspx (accessed on 23 July 2012). (In Portuguese).
28. ISO. *Buildings and Construction Assets—Service Life Planning—Part 5: Life-Cycle Costing*; International Organization for Standardization: Geneva, Switzerland, 2008.
29. Silvestre, J.D.; Brito, J.; Pinheiro, M.D. Life cycle impact "cradle to cradle" of building assemblies. In *Engineering Sustainability*; Institution of Civil Engineers: London, UK, 2013; Volume 167, pp. 53–63.
30. Manso, A.; Fonseca, M.; Espada, L. *Information about Costs—Productivity Sheets*; Laboratório Nacional de Engenharia Civil: Lisbon, Portugal, 2010. (In Portuguese)
31. Mateus, R.; Bragança, L. Sustainability assessment and rating of buildings: Developing the methodology SBToolPT—H. *Build. Environ.* **2011**, *46*, 2166–2175. [CrossRef]
32. RCCTE. Regulation of the Characteristics of Thermal Behaviour of Buildings. In *Law Decree No. 80/2006*; D.R.I-A Series 67-2468-2513; RCCTE: Lisboa, Portugal, 2006. (In Portuguese)
33. Coelho, A.; Brito, J. Economic analysis of conventional versus selective demolition—A case study. *Resour. Conserv. Recycl.* **2011**, *55*, 382–392. [CrossRef]

34. Silvestre, J.D.; Brito, J.; Pinheiro, M.D. Environmental impacts and benefits of the end-of-life of building materials—Calculation rules, results and contribution to a "cradle to cradle" life cycle. *J. Clean. Prod.* **2014**, *66*, 37–45. [CrossRef]

35. Monteiro, H.; Freire, F. Life-Cycle assessment of a house with alternative exterior walls: Comparison of three assessment methods. *Energy Build.* **2012**, *47*, 572–583. [CrossRef]

36. Pinto, A. Application of Life-Cycle Assessment in the Energetic and Environmental Analysis of Buildings. Ph.D. Thesis, Mechanical Engineering, Universidade Técnica de Lisboa, Lisbon, Portugal, 2008; p. 329. (In Portuguese).

Article

The Application of Courtyard and Settlement Layouts of the Traditional Diyarbakır Houses to Contemporary Houses: A Case Study on the Analysis of Energy Performance

İdil Ayçam [1,*], Sevilay Akalp [2] and Leyla Senem Görgülü [1]

[1] Department of Architecture, Graduate School of Natural and Applied Sciences, Gazi University, Ankara 06570, Turkey; leylasenem.gorgulu@gazi.edu.tr
[2] Department of Architecture, Harran University, Şanlıurfa 63300, Turkey; sevilayakalp@gmail.com
* Correspondence: iaycam@gazi.edu.tr; Tel.: +90-312-582-3602

Received: 31 December 2019; Accepted: 23 January 2020; Published: 27 January 2020

Abstract: Conventional energy use has brought environmental problems such as global warming and accelerated efforts to reduce energy consumption in many areas, particularly in the housing sector. For this purpose, bioclimatic design principles and vernacular architecture parameters have started to be examined in residential buildings nowadays. Thus, the demand for less energy-consuming houses has started to increase. In this study, we aimed to specify the significance of traditional architectural parameters for houses in the hot-dry climatic region of Diyarbakır, Turkey. Within the scope of the study, a case was based on the urban fabric of the traditional houses in Historical Diyarbakir Suriçi-Old Town settlement and the Şilbe Mass Housing Area was discussed. The courtyard types, settlement patterns, and street texture of traditional Diyarbakır houses were modeled by using DesignBuilder energy simulation program for the case study. Annual heating, cooling, and total energy loads were calculated, and their thermal performances were compared. The aim is to create a less energy-consuming and sustainable environment with the adaptation of traditional building form-street texture to today's housing sector. Development of a settlement model, which is based on traditional houses' bioclimatic design for hot-dry region, was intended to be applied in the modern housing sector of Turkey. Moreover, adapting local forms, urban texture, and settlement patterns to today has significant potential for sustainable architecture and energy-efficient buildings. According to this study, the optimum form and layout of traditional houses, which are one of the climate balanced building designs, provide annual energy savings if integrated and designed in today's building construction. As a result of this study, if the passive design alternatives such as building shape, layout, and orientation were developed in the first stage of the design, energy efficient building design would be possible. The study is important for the continuation of traditional sustainable design.

Keywords: traditional Diyarbakır houses; courtyard; settlement; Designbuilder simulation; energy performance

---

1. Introduction

The rapid growth of the world's population has led to a significant acceleration in energy demand and consumption, leading to serious environmental problems such as global warming and climate change [1,2]. Industrialization, the instability between humanity and nature, and globalization are important problems in the building sector [3]. In other words, many sectors such as construction, transportation, infrastructure, industry, agriculture have a significant impact on energy consumption and carbon emissions [4]. Among these sectors, the construction sector that has a large energy consumption network is responsible for more than 40% of global energy use and one-third of global greenhouse gas emissions [5]. It is stated that half of the energy consumed in the buildings is used in heating-cooling and air-conditioning ventilation (HVAC) systems to provide climatic comfort [6]. This situation led to different searches in the construction sector to save energy costs. As a result, passive design strategies have been developed to provide climatic comfort in heating, cooling, and ventilation [7]. As a result of these strategies, it is aimed to provide the necessary comfort conditions of the building with minimum energy consumption by designing the building envelope and its thermophysical properties, orientation, form, material, space organization, and many other parameters in an integrated way.

Traditional building settlements in Turkey and structures that make up these settlements have originated in different ways. In other words, the traditional texture was set up using cognitive rules and unwritten organic street pattern [8]. The different space organizations in these building settlements were formed according to climatic conditions. One of these units, courtyard structure form, is a unit that helps to reduce annual energy consumption with passive cooling technique [9]. The courtyard structure form is based on thermal performance, shading, and natural ventilation. Many research studies have been conducted on the climate performance of the courtyard form in order to address the thermal, shading, daylight, and airflow characteristics of low buildings in different climates [10]. The courtyard building form, which provides control of climatic elements such as sun and wind, causes a temperature difference between the inner and outer surface of the building shell. As a result, it was found that while the temperature of the semi-enclosed courtyard form was higher than the outdoor temperature both in winter and at night, it had a lower value compared to the outdoor temperature due to the shadow effect it created during the summer season [11]. Therefore, in the hot-dry climate zone where the hottest season lasts longer than the coldest one, the courtyard structure form appears as a space organization that provides thermal comforts by using the passive air conditioning elements together. When the traditional street texture in the hot arid climatic zones is taken into consideration, the density of the settlement is often encountered. Within the scope of the study, the street texture of the historical city walls of Diyarbakir and the traditional courtyard houses that make up this texture are discussed. In terms of Fathy, settlement texture has two features, respectively, wide courtyards and tight winding streets [12]. The aim of this study is to create a sustainable and less energy-consuming environment by ensuring the integration of traditional building forms and settlement orders, especially in the sector of mass housing into today's housing sector. As the study area, the historical Suriçi-Old Town texture and the closest Şilbe mass housing settlement unit to this region were discussed. The reason for choosing a mass housing settlement unit in addition to the historical urban texture in the scope of the study is that the housing section Housing Development Administration (TOKI) produces numerous houses and when the basalt, which is suitable for the traditional urban fabric, is used in modelling, the wall section is thick and not quite applicable. In the first step of the study, which was carried out in four steps, literature review and field study on traditional Diyarbakır houses and street texture were conducted. In the second step of the study, assumptions were made with reference to the building envelope properties of the houses in the mass housing units closest to the historical street texture by means of DesignBuilder energy simulation program.

As a result of the literature search, heating-cooling load values of the four most commonly used courtyard forms in the traditional housing texture were calculated and optimum courtyard form was determined. After determining the optimum courtyard building form, the heating and cooling loads due to the shadow effect of the structures in the corner and middle parcels were calculated by taking the street widths of the traditional Suriçi-Old Town texture as a reference. In the conclusion part of the study, by evaluating and comparing the results of the analysis, appropriate settlement alternatives have been proposed for low-rise residential settlements in Diyarbakır Turkey.

2. Materials and Method Analysis

2.1. Characteristics of the Traditional Settlement of Diyarbakır

The province of Diyarbakir that is located in the southeast part of Turkey (Figure 1) belongs to the hot-dry region in the classification of climates, and in this region the summer seasons last longer than winter seasons as can be seen via Figure 2 [13,14]. In addition, July is the warmest and January is the coldest month. The average air temperature is 27.5 °C and the average temperature measurements per month are tabulated in Figure 2 [15].

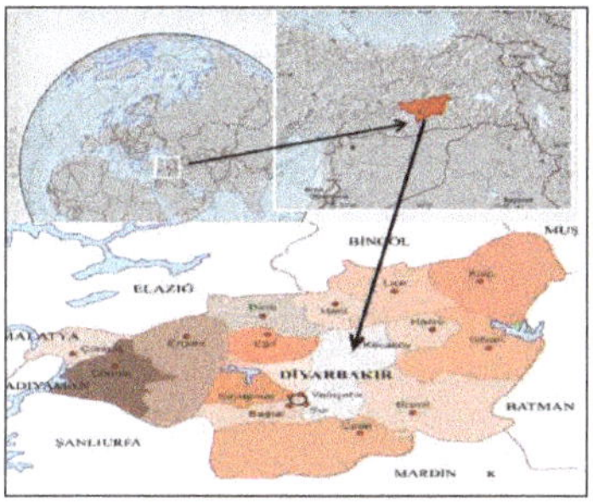

Figure 1. Location of Diyarbakır [15].

	January	February	March	April	May	June	July	August	September	October	November	December
Avg. Temperature (°C)	2.2	4.1	8.3	13.7	18.6	24.7	29.7	29.4	24.4	17.5	10.3	4.7
Min. Temperature (°C)	-2.2	-0.8	2.5	7.3	11.3	16.3	21.3	20.6	15.7	9.8	4.3	0.1
Max. Temperature (°C)	6.6	9	14.2	20.1	26	33.2	38.2	38.2	33.1	25.2	16.4	9.3
Avg. Temperature (°F)	36.0	39.4	46.9	56.7	65.5	76.5	85.5	84.9	75.9	63.5	50.5	40.5
Min. Temperature (°F)	28.0	30.6	36.5	45.1	52.3	61.3	70.3	69.1	60.3	49.6	39.7	32.2
Max. Temperature (°F)	43.9	48.2	57.6	68.2	78.8	91.8	100.8	100.8	91.6	77.4	61.5	48.7
Precipitation / Rainfall (mm)	79	71	76	74	46	8	1	0	3	35	59	78

Figure 2. The average temperature measurements per month for Diyarbakır [14].

The modelling part of the study is composed of the housing units which are the reference units on the scale of the building envelope and are located in the closest distance to the historical city texture. The Şilbe housing estate consists of three stages with different orientations and square meters. The relation with the historical city wall is given in Figure 3.

In this study, we aimed to analyze the effects of climate-based building design features of the historical (Suriçi) residential unit on the energy performance of the building where the traditional residential texture of Diyarbakır is located. In this region, we aimed to give direction regarding climate-based design to find a suitable building form and settlement alternatives for the houses built today. Due to the lack of studies throughout the world and Turkey, this study was designed [16].

Figure 3. Satellite view of historical Old Town texture—Şilbe collective housing settlement unit [17,18].

Throughout history, the city of Diyarbakır, which has been constantly exposed to war and invasion, has found a solution to build the city with city walls all around. This led to the formation of the urban texture in a limited area. In this urban texture, which is divided as Suriçi-Old Town and new town, urban settlements took place in the inside of the walls until the 19th century; with the rise in the population, to provide the need of housing, new buildings led to adjacent narrow alleys by being built next to already existing buildings. Another important factor in the formation of the historical old town texture walls is the climate. Located in the hot-dry climate zone, this settlement created an organic street texture by being transformed into a courtyard space organization that allows spatial and adjacent articulation to create shaded surfaces, protected from solar radiation [19]. When the traditional houses in the "Sur" city walls are examined, although the parcels in this organic street texture are formally separated from each other, the building masses in the parcel intersect vertically or at an angle close to the vertical. This way, even in the most formally damaged parcel, the building masses intersect almost vertically. Moreover, the parcel areas of traditional Diyarbakır houses vary between 85 and 1000 m^2. The areas labeled as the small parcels are 85–100 m^2 and the area of big parcels vary between 700 and 1000 m^2 [20]. In addition, the size and forms of the parcel are the major factors in the formation of the existing urban texture and an organic form.

The narrow streets of the traditional street texture have made the structure physically cut off some negative environmental factors. In other words, the high walls created a shadow effect in the courtyards. In addition, living in the courtyard brought by the warm-dry climate region as a user requirement has caused some sound reflections. However, the basalt stone, which is used as a traditional building material, absorbs the sounds due to its porous structure [28]. In the adjoining narrow street texture, the spaces are only the courtyard and a small number of gaps [29]. The courtyard has a significant influence on the formal formation of the traditional urban texture. It is seen that there are different courtyard types in the Middle East Region which forms the basis of today's courtyard type space organization. In this region, where hot and dry climatic conditions prevail, it is seen that the places are located on the north-south axis as the zoning makes it difficult to control the east and west sun [30]. For this reason, it is seen that the openings are at minimum dimensions in the spaces facing the north direction and the windows placed in the north-south directions benefit from cross-ventilation [23]. The traditional residential settlements of the province of Diyarbakir also have the characteristics of the traditional residential settlements in the hot dry climate. It is seen that the design principles regarding solar control, passive cooling, and increasing ventilation are benefitted from in these settlements. This situation has had a direct impact on traditional housing orientation and space organization. In other words, the courtyard pattern thermal performance links to two integrally working strategies: protecting buildings from solar radiation and natural ventilation [31,32].

In the traditional Diyarbakır houses, space was organized with the courtyard building unit being the center. In other words, with an inward-looking design, the courtyard is the focal point of the whole house and the other sections are positioned so that they are shaped according to the courtyard. The climate factor comes to the forefront in the orientation of the building masses around the courtyard.

In hot-dry climatic zones, the fact that the hottest period lasts longer than the coldest period has led to the emergence of a design understanding where measures to reduce solar radiation and to minimize the drying effect of the wind are at the forefront. The most prominent part of traditional Diyarbakır houses is summer spaces in the south wing of the courtyard [20]. Thus, the spaces in the south wing are positioned in the north direction and designed to receive minimal solar radiation during the day. In addition, these units have further importance in terms of the frequency of use. In the building mass directed to the northern wing of the courtyard, the spaces were designed to face south. In houses that do not have a northern wing, the building masses positioned to the east are considered as winter wings. (Table 1). The examples regarding the seasonal orientation of the traditional Diyarbakır houses are depicted through Table 2. In other words, the buildings in this wing are closed or have few transparent surfaces in order not to expose the spaces directed to the south to sunlight too much [34]. Briefly, the spaces in the south of the courtyard and the openings facing the north (courtyard) are planned as summer wings, and the spaces in the north of the courtyard and facing the openings in the south (courtyard) are planned as winter wings which is illustrated in Table 3 [35–40].

Table 1. Examples of space, street pattern in the historical settlement [21–27].

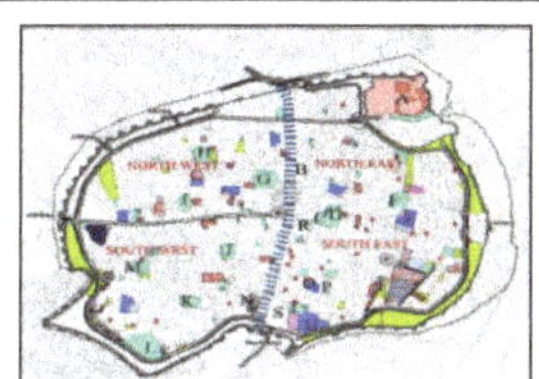 **Location of Suriçi Historical Settlement**		 **Top View of Settlement**
Eyvan	Three sides closed and single arch opening the courtyard.	Example of Eyvan
Kabalatı	Vaulted passageways Kabaltı is enclosed by high walls. Basalt stone is used as building material in Kabaltı.	Examples of Kabaltı
Street Width	Street widths vary between 1.90 and 2.50 m, but it can be seen to increase up to 3.00–4.00 m widths.	Narrow streets in the traditional Suriçi pattern
Rooms	The height of the summer rooms is 4–5.5 cm and the aspect ratio is 4 × 7–3.5 × 6 m; the height of the winter rooms varies between 3 and 3.5 m width and 3 × 4 and 2.5 × 3 m.	Example of traditional room

Table 2. Parts of a typical traditional house in the Diyarbakır old town settlement, Cahit Sıtkı Tarancı house [33].

Part	Image
Site Plan	
Summer	
Winter	
Seasonal Part	

Table 3. The positioning of the seasonal spaces [41].

	Part	Orientation
	Summer Room	
	Winter Room	
	Spring Room	

In traditional Diyarbakir houses, the courtyard and the Eyvan have a great influence on the orientation of the other units. The plan typology of traditional Diyarbakır houses is classified as U, I, and L type according to the wings arranged around the courtyard in the literature [42]. In addition, according to the openings left by the wings around the courtyard, it is possible to rank them as plan types with outer, inner, and middle courtyards. In the L-type plan, the courtyard is surrounded by two adjacent wings. The wing located in the south and whose opening faces north is defined as a summer house. In order to provide privacy in this plan type, exposed areas were separated from the street by high walls. In the U-type plan, all three sides of the courtyard are surrounded by spaces. The biggest advantage of this type of plan is that the wings are positioned as a summer house and winter house in optimum directions, which correspond to the climatic data of Diyarbakır. The U-type plan is the type of residence preferred by families with a high level of economic welfare. In the type I plan, also known as the inner courtyard plan type, the two sides of the courtyard are mutually arranged. The remaining sections were covered by walls and entrance doors. In plan type with a central courtyard, all four sides of the courtyard are surrounded by spaces. The courtyard types mentioned above is presented via Table 4. The major advantage of this type of plan is that it will meet the user requirements in four seasons. The most used plan type in the traditional old town texture is the central courtyard type [42].

Table 4. Orientations used according to housing plan types in Diyarbakır [19].

		Courtyard Typology		
Outer Courtyard Plan	L Type	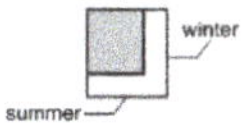 winter / summer	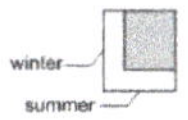 winter / summer	
	U Type	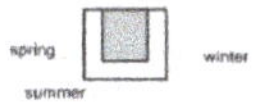 spring / winter / summer	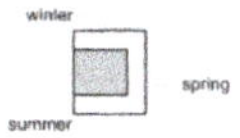 winter / spring / summer	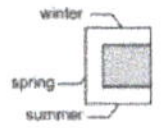 winter / spring / summer
	Mix Type	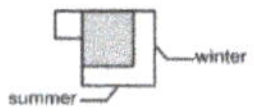 winter / summer	 winter / summer	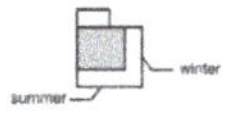 winter / summer
Inner Courtyard Plan			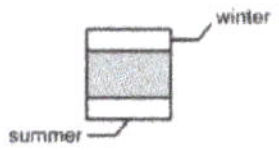 winter / summer	
Center Courtyard Plan			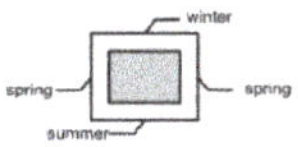 winter / spring / spring / summer	

Courtyard Plan N

2.2. Material and Method

DesignBuilder Energy Simulation program was used in this study. DesignBuilder is a dynamic simulation tool that calculates all building energy, lighting, carbon, and comfort performance analyses [43]. The program is preferred because it has a user-friendly interface and the simulation results are realistic. DesignBuilder simulation program is used in many disciplines because it is reliable software. It can be used actively in architecture, building physics, mechanical engineering, heating and cooling systems modelling. In addition to heating-cooling load modelling, lighting has the ability to calculate daylight and model computational fluid dynamics (CFD) [44].

IWEC (International Weather for Energy Calculations) climate data of the province of Diyarbakır, located in the hot-dry climate region, which was selected as the study area, was introduced to the program. Comfort conditions were determined by making certain constants and assumptions to carry out analyses through the DesignBuilder Energy Simulation Program. Building heating and cooling systems were used as a reference in the mass housing settlements used in the region, the building heating system was introduced as natural gas, and the cooling system was introduced as electric energy to the program. In order to provide climatic comfort, seasonal differentiation was preferred. Furthermore, to provide internal climatic comfort, the indoor air temperature comfort value was set to 21 °C and the heating-setback setting was set to 16 °C in winter. For the summer season, the indoor air comfort value was 25 °C and the cooling setback temperature was 25 °C. In addition, the number of residential users was determined to be four. User activity level 0.9 MET (metabolic equivalent of task) winter garment insulation value was accepted as 1 Clo and summer garment insulation value was accepted as 0.5 Clo.

The residential type was chosen as the construction type. As a floor height, by taking traditional Diyarbakır houses as a reference, a two-storey, 8-m-high summer mass with openings facing north was determined. When it comes to building shell and thermophysical properties, the building shell properties of the collective housing buildings in Diyarbakır were taken as reference. The physical model of the building shell of the courtyard is shown in Figure 4. About the optical properties, since the mass facing the courtyard is more important than the main facade in the traditional Diyarbakır houses, the transparency ratio of the surfaces facing the courtyard was determined to be 35% and the transparency ratio of the facade was 25%. In addition, a single glass of 6 mm thickness was used as the type of glass (Table 5).

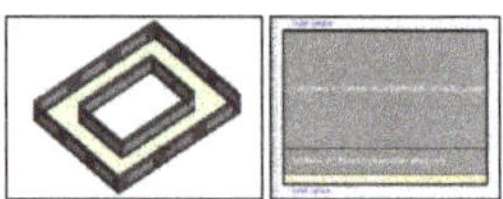

Figure 4. Courtyard and building shell section modeled with DesignBuilder (DB version 5) [45].

Table 5. Building crust and thermophysical properties in investigated houses.

	Properties	**U-Value (W/m^2 K)**
Reinforced Concrete Exterior Wall	Reinforced Concrete Precast Wall (15 cm) Glass Fiber Insulation (3 cm) Gypsum Plastering (1.25 cm)	0.814
Tilling on the Floor	Terrazzo Tile (3 cm) Levelling Concrete (4 cm) Blind Concrete (10 cm) Clinker Filling (20 cm) Concrete Fundament (60 cm) Insulation (3 mm)	0.545
The Ceiling Dividing the Garret	Bitumen Sheet (2 mm) Heat Insulation (2 cm) Bitumen Sheet (2 mm) Reinforced Concrete Wall (15 cm)	0.466

The study was carried out in two stages. In Step 1, a summer mass placed in south with two rooms, a size of 4×12 m^2 and two floors, height of 8 m was determined based on the room sizes in traditional Diyarbakır houses and in the masses with again the same size and height called adiabatic which does not provide any heat exchange with its environment while only providing a shadow effect, 8 m wide courtyard structure forms with L, U, central, and inner courtyards were created (Table 6). The reason why adiabatic mass was preferred in the study is that free convection heat transfer from a horizontal, isothermal fin attached cylinder placed between two nearly adiabatic walls has various applications in industry [46]. Then, the heating and cooling loads due to the shadow effect of the most commonly used forms in traditional Diyarbakır houses were compared and the optimum building form was determined.

Table 6. Courtyard alternatives by using adiabatic building mass.

Courtyard Alternatives by Using Adiabatic Building Mass		
	Reference Building Form (12 × 4 × 8 m)	
	Adiabatic Building Form (12 × 4 × 8 m) (Only Shadow Effect—No Heat Transfer)	
Central Type Courtyard Form	Inner Courtyard Form	L Type Courtyard Form
	U Type Courtyard Form	

In the second step of the study, heating-cooling loads due to the shadow effect of different street widths and orientations were analyzed with reference to the 20 m wide, 16 m long, and 8 m high central courtyard building form (Figure 5). In other words, due to the shadow effect created, the thermal performances of the referenced central courtyard structure form with 3–6 m wide adiabatic building masses in different settlements were analyzed to determine the optimum type of street texture (Table 7).

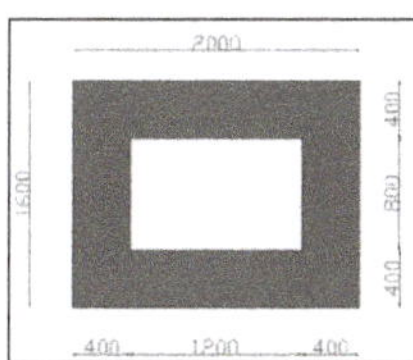

Figure 5. Reference courtyard form and dimensions (m).

Table 7. Different settlement layouts with streets with widths of 3 and 6 m.

	B1	**B2**	**B3**
N	Settlement texture consist of three blocks (3 m street width)	Settlement texture consist of four blocks (3 m street width)	Settlement texture consist of two blocks (3 m street width)
	B4	**B5**	**B6**
Shading Device Type and Position	Settlement texture consist of three blocks (6 m street width)	Settlement texture consist of four blocks (6 m street width)	Settlement texture consist of two blocks (6 m street width)

2.3. Analysis of Simulation

The courtyard building forms, which are frequently used in traditional Diyarbakır houses, were formed by adiabatic masses. In Diyarbakır, which is located in the hot-dry climate region, it was determined that the annual cooling load expenses are higher than the heating load expenses in all forms of construction. For this reason, it was obtained as a result of the analysis that the annual cooling load in the middle courtyard plan type decreased compared to the reference building. Moreover, when the examined building forms were examined, it was found that the cooling load expenses due to the shadow effect reduced. In addition, the building form that would minimize the total cooling load per year was found to be the plan type with a central courtyard. When total annual heating load expenses are compared, annual heating loads increase in inner courtyard building form, while other building forms decrease in values compared to the reference building. As a result of the analysis, it was determined that the optimal structural form was the central courtyard plan type among the heating load values, and also the L type courtyard type had similar values. When the total yearly heating and cooling load values are compared with the reference building, in other words, when the reference building which is not exposed to any shadow effect and the shadow forms with adiabatic building masses are compared, it is found that the optimum form regarding energy performance is the central courtyard building type (Figure 6).

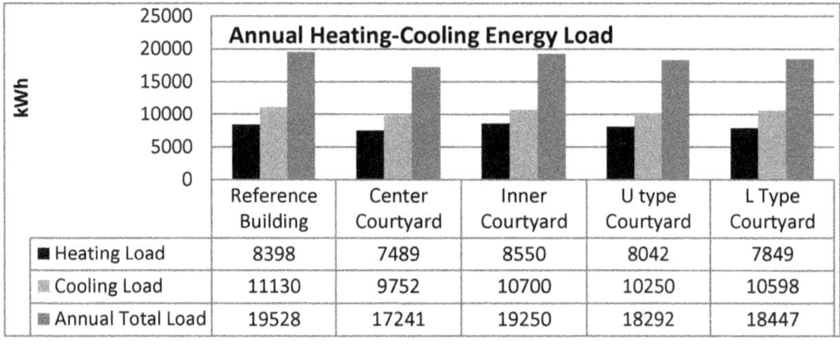

	Reference Building	Center Courtyard	Inner Courtyard	U type Courtyard	L Type Courtyard
■ Heating Load	8398	7489	8550	8042	7849
▨ Cooling Load	11130	9752	10700	10250	10598
▨ Annual Total Load	19528	17241	19250	18292	18447

Figure 6. The effect of change of courtyard form on heating-cooling load values.

The plan of the central courtyard, which was previously determined as the optimum building form, was selected and the annual cooling-heating load values of different street widths and settlements were compared. In other words, the reference building form with the central courtyard, which is not surrounded by any building mass, was compared with the annual heating-cooling load values due to the shadow effect of the building island with a different layout. In the first stage, the street widths were fixed everywhere, 3 m as accepted, and three different settlements were compared with the reference building. The effect of annual heating-cooling load values on heat gain and losses due to shadow effect of the B3 corner parcel surrounded by the structures on two sides, middle parcel B1 surrounded by the structures on three sides, and B2 inner middle parcel surrounded by the structures on four sides were examined. As a result of the analyses, if the cooling loads connected to the B1, B2, B3 layout are compared, the optimal layout is determined as B2 layout surrounded by the structure on four sides provided that street widths are kept constant everywhere. When the heating loads due to the shade effect of the settlement layout were compared, it was found that the shade effect had a negative effect on energy costs. In other words, an increase in the heating loads due to the shadow effect was observed in all three settlements. However, B2 layout was determined as the layout with the highest heating load value due to the shadow effect on the four sides.

When the total annual heating and cooling load values are compared with the reference building (Figure 7), the optimal layout is determined as the B1 layout, which is surrounded by structures on three sides. In order to examine the effect of the same settlement on different street widths, the settlement islands of B4, surrounded by the buildings on three sides, B5 surrounded by the buildings on four sides and B6, which is a corner parcel surrounded by the buildings, were created. When the annual cooling load values of the B6 corner parcel surrounded by two sides structure, B4 middle parcel surrounded by three sides structure and B5 inner middle parcel layouts surrounded by four sides structure are compared, the optimal layout is determined to be B5 settlement layout surrounded by four sides structures. In other words, it was determined that the cooling load values of the settlement arrangement B5 having the maximum shade effect had the least value. When the heating loads were compared, it was found that annual heating loads increased in all settlements, but this increase was less than other combinations in the corner parcel B6 settlement layout, which was surrounded by the structures on both sides. It was determined that the settlement with the highest increase in annual heating loads was the B5 settlement plan surrounded by structures on four sides which was exposed to the maximum shade effect. When the total annual heating-cooling load values of different settlements connected to the street texture with a width of 6 m are compared, it is found that optimal settlements are B6 settlement layout surrounded by the structures on both sides.

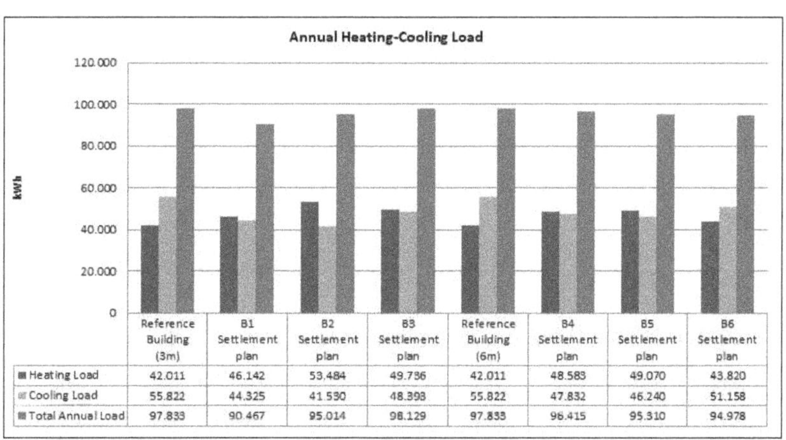

	Reference Building (3m)	B1 Settlement plan	B2 Settlement plan	B3 Settlement plan	Reference Building (6m)	B4 Settlement plan	B5 Settlement plan	B6 Settlement plan
■ Heating Load	42.011	46.142	53.484	49.736	42.011	48.583	49.070	43.820
▨ Cooling Load	55.822	44.325	41.530	48.393	55.822	47.832	46.240	51.158
■ Total Annual Load	97.833	90.467	95.014	98.129	97.833	96.415	95.310	94.978

Figure 7. The effect of change of courtyard form on heating-cooling load values.

3. Results

As a result of the analyses performed through the Design-Builder energy simulation program, changes in the annual heating-cooling load values of the building form and street texture were determined. The mentioned results are examined under the following headings:

3.1. Annual Energy Gain-Loss Changes of Building Form

The heating-cooling energy gains and losses of different building forms formed with the reference structure and adiabatic masses were compared. As a result of the analyses, it is seen that the plan type with the central courtyard has a gain of 12% cooling and 14% heating load annually compared to the reference building. In other words, plan type with a central courtyard was determined as the most efficient form of energy compared to other building forms. When other building forms are examined, it is seen that the cooling load in U courtyard type has increased by 4% and in L courtyard form by 6%. However, in the inner courtyard plan type, cooling loads increase by 2% due to insufficient shadow effect. When the four building forms are considered, it is seen that the shadow effect has a decisive effect on the cooling loads for the province of Diyarbakır, where there is a hot-dry climate. Again, when comparing the annual heating load losses and gains of the reference building with different buildings' forms, it was found that 14% of the annual gain was achieved in the plan type of the central courtyard. Among other building forms, it was determined that the building form, which has the minimum gain among the annual heating load values, was the inner courtyard plan type.

Based on the total heating-cooling load values of the reference building, the annual total heating-cooling load percentage gains of other building forms were calculated, and it was determined that the plan type with the central courtyard among these building forms was the optimum building form. This building form is followed by U courtyard type with 7% and L courtyard type with 6%. It is determined that the inner courtyard plan type does not provide significant gains for the hot-dry climate zone in terms of energy costs, in other words, it is a form of low adaptation form for this climate zone. Furthermore, when Table 8 is examined, it is found that although the shading effect and cooling load values are more important than other parameters in the hot-dry climate zone, heating loads as well have great importance on annual gains in comparison of total heating-cooling load values.

Table 8. Annual energy gain-loss change percentages of building form.

Street Pattern	Cooling Load Exchange (%)	Heating Load Exchange (%)	Annual Total Heating-Cooling Load Exchange (%)
Center Courtyard	+12	+14	13
Inner Courtyard	−2	+4	1
U type Courtyard	+4	+8	7
L type Courtyard	+6	+5	6

3.2. Annual Energy Gain-Loss Percentages due to the Shadow Effect of the Different Settlement Layouts

The heating-cooling load gain percentages are given due to the shade effect of different layouts and street widths. The annual cooling load gains of different settlements surrounded by three sides (B1), four sides (B2), and two sides (B3) buildings with a street width of 3 m are given in the table. Among the alternative settlements, it was found that the B2 settlement, which was most exposed to shade effect and surrounded by building masses on all four sides, had the highest cooling load gain with a rate of 34%. It is determined that the B1 settlement which is surrounded by structures on three sides saves 25% and the B3 settlement which is surrounded by structures on two sides saves 15% cooling loads (Table 9). When the reference building and annual heating load values are compared, heat losses were determined in alternatives including all settlements. However, among the total annual heating expenses, the highest loss was found in the B3 settlement which is surrounded by the structures on two sides and corner parcel. Among the alternatives with a street width of 3 m, the total annual heating-cooling energy gains were determined in the B1 settlement plan surrounded by the buildings on three sides, while it was found there was no significant difference between the B3 settlement plan surrounded by the buildings on the two sides and the reference building. In other words, there is no significant difference in terms of annual energy loss-gains between the reference building on a single building scale and the settlement plan B3, which is surrounded by buildings on both sides. In the case where the street width is fixed everywhere and 6 m, it is determined that the settlement plan which shows a gain in the annual cooling load values is the B4 settlement which is surrounded on three sides. The B6 settlement plan in the corner parcel was the least profitable among the alternatives.

Table 9. Annual energy gain-loss percentage changes due to shadow effect of different settlement layouts.

Street Pattern	Cooling Load Exchange (%)	Heating Load Exchange (%)	Annual TotalHeating-Cooling Load Exchange (%)
B1 Settlement Plan	+25	−9	8
B2 Settlement Plan	+34	−11	3
B3 Settlement Plan	+15	−16	0.3
B4 Settlement Plan	+17	−14	1
B5 Settlement Plan	+13	−14	4
B6 Settlement Plan	+10	−4	4

When annual heating loads were compared, heat losses due to shadow effect were observed among all alternatives. Among these heat losses, the alternative having the least value was found to be the corner parcel B6 settlement layout surrounded by the structures on both sides. It was determined that the heating load losses of the B5, surrounded by the structures on all four sides and B4 buildings, surrounded by the structures on three sides were equal. When the total annual heating and cooling energy gains were compared, it was determined that the optimal alternative with a width of 6 m street was the B5 settlement layout surrounded by the structures on four sides and the corner parcel B6 surrounded by the structures on two sides. If the reference building and six different layout plans are compared together, the B1 settlement layout, which is surrounded by three-side buildings with a street width of 3 m, was identified as the building island that uses energy in the most efficient way. In addition, if the street width was increased, an increase in the cooling loads due to the decrease in shade effect was observed in alternatives with the same layout (B2-B5, B1-B4, B3-B6). The B1 alternative, which is surrounded by structures on three sides with an optimal layout of 3 m of street width, is no longer the optimal layout if the street width is increased to 6 m with the condition that the layout remains constant. In other words, while the optimal layout with a street width of 3 m is B1, which is surrounded by structures on three sides when the street width is increased to 6 m, the corner parcel B6 surrounded by structures on two sides is determined to become the optimum layout. Therefore, it is concluded that street widths have a direct impact on the settlement patterns and affect annual expenses even if the parcel layout remains constant. In conceptual terms, street widths for the hot-dry climate zone have a direct impact on annual energy losses and gains. While, for hot dry climatic zones where the warmest season lasts longer than the coldest season, the B2 settlement layout, which is surrounded by the structures on all four sides and exposed to the shadow effect the most, is expected to be determined as the optimal layout with annual cooling energy gain, determining the B1 layout, which is surrounded by structures on three sides, as the optimal layout proves that energy-efficient design is an integrated design approach. In other words, in terms of bioclimatic based energy-efficient design, total annual energy costs optimization should be evaluated in the correct way.

Although it is seen that it is a priority to take measures to minimize cooling loads in hot dry climate zones, when the B1 settlement layout which is surrounded by three sides structures with optimal layout and B2 settlement layout which is surrounded by four sides structures with the least value regarding cooling costs are compared, the decrease in the annual cooling load value revealed the fact that it will not give optimal results concerning annual total energy expenses. In other words, the design concept that can keep the heating load at the optimal level, as well as the measures to minimize the cooling load in the residential settlements in Diyarbakır province, should be considered together. As a result of all these analyses, the structure form had a significant effect on the heating-cooling load values. Therefore, the central courtyard type was the optimum form. In the hot-dry climate zone, the shadow effect is an important climatic element. The street widths of the traditional city wall were taken as a reference. As a result of analyses, it was determined that the annual heating-cooling load values of all alternatives with a width of 3 m were less than 6 m. As a result of this study, heating-cooling load values increased with increasing distance between structures. The results also differed in different building distances with the same layout. The optimum layout with a width of 3 m is B2, which is surrounded by structures on four sides, whereas the B1 with optimal layout is 6 m when the width of the street is 6 m. Therefore, the construction distances have an effect on the annual heating-cooling load values depending on the settlement patterns. In conclusion, if the building distances switch, the annual heating-cooling load values change in proportion to the decrease in shadow effect even if the layout remains the same. In hot-dry climatic zones, energy load changes due to shade effect affect formal and settlement patterns.

4. Discussion

With this study, suitable building form and settlement pattern samples were determined for low-rise residential settlements to be built in this region and it is aimed at increasing the number of examples of climate-based residential settlements and to transfer traditional residential textures to today's designs.

Design solutions to minimize cooling costs are needed in the city of Diyarbakır, which is located in a hot and dry climate. In the first step of the design, with the environmental data, by looking at the building design parameters such as location, structure form, and direction of buildings with respect to each other, significant energy savings can be achieved by considering a bioclimatic design method which benefits from passive systems. As a result of analyses, shadow climate element has an important place in energy efficient structure design in hot-dry climate regions. In other words, when the street widths of 3 and 6 m are compared, it is concluded that the load values are less than 3 m. The shadow effect can affect the annual energy loads not only on the scale of settlement, but also on the formal scale. Therefore, it is necessary to take design measures to create shadow effects in traditional courtyard type buildings which are examples of climate balanced design. If we adapt all these results to modern houses, we can achieve annual energy savings through the measures taken during the design phase.

In this context, the construction forms of traditional Diyarbakır houses were compared regarding energy costs. The optimum construction form was chosen in terms of heating-cooling load values. Then, the effect of shadow effect on energy loads in a hot dry climate region was examined by taking optimum building form as a reference, simulating different settlement layouts, and street widths of traditional Old Town texture. As a result, it was concluded that the same settlement layouts did not show the same thermal performance at different street widths (3 and 6 m). While the width of the street is 3 m, the settlement layout of B1, which is surrounded by structures on three sides with optimal layout, is expected to be the optimal layout when the width of the street is 6 m as well; the determination of B5, which is surrounded by structures on four sides, and B6, which is surrounded by structures on two sides as the optimal settlement units supported this opinion. According to result of the analysis, it was concluded that heating loads, besides the cooling loads in the hot-dry climate zone, have an effect on the total annual heat gains. In other words, in the case where the street width is 3 m, while it is expected that the settlement layout B2 which is exposed to the most shadow effect among the settlements B1, B2, and B3 to be the optimal settlement unit, the fact that the settlement of B1 which is surrounded by buildings on three sides has less energy expenditure annually explains this situation. With this study, it is understood that design measures should be taken to minimize the cooling loads, especially in the hot-dry climate zone, as well as measures to optimize the heating loads. In addition, it was determined that energy losses and gains may be different if the width of the street is different in the same street layout. As a result of this study, it is possible to create a sustainable environment that consumes less energy by providing the integration of traditional building form and street texture to today's housing sector. If housing sector wants to save energy, traditional housing types and associated climate elements should be analyzed correctly. Moreover, a balanced design method should be developed with climate. It is emphasized that if the results obtained above are integrated into today's housing sector, more energy efficient structures will be produced, and a sustainable environment will be possible.

Author Contributions: Conceptualization, İ.A. and S.A.; methodology, İ.A. and S.A.; validation, İ.A.; formal analysis, İ.A. and S.A.; investigation, İ.A. and S.A.; resources, S.A.; data curation, S.A.; writing—original draft preparation, İ.A., S.A., and L.S.G.; writing—review and editing, İ.A., S.A., and L.S.G.; visualization, S.A.; supervision, İ.A. All authors have read and agreed to the published version of the manuscript.

Funding: This research received no external funding.

Conflicts of Interest: The authors declare no conflict of interest.

References

1. Du, K.; Calautit, J.; Wang, Z.; Wu, Y.; Liu, H. A review of the applications of phase change materials in cooling, heating and power generation in different temperature ranges. *Appl. Energy* **2018**, *220*, 242–273. [CrossRef]
2. Zhou, D.; Zhao, C.Y.; Tian, Y. Review on thermal energy storage with phase change materials (PCMs) in building applications. *Appl. Energy* **2012**, *92*, 593–605. [CrossRef]
3. Hatipoglu, H.K. Understanding social sustainability in housing form the case study "Wohnen Mit Uns" in Vienna and adaptibility to Turkey. *Iconarp Int. J. Archit. Plan.* **2017**, *5*, 87–109. [CrossRef]
4. Coma, J.; Chàfer, M.; Pérez, G.; Cabeza, L.F. How internal heat loads of buildings affect the effectiveness of vertical greenery systems? An experimental study. *Renew. Energy* **2019**, *142*. [CrossRef]
5. International Energy Outlook. 2016. Available online: https://www.eia.gov/outlooks/ieo/pdf/0484(2016).pdf (accessed on 13 October 2019).
6. Young, B.A.; Falzone, G.; Wei, Z.; Sant, G.; Pilon, L. Reduced-scale experiments to evaluate performance of composite building envelopes containing phase change materials. *Constr. Build. Mater.* **2018**, *162*, 584–595. [CrossRef]
7. Soflaei, F.; Shokouhian, M.; Abraveshdar, H.; Alipour, A. The impact of courtyard design variants on shading performance in hot-arid climates of Iran. *Energy Build.* **2017**, *143*, 71–83. [CrossRef]
8. Öz, K.; Özen Yavuz, A. Shape grammar analysis and comparison of the traditional and new urban textures in Sivrihisar, Eskişehir. *Gazi Univ. J. Sci. Part B Art Humanit. Des. Plan.* **2018**, *6*, 113–124.
9. Al-Azzawi, S. Indigenous courtyard houses: A comprehensive checklist for identifying, analysing and appraising their passive solar design characteristics Regions of the hot-dry climates. *Renew. Energy* **1994**, *5*, 1099–1123. [CrossRef]
10. Al Masri, N.; Abu Hijleh, B. Courtyard housing in midrise building: An environmental assessment in hot-arid climate. *Renew. Sustain. Energy Rev.* **2012**, *16*, 1892–1898. [CrossRef]
11. Sinou, M. *Design and Thermal Diversity of Semi-Enclosed Spaces*; Melrose Books: Ogun State, Nigeria, 2007.
12. Fathy, H. *Natural Energy and Vernacular Architecture: Principles and Examples with Reference Hot-Arid Climates*; University of Chicago Press: Chicago, IL, USA, 1986.
13. Berköz, E.; Küçükdoğu, M.; Yılmaz, Z.; Kocaaslan, G.; Ak, F. *Energy Efficient Building and Settlement Design*; TUBITAK-İNTAG 201; Research Report; Istanbul Technical University: Istanbul, Turkey, 1995.
14. Location of Diyarbakır. Available online: http://www.maphill.com/turkey/diyarbakir/location-maps/gray-map/free/ (accessed on 18 October 2019).
15. Climate Data. Available online: https://en.climate-data.org/asia/turkey/diyarbak%C4%B1r/diyarbak%C4%B1r-285/#climate-graph (accessed on 28 November 2019).
16. Sağıroğlu, Ö. Characteristics and construction techniques of Akseki Bucakalan Village Rural Dwellings. *Int. J. Archit. Herit.* **2017**, *11*, 433–455. [CrossRef]

17. Location of Work Area. Available online: http://cografyaharita.com/haritalarim/4l_diyarbakir_ili_haritasi.png (accessed on 20 November 2019).
18. Google Map. Available online: https://www.google.com/maps/search/suri%C3%A7i+%C5%9Filbe/@37.9511218,40.2020973,12.22z (accessed on 27 November 2019).
19. Erginbaş, D. *Diyarbakir Houses*; Pulhan Press: Istanbul, Turkey, 1953.
20. Dalkılıç, N.; Bekleyen, A. *Physical Traces of the Past Reflected in the Present: Traditional Diyarbakır Houses*; Diyarbakır Governorship Publication: Diyarbakir, Turkey, 2011.
21. Akin, C.T.; Koca, C. Modelling transportation axes in Suriçi (Diyarbakir, Turkey) and determining their relationship to social areas allocated for public use. *J. Asian Archit. Build. Eng.* **2017**, *16*, 333–339. [CrossRef]
22. Kuban, D. *Problems of Our Art History: Anatolian Turkish Art, Architecture, Essays on the City*; Çağdaş Publication: Istanbul, Turkey, 1975.
23. Dağtekin, E.; Kakdaş Ateş, D.; Oğur, D. New housing design approaches in Diyarbakir street. *J. Int. Soc. Res.* **2018**, *11*. [CrossRef]
24. Google Earth. Available online: https://www.google.com/maps/@37.912263,40.2289968,14.14z (accessed on 9 November 2019).
25. Özyılmaz, H.; Sahil, S. The reflection of changing social structure: Diyarbakir Example. *Megaron* **2017**, *12*, 531–544.
26. Yıldırım, M. Shading in the outdoor enviroments of climate-friendly hot and dry historical streets: The passageways of Sanlıurfa, Turkey. *Environ. Impact Assess. Rev.* **2020**, *80*, 106318. [CrossRef]
27. Oğuz, G.P.; Halifeoğlu, F.M. Construction tecniques and material protection problems in traditional Diyarbakir Houses. *Dicle Univ. J. Eng.* **2017**, *8*, 345–358.
28. Sözen, M.Ş.; Gedik, G.Z. Evaluation of traditional architecture in terms of building physics: Old Diyarbakır houses. *Build. Environ.* **2007**, *42*, 1810–1816. [CrossRef]
29. Gabriel, A. *Diyarbakır City Walls*; (Voyages archeolojique dans la Turquie Oriental); Gabriel, A., Translator; Diyarbakır Promotion, Culture and Solidarity Foundation: Diyarbakır, Turkey, 1940.
30. Bekleyen, A. *Adapting the Old to New: Interpreting the Originality of Local Architecture in Contemporary Architecture*; Construction Experiences in Historic Environment; Birsen Publication: Istanbul, Turkey, 2018.
31. Al-Hemiddi, N.A.; Al-Saud, K.A.M. The effect of a ventilated interior courtyard on the thermal performance of a house in a hot–arid region. *Renew. Energy* **2001**, *24*, 581–595. [CrossRef]
32. Muhaisen, A.S. Shading simulation of the courtyard form in different climatic regions. *Build. Environ.* **2006**, *41*, 1731–1741. [CrossRef]
33. Baran, M.; Yıldırım, M.; Yılmaz, A. Evaluation of ecological design strategies in traditional houses in Diyarbakir, Turkey. *J. Clean. Prod.* **2011**, *19*, 609–619. [CrossRef]
34. Tuncer, O.C. *Diyarbakir Houses*; Diyarbakir Metropolitan Municipality Culture and Art Press: Diyarbakir, Turkey, 1999.
35. Dalkılıç, N.; Aksulu, I. Traditional housing architecture in Diyarbakir. *Gazi Univ. J. Art* **2001**, *2*, 53–69.
36. Abdulac, S. *Traditional Housing Design in the Arab Countries*; Aga Khan Program for Islamic Architecture: Cambridge, MA, USA, 1982.
37. Ahani, F. Natural light in traditional architecture of Iran: Lessons to remember. *Light Eng. Archit. Environ.* **2011**, *121*, 25–36.
38. Foruzanmehr, A. Summer-time thermal comfort in vernacular earth dwellings in Yazd, Iran. *Int. J. Sustain. Des.* **2012**, *2*, 46–63. [CrossRef]
39. Herdeg, K. *Formal Structure in Islamic Architecture of Iran and Turkistan*; Rizzoli International Publication: New York, NY, USA, 1990.
40. Khajehzadeh, I.; Vale, B.; Yavari, F. A comparison of the traditional use of court houses in two cities. *Int. J. Sustain. Built Environ.* **2016**, *5*, 470–483. [CrossRef]
41. Saljoughinejad, S.; Sharifabad, S. Classification of climatic strategies, used in Iranian vernacular residences based on spatial constituent elements. *Build. Environ.* **2015**, *92*, 475–493. [CrossRef]
42. Sözen, İ.; Oral, G.K. Outdoor thermal comfort in urban canyon and courtyard in hot arid climate: A parametric study based on the vernacular settlement of Mardin. *Sustain. Cities Soc.* **2019**, *48*, 101398. [CrossRef]
43. Zhang, L. Simulation analysis of built environment based on design builder software. *Appl. Mech. Mater.* **2014**, *580*, 3134–3137. [CrossRef]

44. Riahi Zaniani, J.; Taghipour Ghahfarokhi, S.; Jahangiri, M.; Alidadi Shamsabadi, A. Design and optimization of heating, cooling and lightening systems for a residential villa at Saman city, Iran. *J. Eng. Des. Technol.* **2019**, *17*, 41–52. [CrossRef]
45. *Design Builder Software Version 5.03.007*, Designbuilder Software Limited, Clarendon Court 1st Floor 54/56 London Rd Stroud, Gloucestershire GL5 2AD, UK. 2017.
46. Rezaei, A.A.; Yousefi, T. Free convection heat transfer from a horizontal fin attached cylinder between confined nearly adiabatic walls. *Exp. Therm. Fluid Sci.* **2010**, *34*, 177–182. [CrossRef]

Article

Natural Ventilation Effectiveness of Awning Windows in Restrooms in K-12 Public Schools

Sung-Chin Chung [1], Yi-Pin Lin [1], Chun Yang [1] and Chi-Ming Lai [2,*]

[1] Department of Creative Design, National Yunlin University of Science and Technology, Douliu 640, Taiwan; chungsc@yuntech.edu.tw (S.-C.C.); linyp@yuntech.edu.tw (Y.-P.L.); yc004009@gmail.com (C.Y.)

[2] Department of Civil Engineering, National Cheng Kung University, Tainan 701, Taiwan

* Correspondence: cmlai@mail.ncku.edu.tw; Tel.: +886-627-57-575 (ext. 63136)

Received: 8 April 2019; Accepted: 17 June 2019; Published: 23 June 2019

Abstract: Using computational fluid dynamics (CFD), this study explores the effect of a different number of awning windows and their installation locations on the airflow patterns and air contaminant distributions in restrooms in K-12 (for kindergarten to 12th grade) public schools in Taiwan. A representative restroom configuration with dimensions of 10.65 m × 9.2 m × 3.2 m (height) was selected as the investigated object. Based on the façade design feasibility, seven possible awning window configurations were considered. The results indicate that an adequate number of windows and appropriate installation locations are required to ensure the natural ventilation effectiveness of awning windows. The recommended installation configuration is provided.

Keywords: natural ventilation; air environment; school restroom; awning window; CFD

---

1. Introduction

In Taiwan, restrooms in K-12 public schools (kindergarten (K) and the 1st through the 12th grade (1-12)) are frequently accessed and are open to the public, which hinders their management. In particular, restrooms that have been in service for more than 20 years often exhibit problems, such as poor ventilation, inadequate lighting, outdated layouts and designs and limited toilet spaces. Thus, these restrooms have become a nuisance for school environment management. Poor-quality restrooms create fear of using the restroom among teachers and students, which increases incidents in which teachers and students are affected by acute or chronic diseases that can adversely affect their physical and mental health.

There are few studies on bathroom and restroom ventilation that address the ventilation efficiency of entire bathrooms and restrooms as well as the ventilation efficiency of their components (e.g., toilets and fans). Investigating the ventilation of an entire bathroom and restroom, Tung et al. [1] analyzed a new negative pressure wall-exhaust ventilation system (that differs from the traditional ceiling-exhaust system) installed in a residential bathroom and restroom using a full-scale test. In the test, toilets were deployed in several different patterns and positions. The test results indicated that the restroom ventilation system could take advantage of a negative pressure difference to prevent the escape of restroom malodor to an adjacent room. Increasing the ventilation volume or decreasing the distance between vent and toilet could improve both the indoor pollutant removal rate and the ventilation rate. From the perspective of energy efficiency, an air change rate of 8.5 h^{-1} was the optimal value. Tung et al. [2] analyzed the effectiveness of three ventilation strategies for residential bathroom malodor removal: (1) forced ceiling-supply and wall-exhaust systems, (2) natural window-inlet and forced ceiling-exhaust systems and (3) forced ceiling-supply and ceiling-exhaust systems. The first strategy achieved the best malodor removal. Yang and Kim [3] employed computational fluid dynamics (CFD) to analyze the effect of changing the glass partition shape in an apartment bathroom on the bathroom ventilation. Their study provides a glass partition design for a bathroom (with a shower and restroom) that can maintain excellent household hygiene.

Regarding the effectiveness of bathroom and restroom components, Seo and Park [4] designed a new toilet ventilation system, in which a ring opening for odor suction was included under the toilet seat. Malodor at the source was directly sucked out to reduce the escape of malodor into the restroom. The effectiveness of this ventilation system was validated via testing and CFD simulation. The test and numerical simulation results were essentially consistent. Dual openings at the rear of the toilet with 4 mm^2 × 4 mm^2 odor suction holes represented the optimal design. The ventilation system draws outdoor fresh air into the restroom via door opening. Sato et al. [5] analyzed the amounts of volatile substances of human waste malodor (i.e., feces and urine). The substances were collected via Tenax-TA, and their concentrations were determined by thermal-desorption cold-trap injector/gas chromatography/mass spectrometry (TCT/GC/MS). The results revealed that approximately 90% of malodor substances were fatty acids: acetic acid (65%), propionic acid (15%), butyric acid (6.5%), i-Valeric acid and n-Valeric acid. Approximately 8% of these substances were N-containing compounds: ammonia (6.5%), pyridine, pyrrole, indole, skatole and trimethylamine.

Kim and Yang [6] analyzed the ventilation effectiveness of exhaust fans installed in a residential bathroom in Korea via field measurement. In the paper, construction and design methods were revised to enhance the bathroom ventilation. Yin et al. [7] analyzed the effect of increasing the operational pressure of a residential bathroom ventilation fan on the ventilation effectiveness. The survey objects were more than 80 families who used AC-motor ventilation fans. The results revealed that the performance of the exhaust fan was significantly penalized by an increase in external static pressure. Choi et al. [8] provided statistics from bathroom ventilation fan tests from 2005 to 2013 and interpreted these statistics based on the development trend of residential ventilation standards and guidelines.

Numerous studies have examined natural ventilation for buildings. Comprehensive reviews on natural ventilation studies, including effects of building façade and ventilation opening, ventilation shaft design, shape of louvered windows, apertures and vernacular element, the representative air change rates, applications in multi-story buildings, the effect of thermal energy storage on night ventilation efficiency, night ventilation effectiveness and design, etc., are found in the literature [9–15]. However, studies of the natural ventilation of restrooms, particularly restrooms in K-12 public schools, are limited. In this study, restrooms in these schools in Taiwan are selected as the study subjects. The effect of the installation quantity, position and opening angle of awning windows on the flow pattern and air pollutant (NH$_3$) distribution in restrooms with various wind speeds and directions are analyzed via CFD.

2. Materials and Methods

2.1. Study Subject

Restroom indoor configurations differ due to their position, site area, orientation, and building design. It is very common for public school buildings to have east-west orientation (facing the south and the north) with restrooms at the opposite two ends under the requirement of the building code of Taiwan, as illustrated in Figure 1a. After analyzing restrooms in 140 K-12 public schools in Taiwan, a representative model restroom with the following dimensions is selected as the study subject: 10.65 m × 9.2 m × 3.2 m (height) (Figure 1). In Figure 1c, ① indicates the aisle in the restroom, ② is the squatting toilet area in the women's restroom, and ③ is the urinal area in the men's restroom.

Currently, the windows that are commonly employed in school restrooms include horizontal push windows, awning windows and blinds. In this study, based on Figure 1, a preliminary study of the effect of the window configuration on the natural ventilation performance of a restroom (0.5 m/s north wind) is conducted using CFD (discussed in a subsequent section). The results indicate that an awning window has the best ventilation effect, followed by blinds. The order of ventilation effectiveness is described as follows: awning window > blinds > horizontal push window (data not shown). The awning window has the following advantages: sufficient ventilation area and an adjustable opening angle to control the ventilation area and deflect rain (which is particularly suitable for the typhoon and rainy season in Taiwan). Therefore, the awning window (as illustrated in Figure 2) is selected as the window configuration for the subject.

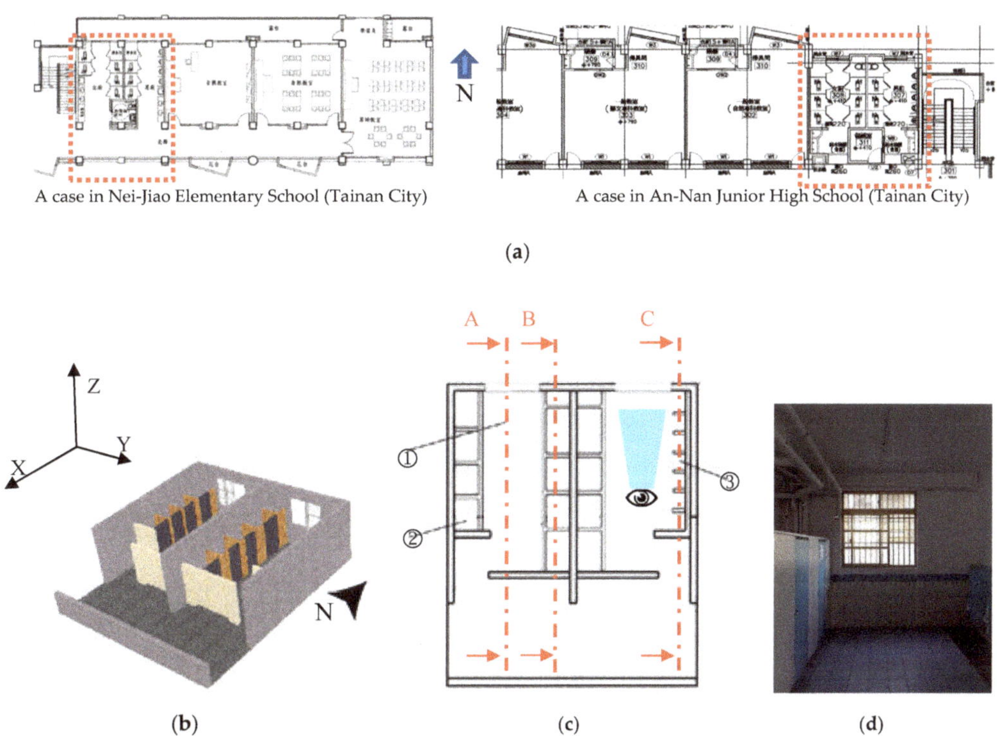

Figure 1. Study subject. (**a**) The investigated model restroom comes from practical cases (red dotted area) (not to scale). (**b**) 3D diagram. (**c**) Plan. (**d**) Field image.

Figure 2. Awning window examples.

The dimensions of the awning windows that are available in the market vary in the range of 0.4 m × 0.4 m–1.4 m × 1.0 m. The dimensions of the awning window that is investigated in this study are 1.0 m × 0.9 m. The position of the restroom window is varied. In this study, the awning window configuration is based on the north façade (Table 1), which is located on a certain floor and has seven window configurations (W1–W7). The windows for men's and women's restrooms are installed symmetrically. Whether a window frame is listed as a CFD simulation object affects the number of grids in the simulation and may significantly increase computing time. Therefore, the glazing dimensions are used as the dimensions of the ventilation opening; i.e., the window frame is ignored here. The installation of four awning windows comprises three designs (the black blocks in Table 1b–d). The installation of eight awning windows consists of three designs (Table 1e–g). In addition, the installation of 12 awning windows has one design (Table 1h). The opening angles θ for the awning windows are 30°, 45° and 60°. The approaching wind speeds are 0.5 m/s, 1 m/s and 2 m/s. In Taiwan, the weather and climate are greatly affected by monsoons. In summer, the prevailing wind is a southwesterly or southerly monsoon, while in winter it is northeasterly or northerly monsoon. Thus, the wind directions set in the CFD simulation include south and north winds. There is no heat source in the restroom, and the toilet doors are closed to simulate the scenario when the restroom is in use. Geometric data for the model under investigation are listed in Table 2.

Table 1. Quantity and position of installed awning windows (north facade).

Windows	Configuration Legend	Windows	Configuration Legend
	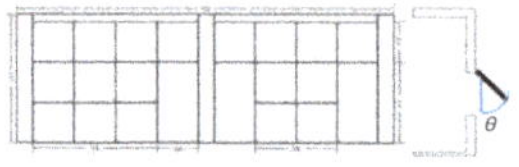 (a) North facade and opening angle.	8	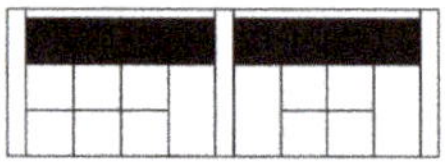 (e) Window configuration W4.
4	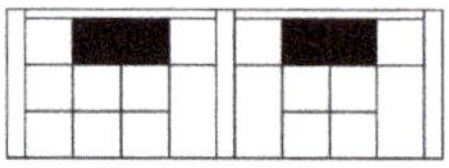 (b) Window configuration W1.	8	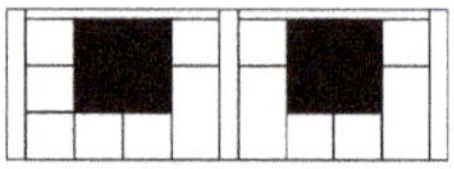 (f) Window configuration W5.
4	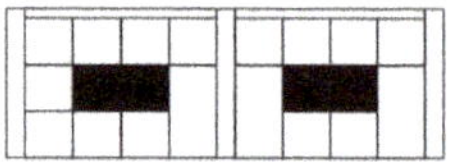 (c) Window configuration W2.	8	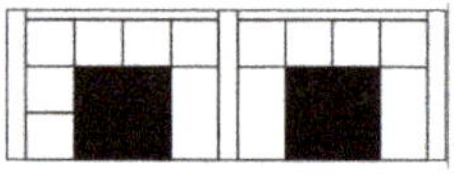 (g) Window configuration W6.
4	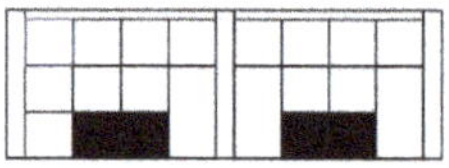 (d) Window configuration W3.	12	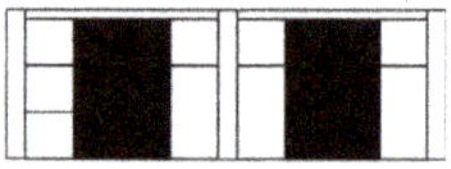 (h) Window configuration W7.

Table 2. Geometric data for restroom under investigation.

Parts of the Model	Geometric Data
Investigated restroom	10.65 m × 9.2 m × 3.2 m (height)
Dimensions of the awning window	1.0 m × 0.9 m (height)
Net ceiling height	3.2 m (Z direction)

The major cause of poor air quality in restrooms is the malodor of feces and urine. The smell of feces and urine is primarily the smell of ammonia. In contrast, sewer malodor consists of mixed chemical substances, such as hydrogen sulfide, methyl mercaptan, trimethylamine, dimethyl disulfide, indole and methyl indole. This study focuses on the restroom flow field and the malodor concentration field. Because the irritating odor of ammonia is believed to be a major contributor to the offensive odor of human waste [5], the NH_3 concentration in the malodor is analyzed. To simulate the very worst condition, the generation rate of unpleasant odors (represented by NH_3) was assumed to be 0.3 L/min (0.2 g (NH_3)/min) in this study. The pollution source area is set to 0.1 m × 0.1 m.

A modified odor removal efficiency (ORE) [2,16,17] was employed to express the ventilation performance of the whole restroom. A higher ORE indicates a lower concentration level, thus indicating better ventilation efficiency for odor removal. The ORE is defined as:

$$\text{ORE} = \frac{C_e - C_0}{C_p - C_0} \tag{1}$$

where C_e is the odor concentration at the exhaust, C_0 is the indoor background concentration and C_p is the average concentration at the height of the breathing zone.

2.2. Numerical Methods

Numerical simulations of the problem that is being investigated are performed via a finite volume method to solve the governing equations with the previously discussed boundary conditions (Table 3). The calculation domain (50 m × 50 m × 3.2 m) is shown in Figure 3a. The commercial CFD code PHOENICS is used to simulate the airflow and NH_3 distributions. The governing equations solved by PHOENICS include a three-dimensional time-dependent incompressible Navier-Stokes equation, a time-independent convection diffusion equation and a k-ε turbulence equation. The formulations of these equations can be found in the PHOENICS manual [18] and in most CFD textbooks; thus, they are not provided here. The empirical turbulence coefficients for the k-ε turbulence equation are assigned as follows: $\sigma_k = 1.0$, $\sigma_\varepsilon = 1.22$, $\sigma_{\varepsilon 1} = 1.44$, $\sigma_{\varepsilon 2} = 1.92$ and $C\mu = 0.09$. These values are widely accepted in CFD k-ε models. To bridge the steep gradients of dependent variables near a solid surface, a general wall function is employed. Iterative calculation continues until a prescribed relative convergence of 10^{-3} is satisfied for all field variables of this problem.

When testing the grid independence of a mesh domain, the NH_3 distribution at the user's squatting position (X = 0.925 m, Y = 1.2 m), which is based on different grid points, is used to calculate the deviation percentages and determine a suitable grid point system for the calculation (Figure 3b). Numerical simulation accuracy depends on the resolution of the computational mesh. A finer grid produces more accurate solutions. In this study, a grid system with approximately 131 × 113 × 52 (769,756) cells is used for numerical simulations. Each cell in the investigated restroom is about 0.1 m × 0.15 m × 0.06 m. An increase in the number of cells provides better information. However, such an increase is accompanied by a significant increase in computational resources.

Table 3. Parameters specified in numerical calculation.

Walls, Ceilings, Doors, Awning Window Glazing	Adiabatic
Outlet planes	Zero static pressure
Quantity and position of installed awning windows	Table 1
Opening angles of awning window	30°, 45°, 60°
Wind directions	North wind, south wind
Wind speeds (with logarithmic velocity profiles, reference height = 10 m)	0.5, 1.0, 2.0 m/s
Toilet door	Closed
Ambient air temperature	25 °C
Pollution source (NH_3)	Volumetric flow rate = 5×10^{-6} m^3/s Mass percentage concentration of NH_3 = 1 (kg NH_3/kg air mixture) Flow area = 0.1 m × 0.1 m
Ambient NH_3 concentration	0 PPMV

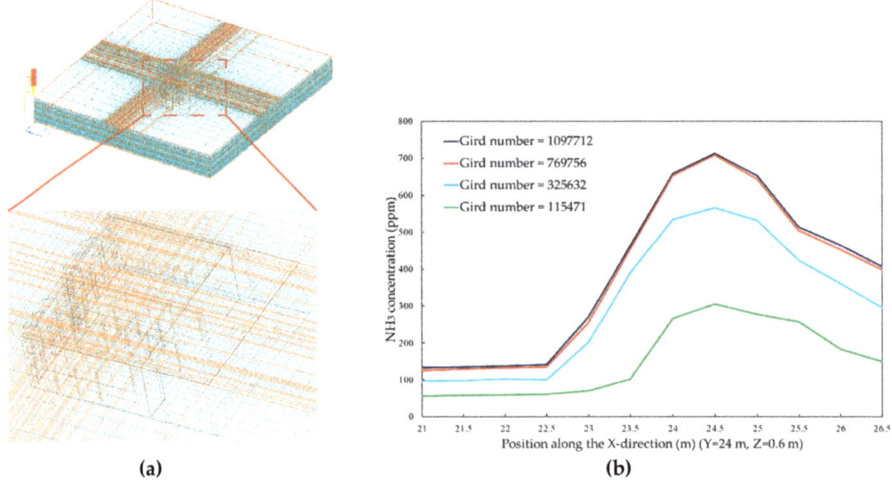

Figure 3. Calculation domain (**a**) and grid independence test results (**b**).

2.3. Model Validation

In this study, a reduced-scale model of the investigated restroom shown in Figure 1 is constructed (Figure 4a). The material used for the model is 3-mm-thick gray hard cardboard and foam core board. The model scale is 1:45. The opening in the model is simply an opening without an installed window. A 4-inch fan is installed at the restroom entrance to simulate a south wind, and the airflow velocity in the model was measured by a multifunction measuring instrument (Testo 435-1) with an anemometer (Testo 0635 1535). There are seven measurement locations: at the two entrances (two locations), the window opening centers (two locations), the aisle centers (two locations) and the central toilet (one point). Each measurement location measures three heights: 1 cm, 3 cm and 5 cm. Next, CFD simulation with the same method mentioned in Section 2.2 is performed based on this reduced-scale model, and the simulation result is compared with the reduced-scale test result. As shown in Figure 4b, the difference between the CFD results and the experimental results is not significant. Thus, the reliability of the simulation results was confirmed.

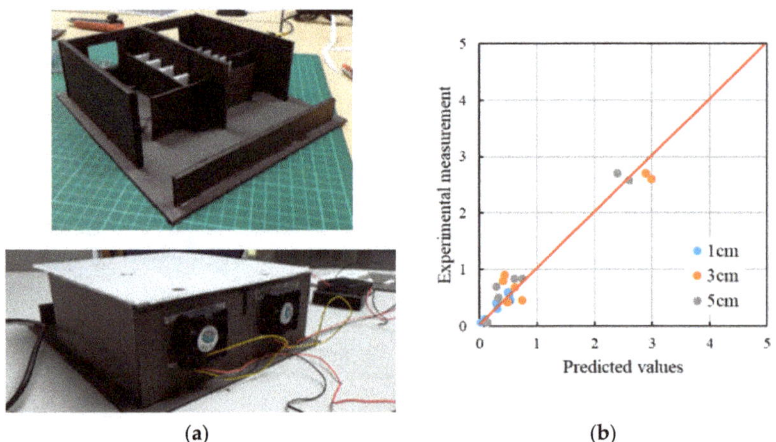

Figure 4. Comparison between the simulation results and the measurements. (**a**) Reduced scale model. (**b**) Predicted values versus simulation results.

3. Results and Discussion

3.1. Case Study: the Effect of Wind Direction

In Taiwan, the perennial wind direction pattern is south in summer and north in winter. In this section, the indoor flow field and NH_3 concentration distribution for different outdoor wind directions are investigated using a case study with the following settings: awning windows installed at the center of the exterior north wall (Table 1c, window configuration W2), an opening angle θ of 45° and an outdoor wind speed of 1 m/s.

Figure 5a shows the flow field in the aisle in the women's restroom (Figure 1c, section A). A north wind flows from the left side of the diagram toward the awning window. Guided by the inclined window surface, outdoor air flows toward the indoor ceiling at an angle pointing to the upper right (Figure 5a, symbol ❶). Then, it flows along the ceiling to the exterior door in the south (right side of the diagram; ❷). This flow becomes the main stream that drives indoor air circulation (❸).

Figure 5b shows the flow field in the toilet area in the women's restroom (Figure 1c, section B). Under the main stream ❹, three air circulations are formed. The clockwise circulations ❺❻ form in the toilet area. Their flow speeds are fast enough to carry NH_3 to the top. After these flows merge into the main stream, they flow to the exterior door in the south. Circulation ❼ is a large-scale clockwise circulation. This circulation flows to the toilet area from the main stream, carries away NH_3 and flows toward the top left ❽. Then, it merges into the main stream and flows toward the exterior door in the south (right side of diagram). Because circulation ❼ is weak, and the circulation follows a particular pattern, the flow in space ❻ at the bottom of ❽ is weak and unable to carry away NH_3, which produces a higher NH_3 concentration than the concentration in the other toilet areas (Figure 5c).

Figure 5d shows the flow field in the urinal area in the men's restroom (Figure 1c, section C). Driven by main stream A and affected by the urinal partition board, the clockwise flows BCD (please refer to the symbols on Figure 5d) merge with the counter-clockwise flow E. As flow BCD flows through the urinal area, the NH_3 concentrations in the three individual urinal areas are low (Figure 5e). Affected by the flow structure, the air in space F stagnates, which results in a higher NH_3 concentration than in other locations.

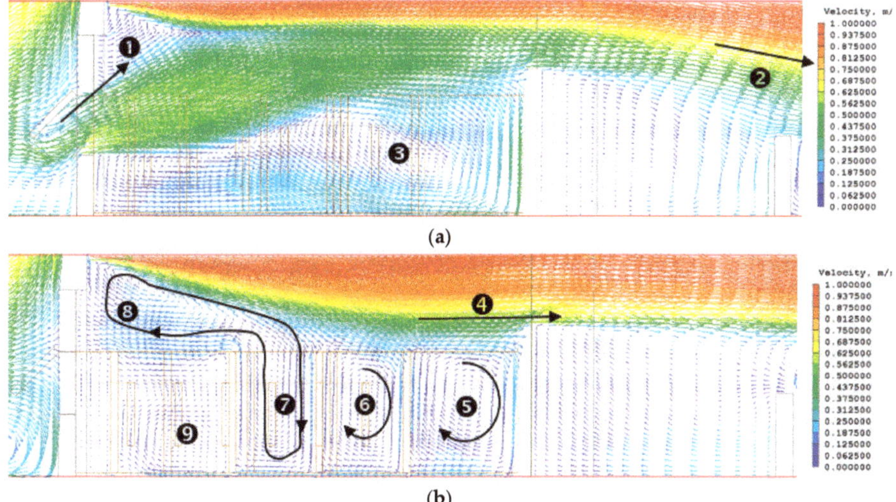

(a)

(b)

Figure 5. *Cont.*

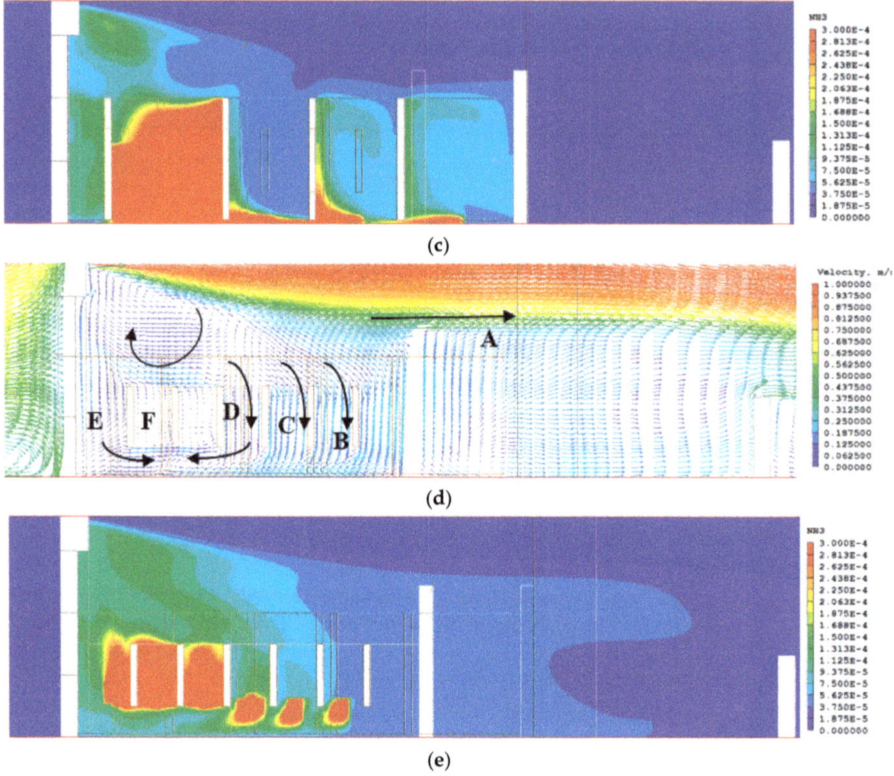

Figure 5. Indoor flow fields and NH_3 concentration (PPMV) distributions for north wind (window configuration: W2 (Table 1c); the opening angle θ is 45°; the outdoor wind speed is 1 m/s). (**a**) Flow field in aisle in the women's restroom (Figure 1c, section A). (**b**) Flow field in toilet area in the women's restroom (Figure 1c, section B). (**c**) NH_3 concentration distribution in toilet area in the women's restroom (Figure 1c, section B). (**d**) Flow field in urinal area in the men's restroom (Figure 1c, section C). (**e**) NH_3 concentration distribution in urinal area in the men's restroom (Figure 1c, section C).

The preceding analysis and Figure 6 reveal that the toilet area and the urinal area near the exterior wall have poor ventilation due to the flow field structure. Figure 6a shows that when users defecate in a squatting position, the NH_3 concentration is high at the breathing zone height, in the area marked with a red dotted line (Z = 0.6 m). When men urinate in a standing position, Figure 6b shows that the NH_3 concentration is high at the breathing zone height, in the area marked with a red dotted line (Z = 1.5 m).

The Annex I 109.03 of "Information notices on occupational diseases: a guide to diagnosis (European Commission, 2009)" [19] indicates that the odor threshold of NH_3 is about 20 ppm; exposure levels of NH_3 that surpass 50 ppm will result in immediate irritation to the nose and throat; exposure level to 250 ppm is bearable for 30–60 min; and exposure level to 300 ppm is considered to be immediately dangerous to life and health. It is good to define an acceptable level of NH_3 from which the ventilation performance of each case could be evaluated. However, the recommended values above cannot be well applied in quasi-steady-state problems raised in this study that urinating or defecating is within a limited time-period. More observation and discussion are needed. Besides, the generation rate of unpleasant odors set in this study presents the very worst condition; if a referenced threshold level was used and linked to our simulations, the results would be misleading. Such constraints limit this study to a relative comparison among cases.

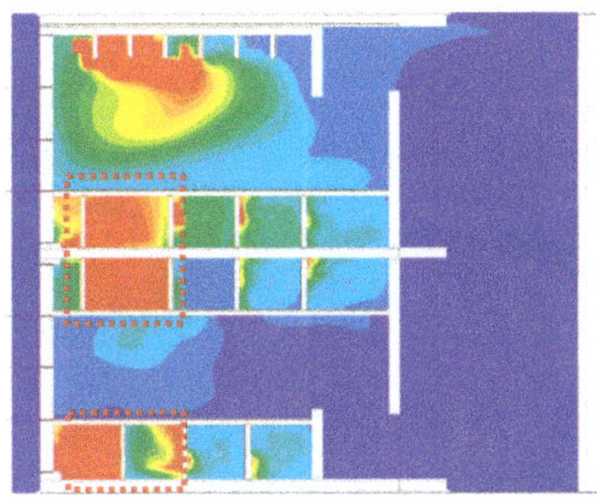
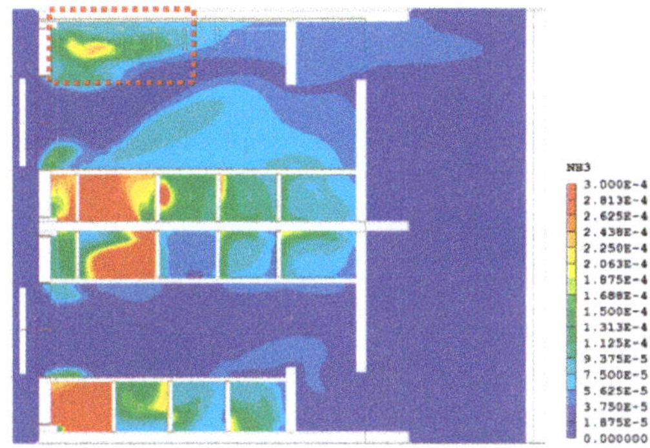

(**a**) Height Z = 0.6 m (height of the squatter's breathing zone).

(**b**) Height Z = 1.5 m (height of the standee's breathing zone).

Figure 6. NH$_3$ concentration (PPMV) distributions at breathing zone heights for (**a**) squatting position and (**b**) standing position in north wind (window configuration: W2; the opening angle θ is 45°; the outdoor wind speed is 1 m/s).

Figure 7a shows flow field in the aisle in the women's restroom (Figure 1c, section A). Affected by the partition wall (❶), the south wind flows into the room via the south exterior door (right side of the diagram) and then flows toward the top left ceiling (❷). Subsequently, it flows outdoors via the north window (left side of the diagram). This flow becomes the main stream that drives indoor circulations (❸) at the bottom zone.

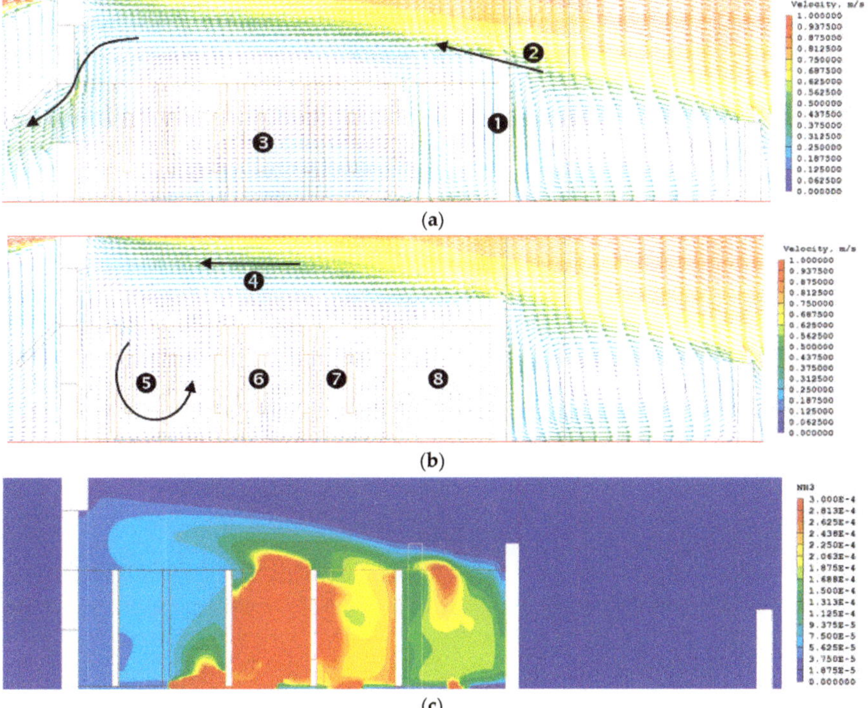

(**a**)

(**b**)

(**c**)

Figure 7. *Cont.*

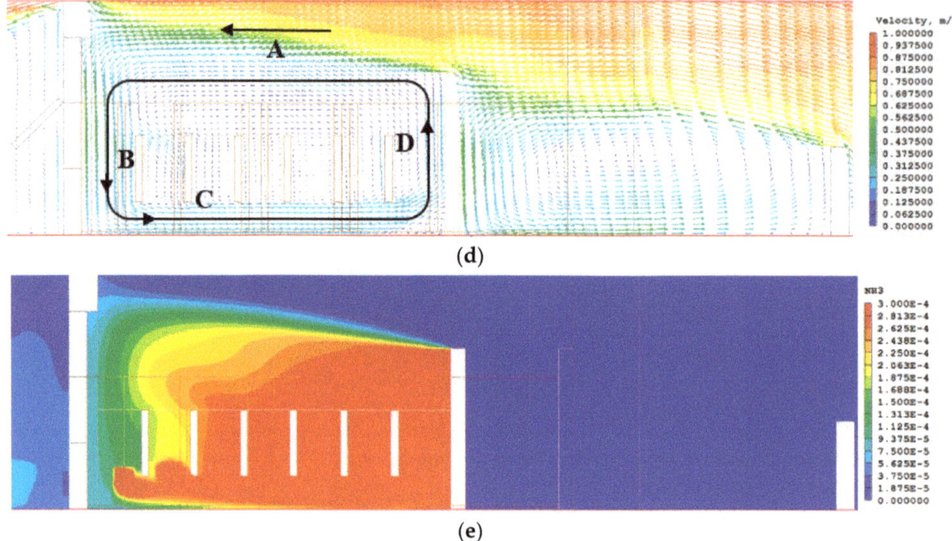

(d)

(e)

Figure 7. Indoor flow field and NH_3 concentration (PPMV) distribution for south wind (window configuration: W2; the opening angle θ is 45°; the outdoor wind speed is 1 m/s). (**a**) Flow field in aisle in the women's restroom (Figure 1c, section A). (**b**) Flow field in toilet area in the women's restroom (Figure 1c, section B). (**c**) NH_3 concentration distribution in toilet area in the women's restroom (Figure 1c, section B). (**d**) Flow field in urinal area in the men's restroom (Figure 1c, section C). (**e**) NH_3 concentration distribution in urinal area in the men's restroom (Figure 1c, section C).

Figure 7b shows the flow field in the toilet area in the women's restroom (Figure 1c, section B). The main stream ❹ flows into the room via the south exterior door (right side of the diagram). Near the north wall (left side of the diagram), part of the flow flows out the window, but the majority of the flow moves down along the wall. When the flow enters the toilet, counter-clockwise air circulation ❺ is formed, which results in a low NH_3 concentration in the restroom (Figure 7c). Because the main stream is close to the ceiling and the flow speed is low, stagnant air and high NH_3 concentrations result in the spaces at the bottom ❻❼❽.

Figure 7d shows the flow field in the urinal area in the men's restroom (Figure 1c, section C). Main stream A flows into the room via the south exterior door (right side of the diagram). Near the north wall (left side of the diagram), part of the flow flows out the window, but the majority of the flow flows downward along the wall surface (B). Affected by the partition board, flow B turns right and exhibits a pattern of horizontal flow (C). Then, it turns upward at the partition wall (D) and forms a major counter-clockwise circulation (BCD). This large circulation surrounds the urinal area, which results in a high NH_3 concentration in this area (Figure 7e).

The preceding analysis and Figure 8 show that as a result of the flow structure, the toilet areas and urinal area near the indoor partition wall have inferior ventilation. Figure 8a shows that when users defecate in a squatting position the NH_3 concentration is high at the breathing zone height, in the area marked with a red dotted line (Z = 0.6 m). When men urinate in a standing position, the NH_3 concentration is high at the breathing zone height, in the area marked with a red dotted line (Z = 1.5 m) (Figure 8b). A possible solution is to add a vertical guiding board to the ceiling in the area with inferior ventilation (red blocks in Figures 6 and 8). In this manner, part of the main stream flow is guided into an area with poor ventilation, and the NH_3 is carried away. Although this issue is not the focus of this study, it warrants for future investigation.

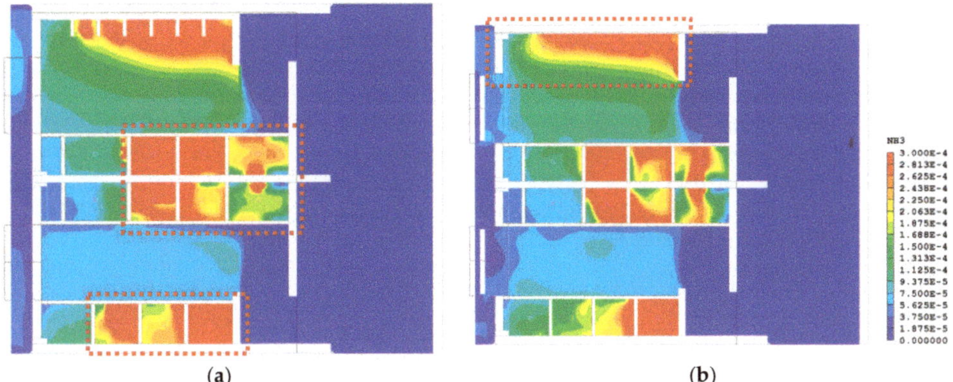

Figure 8. NH_3 concentration (PPMV) distributions at heights of squatter and standee breathing zones for south wind (window configuration: W2; the opening angle θ is 45°; the outdoor wind speed is 1 m/s). (**a**) Height Z = 0.6 m (height of the squatter's breathing zone). (**b**) Height Z = 1.5 m (height of the standee's breathing zone).

3.2. Findings and Design Recommendation

The previous section explains how the ventilation performance of each restroom is analyzed via flow pattern observation and the NH_3 concentration distribution. Due to page length limitations, this paper cannot elaborate the restroom ventilation effectiveness of each window configuration. Instead, the modified odor removal efficiency (ORE) was used to investigate the ventilation effectiveness in each case. Figure 9 shows the ORE for the cases with 0.5 m/s south wind (commonly seen in spring/summer) and 2.0 m/s north wind (autumn/winter). This figure reveals that the window configuration (W1–W7) has a major impact on the ventilation of the restrooms; this impact significantly exceeds the effect of other parameters (wind speed, wind direction and opening angle). When window configuration W2 is employed, an arbitrary opening angle (30°–60°) under the two wind conditions is acceptable; opening angle 45° is proposed. When window configuration W5 is adopted, the 30° and 60° opening angle configurations are recommended for spring/summer and autumn/winter respectively (marked in light yellow and light blue in Figure 9).

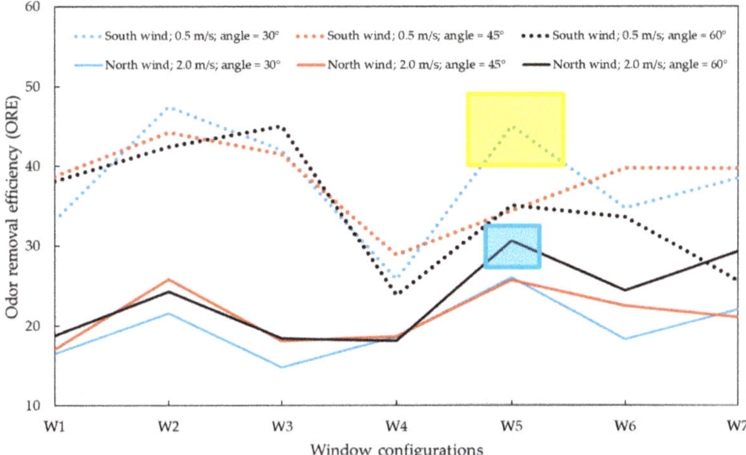

Figure 9. Odor removal efficiency for the cases with 0.5 m/s south wind and 2.0 m/s north wind.

Design recommendations for the installation quantity and position of awning windows are listed in Table 4. K-12 school restrooms could adopt window configuration W2 for the north walls with an opening angle of 45° for all seasons. Window configuration W5 is also recommended. However, here, different seasons require different opening angles. In spring and summer with low-speed wind, the window should be opened 30°. Such a configuration could also prevent rain from entering the room during the rainy season in summer. In autumn and winter with stronger wind, the window should be opened 60°. However, the cost of the W5 window configuration is high. In addition, some of the windows are located in high positions and are thus difficult to open. Therefore, window configuration W5 is not our first recommendation.

Table 4. Recommended awning window configuration and opening angle.

Recommendation	Window Configuration	Season	Opening Angle
First recommendation	W2 (Window number: 4)	All	45°
Second recommendation	W5 (Window number: 8)	Spring and summer	30°
		Autumn and winter	60°

4. Conclusions

In this study, the restrooms in K-12 public schools in Taiwan are selected as the study subjects. The effect of awning window quantity and installation position on airflow pattern and air contaminant (NH_3) distribution in the representative restroom is analyzed using CFD. The research findings are summarized as follows.

1. In autumn and winter, the north monsoon wind flows into the awning window from the north. Guided by the inclined window surface, it flows toward the indoor ceiling at an angle that is oriented upwards. It then flows to the south exterior door along the ceiling. In spring and summer, a south monsoon occurs, and a reverse indoor flow pattern is observed. This north-south flow becomes the main stream that drives restroom indoor circulation at the bottom (and along both sides).

2. In the toilet areas in the men's and women's restrooms and the urinal area in the men's restroom, if the air flow into these areas forms a circulation pattern in the main stream and the flow speed is sufficiently high, the air circulation will carry air pollutants from the areas to the top of the space, where they flow out of the restroom after merging with the main stream. If these conditions are not satisfied, the air in the areas will stagnate, and the air pollutant concentration will be high. Under a south wind in summer, the flow at the bottom of the main stream is affected by the urinal partition board and partition wall and forms a large circulation surrounding the urinal area. This circulation causes poor ventilation in this area.

3. Based on the ventilation performance analysis using the modified odor removal efficiency (ORE), we suggest that K-12 school restrooms use window configuration W2 in their north walls (as shown in Table 4). The opening angle should be set to 45° for all seasons.

The modeling and simulation results are limited to the urban and building morphologies chosen with the specific wind environments. At other conditions, recommendations given in the paper may not be applicable. The surroundings would greatly affect the magnitude and direction of the approaching wind. The interior partition design and layout would also affect the ventilation performance. Although investigation on other themes (surroundings, outside environments, interior partitioning, buoyancy effect, link to IAQ studies, etc.) are not what we are exploring in this study but is worthy of further consideration.

Author Contributions: Conceptualization, S.-C.C. and C.-M.L.; methodology, C.-M.L.; software, Y.-P.L. and C.Y.; validation, Y.-P.L. and C.Y.; formal analysis, S.-C.C., C.Y., and Y.-P.L.; investigation, S.-C.C. and C.-M.L.; resources, S.-C.C.; data curation, S.-C.C. and C.-M.L.; writing-original draft preparation, C.-M.L.; writing-review and editing, S.-C.C., Y.-P.L. and C.-M.L.; visualization, C.Y.; supervision, S.-C.C. and C.-M.L.; project administration, S.-C.C.; funding acquisition, S.-C.C.

Conflicts of Interest: The authors declare that there is no conflict of interest.

References

1. Tung, Y.-C.; Hu, S.-C.; Tsai, T.-Y. Influence of bathroom ventilation rates and toilet location on odor removal. *Build. Environ.* **2009**, *44*, 1810–1817. [CrossRef]
2. Tung, Y.-C.; Shih, Y.-C.; Hu, S.-C.; Chang, Y.-L. Experimental performance investigation of ventilation schemes in a private bathroom. *Build. Environ.* **2010**, *45*, 243–251. [CrossRef]
3. Yang, J.-H.; Kim, O. Improvement of ventilation efficiency by changing the shape of glass partition in bathroom of apartment house. *Indoor Built. Environ.* **2017**, *26*, 1274–1291. [CrossRef]
4. Seo, Y.; Park, I.S. Study for flow and mass transfer in toilet bowl by using toilet seat adopting odor/bacteria suction feature. *Build. Environ.* **2013**, *67*, 46–55. [CrossRef]
5. Sato, H.; Hirose, T.; Kimura, T.; Moriyama, Y.; Nakashima, Y. Analysis of malodorous volatile substances of human waste: Feces and urine. *J. Health Sci.* **2001**, *47*, 483–490. [CrossRef]
6. Kim, O.; Yang, J.-H. On the design to improve the bathroom exhaust performance in multi-unit residential buildings. *J. Asian Arch. Build. Eng.* **2016**, *15*, 349–356. [CrossRef]
7. Yin, P.; Pate, M.B.; Sweeney, J.F. The impact of operating pressure on residential bathroom exhaust fan performance. *J. Build. Eng.* **2016**, *6*, 163–172. [CrossRef]
8. Choi, W.; Pate, M.B.; Sweeney, J.F. Study of bathroom ventilation fan performance trends for years 2005 to 2013—Data analysis of loudness and efficacy. *Energy Build.* **2016**, *116*, 468–477. [CrossRef]
9. Aflaki, A.; Mahyuddin, N.; Mahmoud, Z.A.; Baharum, M.R. A review on natural ventilation applications through building façade components and ventilation openings in tropical climates. *Energy Build.* **2015**, *101*, 153–162. [CrossRef]
10. Zhai, Z.; Mankibi, M.E.; Zoubir, A. Review of natural ventilation models. *Energy Procedia* **2015**, *78*, 2700–2705. [CrossRef]
11. Jomehzadeh, F.; Nejat, P.; Calautit, J.K.; Yusof, M.B.M.; Zaki, S.A.; Hughes, B.R.; Yazid, M.N.A.W.M. A review on wind catcher for passive cooling and natural ventilation in buildings, Part 1: Indoor air quality and thermal comfort assessment. *Renew. Sustain. Energy Rev.* **2017**, *70*, 736–756. [CrossRef]
12. Nomura, M.; Hiyama, K. A review: Natural ventilation performance of office buildings in Japan. *Renew. Sustain. Energy Rev.* **2017**, *74*, 746–754. [CrossRef]
13. Omrani, S.; Garcia-Hansen, V.; Capra, B.; Drogemuller, R. Natural ventilation in multi-storey buildings: Design process and review of evaluation tools. *Build. Environ.* **2017**, *116*, 182–194. [CrossRef]
14. Oropeza-Perez, I.; Østergaard, P.A. Active and passive cooling methods for dwellings: A review. *Renew. Sustain. Energy Rev.* **2018**, *82*, 531–544. [CrossRef]
15. Solgi, E.; Hamedani, Z.; Fernando, R.; Skates, H.; Orji, N.E. A literature review of night ventilation strategies in buildings. *Energy Build.* **2018**, *173*, 337–352. [CrossRef]
16. Brohus, H. Personal exposure to contaminant sources in ventilated rooms. Ph.D. Thesis, Aalborg University, Aalborg, Denmark, 1997.

17. Qian, H.; Li, Y.; Nielsen, P.V.; Hyldgaard, C.E.; Wong, T.W.; Chwang, A.T. Dispersion of exhaled droplet nuclei in a two-bed hospital ward with three different ventilation systems. *Indoor Air* **2006**, *16*, 111–128. [CrossRef] [PubMed]
18. Spalding, D. *The Phoenics Encyclopedia*; Cham Ltd.: London, UK, 1994.
19. European Commission. *Information Notices on Occupational Diseases: A Guide to Diagnosis*; European Commission: Luxembourg, 2009.

Article

Experimental Studies Involving the Impact of Solar Radiation on the Properties of Expanded Graphite Polystyrene

Paweł Krause and Artur Nowoświat *

Faculty of Civil Engineering, Silesian University of Technology, Gliwice 44–100, Poland; pawel.krause@polsl.pl
* Correspondence: artur.nowoswiat@polsl.pl

Received: 27 November 2019; Accepted: 20 December 2019; Published: 22 December 2019

Abstract: This article presents the research studies aimed at identifying the behavior of expanded polystyrene with the addition of graphite in the conditions of exposure to solar radiation. For this purpose, a series of in situ tests and laboratory studies were carried out. Three types of material were tested, i.e. expanded polystyrene (EPS) (white polystyrene), polystyrene with the addition of graphite (gray polystyrene) and two-layer polystyrene (gray bottom layer and white top layer). Temperature distributions on the surfaces of the panels in field and laboratory conditions were determined. The distributions of temperature were recorded at varied wind impact (field conditions and laboratory conditions) and at varied impact of solar radiation (laboratory conditions). Based on the conducted experiments, differences in temperature distribution on the surfaces of the tested panels were determined. In addition, geometric changes and deformation levels of the tested white and gray expanded polystyrene panels exposed to artificial sun radiation were determined in laboratory conditions.

Keywords: expanded polystyrene (EPS); gray polystyrene; artificial sun; thermographic measurement; temperature distribution

1. Introduction

The subject of the thermal insulation of buildings is very important because, according to Berardi [1], energy consumption in buildings accounts for 30% of the total final energy consumption in the world. In addition to environmental protection, it is very important to meet the expectations of ordinary building users, who in general are not experts in the field of energy saving [2]. In building envelope insulation systems, we very often apply such insulation materials as expanded polystyrene (EPS) and extruded polystyrene (XPS) [3], or their modifications [4–7]. Such systems can be used on building envelopes as insulation from the inside, which is often applied in buildings having historical character [8]. When using EPS as insulation from the inside, we consider, among others, the problems of the thermal accumulation of building envelopes, which have impact on their energy performance [9]. The second and more frequent application of such systems involves the insulation from the outside [10]. Also for that type of application, various physical processes are taken into consideration [11]. One of such important problems is the impact of the moisture content of polystyrene used in thermal insulation systems on thermal conductivity [12,13]. Another important issue related to the thermal insulation of buildings is the distribution of temperature on the outer wall surface [14–16], which often depends upon various perturbations [17]. Another application of expanded polystyrene (EPS) involves its use in ready-made wall systems in the form of lightweight panels [18]. The most commonly applied system for the thermal insulation of buildings is the system for wall insulation from the outside, such as External Thermal Insulation Composite System (ETICS) [19].

Frequently, graphite is applied for expanded polystyrene composites [20], which improves the capacity of energy storage [21]. The described problems involving the impact of temperature on polystyrene with thermal additives have been discussed by various research teams [22–24]. There is relatively little information on the scale of the issue under consideration. Based on the Polish literature, we can find information that, among others, relates how rapid changes of temperature on the surface of 'black polystyrene panels' contribute to their damage on building facades, in particular from the south. In other works in Poland, the issue involving the impact of solar radiation was investigated both in laboratory and natural conditions, as well as in numerical simulations. It was found that the temperature on the surface of polystyrene under intense solar radiation may surpass the softening point. In addition, it may result in the deformation of the thermal insulation boards, which can bring about the detachment of the polystyrene boards from the wall surface. Exposing the material to excessive, direct solar radiation may damage polystyrene. The said problem was pointed out by Kussauer and Ruprecht [23] in their monograph.

The parameters of such composites may depend on the method of their development [25]. Minh-Phuong Tran et al. demonstrated that adding various types of graphite particles to polystyrene during the production process reduces its thermal conductivity [26]. However, it should be noted that when exposed to solar radiation, gray polystyrene, i.e., with the addition of graphite, behaves differently to white polystyrene, i.e., without such addition. Due to the fact that polystyrene with the addition of graphite is a popular material reducing energy consumption in buildings, and at the same time is sensitive to the impact of solar radiation, the objective of the work is to assess the impact of solar radiation intensity on temperature distribution on the surface of various polystyrenes and on the development of structural and geometric imperfections of thermal insulation materials. The objective formulated in that way allows us to identify the problems related to the impact of solar radiation intensity on unprotected polystyrene, both with and without the addition of graphite.

The implementation technology in the ETICS system for fixing thermal insulation to the wall requires the mortar setting time of over 72 h.

2. Methodology

The influence of solar radiation intensity on unprotected thermal insulation boards was examined. It takes place during construction works, before the successive stages of the ETICS system implementation.

The study involved three 0.15 m-thick expanded polystyrene (EPS) boards of the dimensions 0.5 m × 1.0 m marked with letters A, B, C. The dimensions of the tested elements reflect on the scale 1:1 the actual dimensions of the thermal insulation boards used for insulation. Therefore, it was not necessary to plan scale studies.

Panel A—composite polystyrene consisting of two layers: the main layer in the form of gray, expanded polystyrene, and an additional layer 0.006 m thick in the form of white expanded polystyrene (a layer protecting the gray layer against the impact of solar radiation)

Panel B—polystyrene board made entirely of expanded polystyrene with the addition of graphite (gray polystyrene)

Panel C—expanded polystyrene board without the addition of graphite (white polystyrene) with raised technical parameters.

When testing the distribution of temperature on the surface of a material, it is very important to test its emissivity. These tests are important, because incorrectly determined emissivity does not allow us to determine the real distribution of temperature.

Therefore, in the Section 2.2.1, the methodology of the emissivity test was presented. However, to make the remaining part of the work clearer, except for the methodology in the sub-chapter 2.2.1, also the values of the adopted emissivity of the tested panels A, B, C were provided.

2.1. In Situ Research

2.1.1. Determining the Emissivity of the Panels

Since there are no unequivocal emissivity values for individual polystyrene boards (no declarations in this respect offered by manufacturers), a testing study was carried out. The emissivity of the expanded polystyrene boards (panel B) was determined using a comparative method with respect to the reference tape of the known emissivity of 0.96. The emissivity of the panel B determined on that basis was 0.95, and basing on the literature [27], for white polystyrene (panel C) the emissivity is assumed to be 0.60. In order to determine the impact of the adopted emissivity value on the temperature distribution on the surface for the white polystyrene, a thermographic measurement was made. Two measurement points Sp1 and Sp2 were selected, at which the emissivity values of 0.60 and 0.95 were assumed, respectively. The results are presented in Figure 1.

Figure 1. Test thermogram of white polystyrene board with two measurement points.

The temperature values obtained on the surface of white polystyrene for two different measurement points of different surface emissivity differ from each other by 0.2 °C. The difference is lower than 1% of the measured values. It bespeaks of the fact that the change of emissivity for white polystyrene (panel C) under solar radiation conditions does not significantly affect temperature distribution on the surface of the polystyrene board. Therefore, there was no need to determine the exact emissivity value for the white board, which is why, for the purpose of comparison with panels A and B, the emissivity value of 0.95 (value determined for panel B) was adopted for all three panels.

2.1.2. Field Study

For the tests in real conditions, the following equipment was applied:

- Thermal imaging camera Flir E95 of the temperature measurement range of −20 °C–1500 °C, resolution of 161,472 pixels, thermal sensitivity for the 42° × 32° lens of <30 mK, spatial IFOV resolution for 42° × 32° lens of 2.41 mrad/pixel and fractional spectral sensitivity of 7.5–14 µm.
- Solar data logger of the measuring range of 0–1999 W/m², resolution of 1 W/m², operating temperature of 0 °C–50 °C, storage temperature of −10 °C– 60 °C, operation in the relative air humidity of 10%–90% and storage in the relative air humidity of 10%–80%.

- Electronic thermometer for measuring temperature inside the polystyrene boards equipped with a stainless steel probe skewer of the length of 0.15 m and temperature measurement range of −50 °C–300 °C.

The tests were carried out at two test stands for in situ studies, which were located at the place having the geographical coordinates of: 18°54′E 50°10′N (i.e., Mikołów, Silesa, Poland).

1. Stationary stand—polystyrene boards glued to the wall with polyurethane adhesive (Figure 2a).
2. Mobile stand—a support structure made of 0.018 m-thick boards, wooden slats, hinges and strings, was prepared for each of the polystyrene boards. The structure enabled us to position the tested polystyrene board vertically to the base surface (Figure 2b).

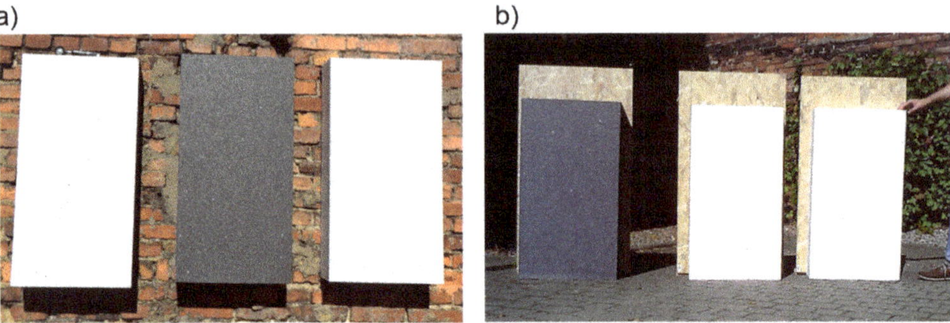

Figure 2. Stand for the in situ tests. (**a**) stationary stand, (**b**) mobile stand.

The tests comprised the registration of temperature distribution on the surfaces of the tested polystyrene panels, and they were carried out for four panels of each of the three types.

The results were recorded at both measuring stands. As part of the measurements, the maximum temperatures, obtainable on a given day, on the surface of the analyzed polystyrene panels, were determined. During the tests, the position of the polystyrene panels on the mobile stand was being adjusted in such a way as to ensure the highest possible load of the tested panels with solar radiation. Simultaneously with the thermographic measurements, the measurements of solar radiation intensity (W/m^2) were carried out. The studies were conducted in the period from 12 September 2018 to 22 September 2018.

2.1.3. Meteorological Measurements

The parameters of the external climate were obtained on the basis of continuous measurements carried out at the local meteorological station, located about 1 km from the place of measurements. The recording of external environment conditions was carried out using the wireless meteorological station VantagePro2 Davis Instruments, powered by a PV cell. The weather station comprised sensors of temperature, air humidity, an anemometer and Pirani meter.

Climate parameters are presented in the graphs. All results were obtained on the basis of continuous measurements carried out at the local meteorological station located in the town where the measurements of the polystyrene panels were carried out.

Figure 3 presents the results for air temperature.

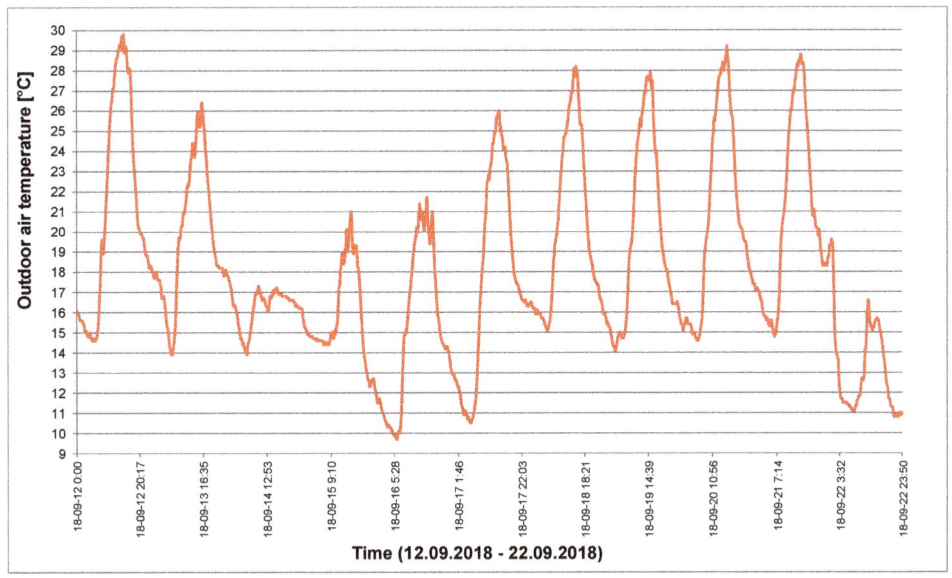

Figure 3. Outdoor air temperature distribution obtained on the basis of measurements carried out in the period from 12 September 2018 to 22 September 2018.

Also the measurements of average air humidity and wind speed were carried out. The results are presented in Figures 4 and 5, respectively.

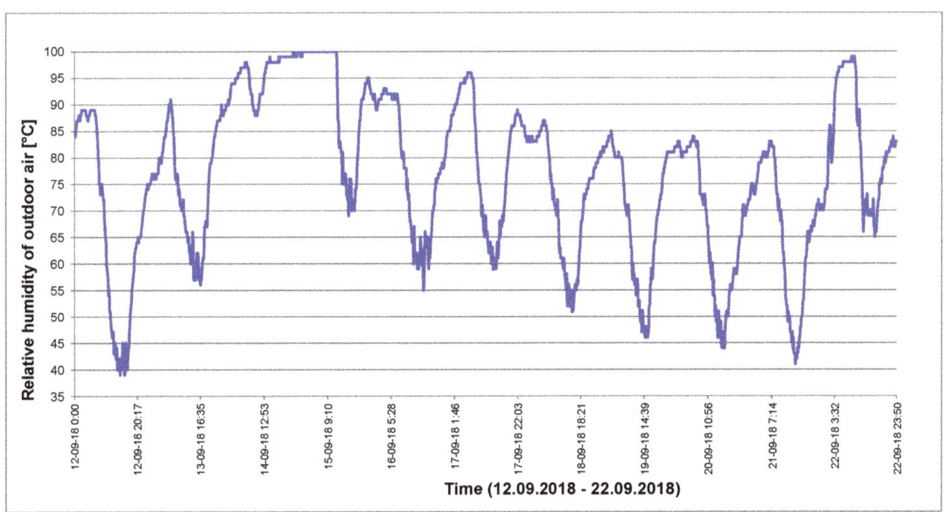

Figure 4. Relative humidity distribution of the outdoor air obtained on the basis of measurements carried out in the period from 12 September 2018 to 22 September 2018.

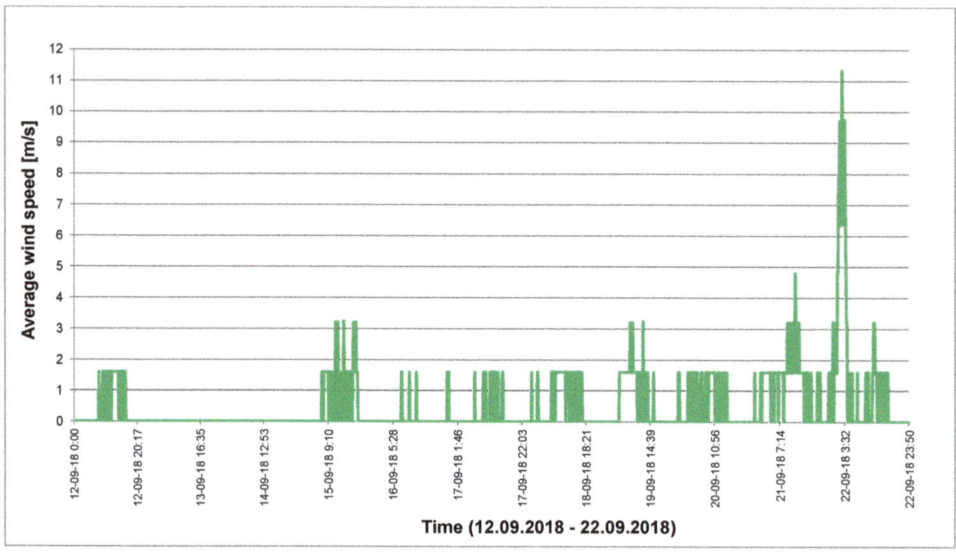

Figure 5. Distribution of average wind speed obtained on the basis of measurements carried out in the period from 12 September 2018 to 22 September 2018.

The last series of results involved the intensity of solar radiation Figure 6.

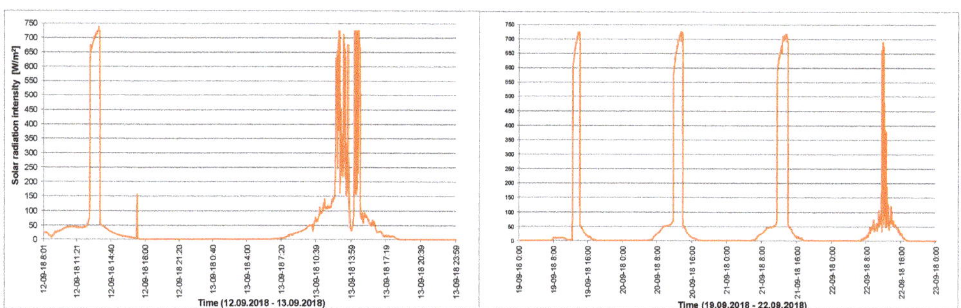

Figure 6. Distribution of solar radiation intensity.

2.2. Laboratory Tests

The laboratory tests of the temperature distribution of the insolated polystyrene panels were carried out in the laboratory hall of the Solar Systems Testing Center located in the laboratory hall of the Euro-Center Science and Technology Park in Katowice in Poland (very close to Mikołów, Silesa). The tests comprised the determination of maximum temperatures at different levels of solar radiation and at different wind speeds. The tested panels had the same geometrical dimensions as the ones tested in situ.

2.2.1. Solar Radiation Stimulator

A solar radiation stimulator (artificial sun) is a stationary complex system of devices that enables the generation of radiation in laboratory conditions which reflects the emission of solar radiation. The simulator consists of a field of lamps, artificial sky, temperature sensors and data generation devices. The main element of the research station is made up of eight metal halide lamps of the total power of 32 kW, enabling the simulation of natural solar radiation. The design of the device allows precise positioning of the tested object and the field of lamps, ensuring angle adjustments in the range of 0°–90°. In that way we can imitate the actual 'journey' of the sun on the horizon (Figure 7).

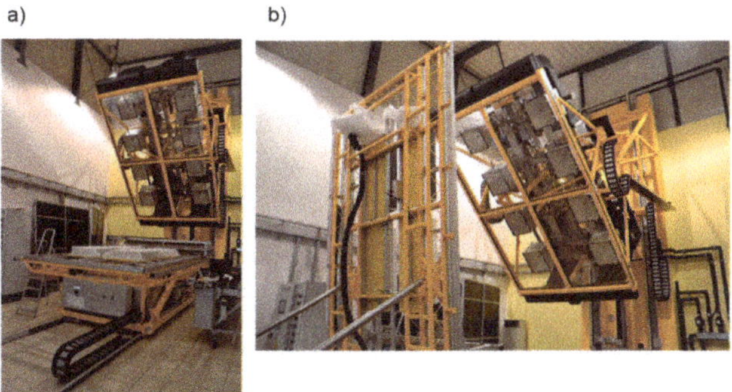

Figure 7. Solar radiation simulator. (**a**) location of samples in the horizontal plane, (**b**) location of samples in the vertical plane.

2.2.2. Measurement in Laboratory Conditions

Along with the laboratory tests with the use of a solar radiation simulator, the measurements of microclimate parameters prevailing inside the laboratory room were carried out simultaneously. The loggers measuring temperature and relative humidity were located in the vicinity of the work table of the solar radiation simulator (Figure 8).

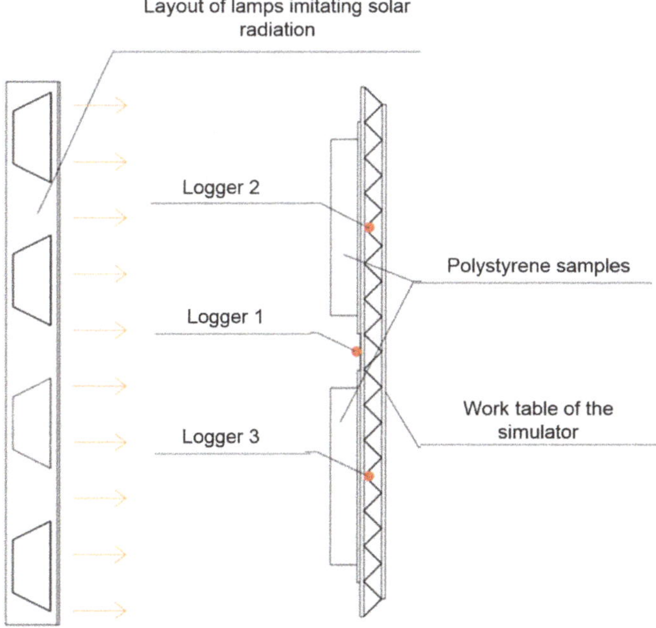

Figure 8. Schematic distribution of loggers on the test stand in the laboratory.

The logger No.1 was placed between the tested polystyrene panels (in the middle of the simulator's work table) from the side of the lamps (artificial sun). The logger No.1 was covered with a layer of aluminum foil from the side of the lamps, which allowed us to eliminate the direct influence of solar radiation on the recorded results of the room climate. The recorders nos. 2 and 3 were placed in the plane of the tested polystyrene samples under the simulator's work table. The logger No. 2 was located in the upper part of the work table, while the logger No. 3 in its lower part. The measurement time step was set at 10 s. Figures 9 and 10 present the measurement results.

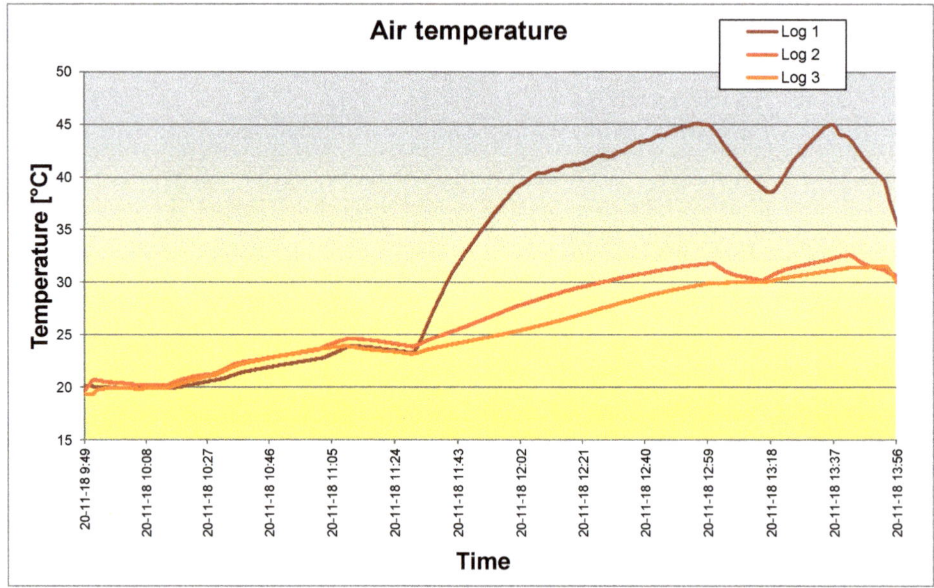

Figure 9. Indoor air temperature of the laboratory recorded by the three loggers.

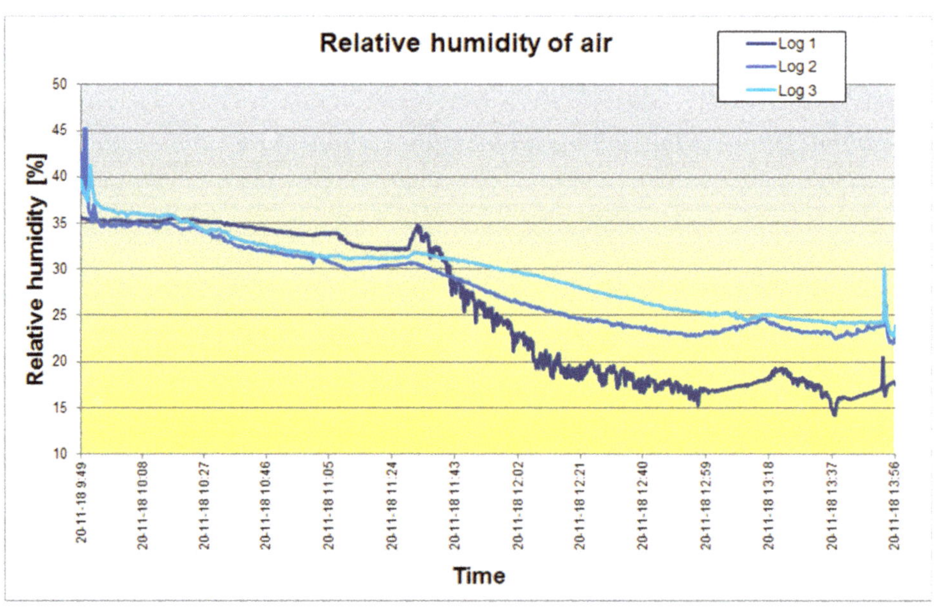

Figure 10. Relative humidity of indoor air of the laboratory recorded by the three loggers.

The tests in the laboratory conditions comprised the recording of temperature distribution on the surfaces of the tested polystyrene boards using a thermal imaging camera in two test series. The tested polystyrene boards were subjected to the impact of solar radiation generated by the simulator.

In the first series of tests, the measurements of temperature distribution on the surfaces of EPS panels were carried out for three different values of solar radiation intensity, i.e. 640 W/m^2, 950 W/m^2 and 1008 W/m^2. For each of these intensities, the actual position of the sun above the horizon was simulated by setting the normal to the lamps plane at an appropriate angle to the horizontal surface. Those positions were, respectively, 640 W/m^2 → 43.28°, 950 W/m^2 → 21°, 1008 W/m^2 → 16.28°.

The intensity at the level of 1008 W/m^2 corresponds to the maximum value of solar radiation intensity onto the vertical plane turned to the south (azimuth of 180°) selected from all Polish climate bases. The value of 950 W/m^2 corresponds to the maximum value of solar radiation intensity onto the vertical plane turned to the south in the period of March–November for the Katowice climate station, i.e. for the climate corresponding to the in situ measurements. The value of 640 W/m^2 corresponds to the maximum value obtained from the initial measurements carried out by the authors of this study at the exact location of the in-situ tests.

In the second series, the tests were carried out with the same position of solar radiation simulator lamps, turned perpendicular to the surface of the tested panels. The average radiation intensity for such a position was adopted at 1104 W/m^2.

3. Results and Discussion

3.1. In Situ Tests

The in situ tests were principally based on thermographic measurements. On the images of the investigated objects, we applied the measuring elements in the form of measuring points (Sp) and rectangular areas (Bx). An exemplary thermogram from the measurements is presented in Figure 11.

Tables 1 and 2 summarize the results of thermographic tests from the in situ measurements for two types of test stands.

Table 1. Temperatures read in the measuring fields of the thermogram made for the stationary stand. The thermograms were obtained in weather conditions close to windless conditions, with a low average wind speed of up to 1.6 m/s. The maximum recorded values of solar radiation intensity were within the range of 716.7 W/m^2–738.4 W/m^2.

Panel Type	T_{MAX} (°C)	T_{MIN} (°C)	T_{AVR} (°C)
Panel B	77.4	59.9	72.5
Panel A	39.7	33.0	34.2
Panel C	34.8	29.0	29.8

Table 2. Temperatures read in the measurement fields of the thermogram made for the mobile stand. The thermograms were obtained in weather conditions close to windless conditions, with a low average wind speed of up to 1.6 m/s. The maximum recorded values of solar radiation intensity were within the range of 716.7 W/m^2–738.4 W/m^2.

Panel Type	T_{MAX} (°C)	T_{MIN} (°C)	T_{AVR} (°C)
Panel B	68.7	56.6	63.6
Panel A	36.7	29.6	30.7
Panel C	33.6	27.2	27.9

The results of the carried out in situ thermographic tests demonstrate that in the case of direct impact of solar radiation, the addition of graphite to expanded polystyrene (EPS) panels (panel B) brings about much higher temperature values on the surface as compared to the A and C panels. The quality of the results is not dependent on the type of test stand. Quantitative differences in the results are due to different intensities of solar radiation. For the meteorological conditions during the realization of the measurements, with the emissivity value of the panel being 0.95, we recorded temperatures which did not exceed the softening temperature of the panel, this being 80 °C. The differences of temperature on the surfaces of the remaining panels did not differ significantly, and these had values far different from the critical value i.e. the softening point.

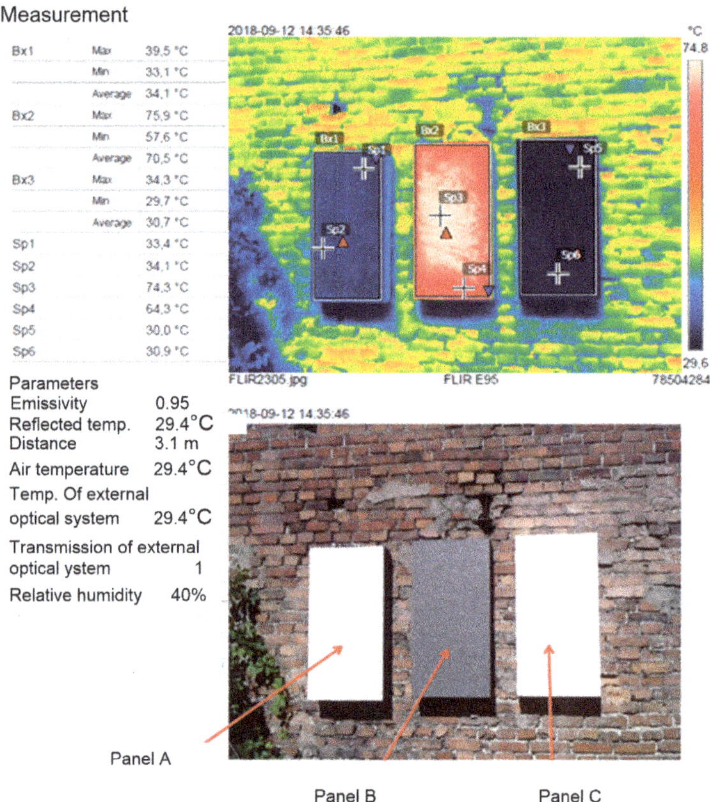

Figure 11. Temperature measurements made for three polystyrene boards. Average wind speed: 1.6 m/s.

Simultaneously with the field thermal imaging tests of the insolated panels, temperature measurements were carried out inside the samples placed on the mobile stand, and the following results were obtained:

Panel B 50.1 °C, Panel A 36.4 °C, Panel C 31.2 °C

The thermographic measurement is confirmed in that case. The highest surface temperature also corresponds to the highest internal temperature of the panel.

3.2. Laboratory Tests

Based on the measurements of solar radiation intensity for the adopted measuring grid, the images presented in Figures 12 and 13 were obtained, using the software dedicated for the test stand.

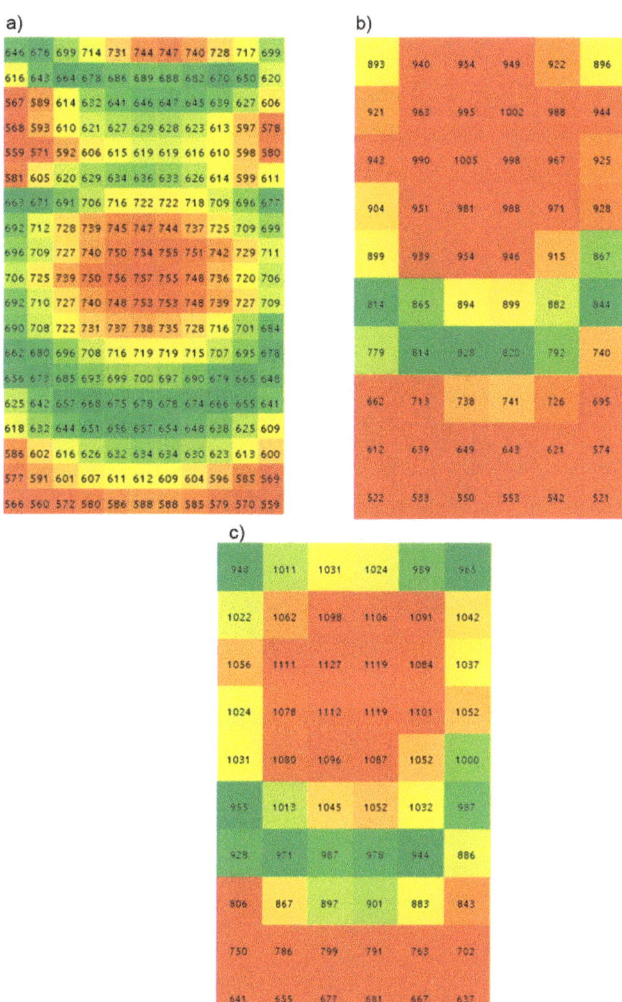

Figure 12. Distribution of solar radiation intensity on the surface comprising the tested samples. (**a**) distribution for artificial sun slope of 43.28°—minimum intensity 665 W/m^2, maximum 757 W/m^2, medium 665 W/m^2, (**b**) distribution for artificial sun slope of 21.0°—minimum intensity 521 W/m^2, maximum 1005 W/m^2, medium 827 W/m^2, (**c**) distribution for artificial sun slope of 16.28°—minimum intensity 637 W/m^2, maximum 1127 W/m^2, average 955 W/m^2.

The second series of tests consisted of one situation presented in Figure 13.

As demonstrated by the in situ tests, the distributions of temperature for panels A and C were very similar. Therefore, the laboratory tests were limited to panel C, which was being compared to the B-type panel. The results of surface temperatures were collected in Tables 3 and 4.

The results presented in Table 3 imply the same conclusions as for the in situ tests, i.e. expanded polystyrene with the addition of graphite has much higher temperature values on the panel surface than polystyrene without the addition of graphite. In addition, we observe that with the absence of wind for the solar radiation intensity of 950 W/m^2 and higher, the risk of exceeding the softening temperature increases significantly. We must admit that such a high intensity did not occur during the in situ tests, but we can read from the data collected for the area of Poland that such high intensities do occur, which was described in the Section 2.2.2 of this work.

1184	1196	1203	1201	1192	1194	1199	1197	1193	1181	1180	1182	1173	1159
1202	1210	1211	1203	1190	1188	1192	1190	1185	1174	1176	1181	1178	1168
1182	1184	1188	1178	1163	1162	1166	1162	1156	1145	1147	1154	1152	1145
1183	1185	1182	1167	1157	1159	1164	1161	1155	1139	1135	1146	1150	1147
1175	1179	1174	1157	1141	1141	1146	1143	1137	1122	1123	1135	1139	1137
1156	1161	1153	1137	1122	1121	1126	1124	1118	1105	1108	1118	1125	1123
1144	1149	1140	1124	1109	1110	1116	1113	1107	1094	1096	1107	1116	1115
1136	1141	1131	1116	1103	1103	1110	1108	1101	1089	1091	1103	1112	1112
1132	1135	1126	1112	1098	1099	1105	1103	1095	1085	1089	1102	1111	1110
1125	1131	1125	1112	1098	1097	1104	1102	1094	1085	1089	1103	1112	1112
1134	1136	1133	1121	1108	1107	1112	1109	1100	1093	1098	1109	1114	1116
1144	1145	1141	1131	1121	1123	1127	1125	1115	1105	1105	1115	1121	1122
1146	1146	1145	1139	1130	1130	1135	1133	1124	1115	1117	1123	1126	1127
1147	1151	1153	1149	1138	1137	1140	1137	1127	1119	1121	1125	1124	1121
1151	1157	1162	1162	1153	1153	1156	1153	1143	1134	1134	1133	1130	1124
1124	1134	1141	1146	1142	1144	1149	1146	1135	1125	1122	1115	1110	1101
1084	1094	1102	1111	1108	1110	1113	1111	1102	1093	1087	1076	1071	1059
1071	1084	1094	1106	1106	1109	1112	1110	1103	1093	1084	1070	1062	1049
1055	1070	1083	1098	1100	1103	1105	1103	1095	1086	1073	1055	1043	1027
1026	1043	1058	1075	1079	1083	1084	1082	1075	1065	1050	1029	1013	996
986	1005	1021	1039	1044	1047	1048	1045	1037	1027	1011	987	969	949
931	950	969	987	993	996	996	993	987	977	960	934	913	893
867	886	906	926	931	934	934	930	923	913	894	867	842	822

Figure 13. Distribution of solar radiation intensity on the surface comprising the tested samples. Location of lamps perpendicular to the tested samples. Minimum intensity 822W/m^2, maximum 1211 W/m^2, medium 1104 W/m^2.

Table 3. Temperatures as a function of the changes in solar radiation intensity read in the measurement fields of the thermogram made for the laboratory stand (Wind speed v = 0 m/s).

Panel Type	T_{MAX} (°C)	T_{MIN} (°C)	T_{AVR} (°C)
	Intensity ls = 640 W/m^2		
Panel B	71.7	59.2	67.2
Panel C	31.0	26.4	27.0
	Intensity ls = 950 W/m^2		
Panel B	84.4	67.0	80.2
Panel C	32.0	26.1	26.8
	Intensity ls = 1008 W/m^2		
Panel B	93.4	74.6	89.0
Panel C	36.4	29.7	30.4

Tables 3 and 4 plainly demonstrate that wind speeds of 3 m/s and 5 m/s significantly reduced the temperatures on the surfaces of the panels. Even with the maximum solar radiation intensity of ls = 1008 W/m^2, the temperatures dropped below the softening point of the panels.

For the second series of tests, i.e. with the lamps of the solar radiation simulator directed perpendicular to the surface of the tested panels, the results are collected in Table 5.

Table 4. Temperatures as the function of wind speed changes read in the measurement fields of the thermogram made for the laboratory stand. Radiation intensity ls = 1008 W/m^2.

Panel Type	T_{MAX} (°C)	T_{MIN} (°C)	T_{AVR} (°C)
	Wind speed v = 1 m/s		
Panel B	93.2	74.9	88.6
Panel C	35.4	29.9	30.4
	Wind speed v = 3 m/s		
Panel B	81.4	61.0	75.8
Panel C	33.6	29.4	29.8
	Wind speed v = 5 m/s		
Panel B	71.0	54.3	66.3
Panel C	32.5	29.6	29.9

Table 5. Temperatures for the second series of tests.

Panel Type	T_{MAX} (°C)	T_{MIN} (°C)	T_{AVR} (°C)
Panel B	101.9	72.7	91.6
Panel C	44.8	35.8	37.0

As we can see in Table 5, solar radiation perpendicular to the surface of the panels is plainly threatening their durability. This may bring about a complete destruction of expanded polystyrene with the addition of graphite.

Then the comparative tests of internal temperature distribution were carried out for panels A and B.

In order to examine the distribution of temperature inside the panels, two temperature measurement sensors were placed on the lateral planes of the panels. The sensors were arranged in such a way (see Figure 14) to examine the impact of the protective layer of panel A on temperature distribution inside it as compared to panel B.

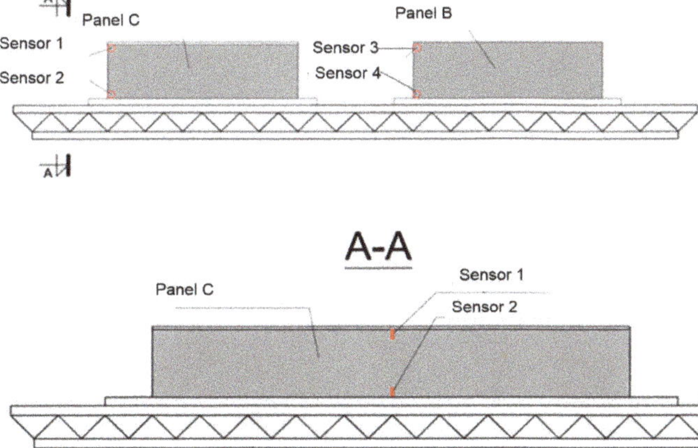

Figure 14. Schematic of the arrangement of sensors in the panels as part of temperature measurements in the samples.

The measurement results are presented in Figure 15.

With the uniform insolation of the panels with solar radiation, the temperature readings indicated that panel B exhibited the susceptibility to strong heating. At the place of sensor 3, the maximum recorded temperature was 69.3 °C. Due to that heating susceptibility of panel B, we decided to carry out additional tests involving the impact of temperature on the deformation of polystyrene boards installed on the surface of building envelopes. The boards were mounted to the base surface using only the spot method. Macroscopic examinations of the panel B demonstrated significant deformations in the form of raised edges of the panel and local degradation of the polystyrene surface in the form of softening. In order to illustrate the resulting deformations and damages of panel B, a measuring staff was applied to it, and the measurements were carried out with the use of a steel tape and a folding measure, as presented in Figure 16.

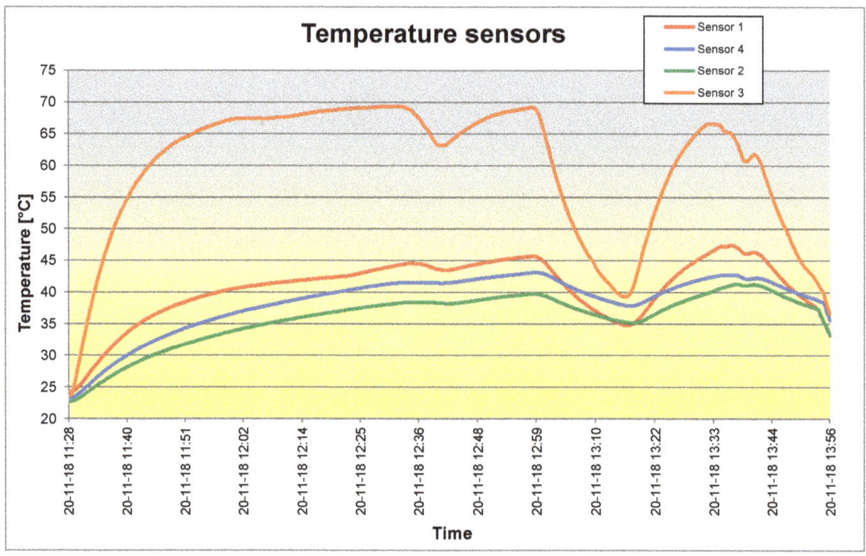

Figure 15. Graph of temperature values on the time axis obtained from the measurements.

Figure 16. Damage to panels B. (**a**) loss of polystyrene granules in effect of exceeding the softening temperature, (**b**) and (**c**) deformation of the panel due to the exposure to artificial sun, (**d**) panel edge raised away from the ground.

The deformations of the panels were confirmed by the measurements using a 3D scanner. The results are presented in Figures 17 and 18.

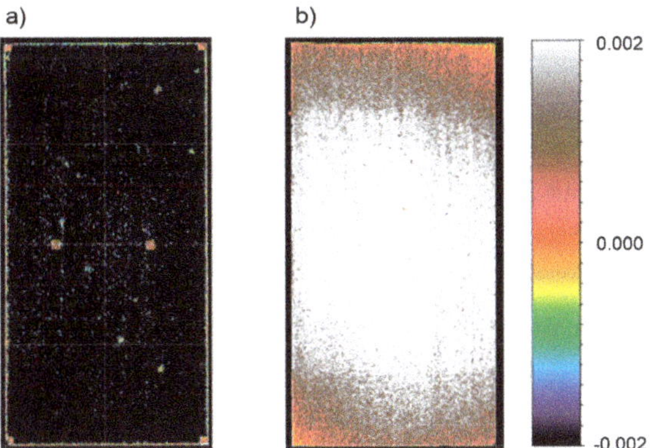

Figure 17. Deformation map of spot-glued panels obtained from 3D scanner readings. (**a**) C-type panel, (**b**) B-type panel.

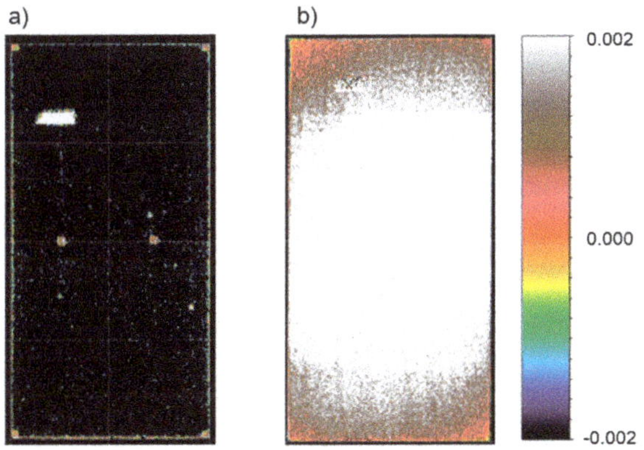

Figure 18. Deformation map of spot-glued panels obtained from 3D scanner readings, (**a**) C-type panel, (**b**) B-type panel.

In the case of C panels, we are faced with point deformations of the tested surfaces. We can presume that these deformations result from the occurrence of trace inclusions of gray polystyrene granules in the structure of polystyrene C. In the B-type panels (gray polystyrene) due to the impact of solar radiation, deformations of entire surfaces of the tested panels were identified.

4. Conclusions

Temperature measurements carried out using different methods demonstrated a different impact of solar radiation intensity on unprotected EPS insulation panels in the ETICS system. Panels A and C are characterized by low sensitivity to the changes of solar radiation intensity as adopted in the study. Panel B, entirely made of polystyrene with the addition of graphite, demonstrated high sensitivity to external factors such as insolation and wind. The temperatures measured on the surface of the panels with the addition of graphite for the intensity of 950 W/m^2 are similar to the softening point of polystyrene declared by the manufacturer. In the case of radiation intensity of 1000 W/m^2 and windless conditions, the softening temperature is exceeded, and destruction of material is taking place in the form of melting traces and deformation of the panels. The conducted research has demonstrated that there is a significant impact of wind influence on lowering the temperature of the polystyrene surface, even by 20 °C for the wind speed of 5 m/s. It should also be noted that the extent and nature of temperature changes on the external surface of the panels depends on the area of impact and exposure time, while the extent of deformations is additionally dependent on the method used to fix the panels to the ground. Thus, adverse environmental conditions, such as windless conditions and strong insolation, foster destructive processes, causing geometrical changes and deformations of elements made of B-panels. By using a composite panel (A-panel), many adverse effects of environmental conditions are avoided compared to B-panels. Further studies should be addressed to determining the environmental conditions ensuring safe realization of thermal insulation works in the ETICS system using B-panels.

Author Contributions: A.N. wrote the article, P.K. methodology, writing-original draft preparation and checked the manuscript and carried out the revision. All authors have read and agreed to the published version of the manuscript.

Funding: Publication supported is a part of the postdoctoral habilitation grant. Silesian University of Technology, no. 03/030/RGH18/0077.

Conflicts of Interest: The authors declare no conflict of interest.

References

1. Berardi, U. A cross-country comparison of the building Energy consumptions and their trends. *Resour. Conserv. Recycl.* **2017**, *123*, 230–241. [CrossRef]
2. Caniato, M.; Gasparella, A. Discriminating people's attitude towards building physical features in sustainable and conventional buildings. *Energies* **2019**, *12*, 1429. [CrossRef]
3. Cai, S.; Zhang, B.; Cremaschi, L. Review of moisture behavior and thermal performance of polystyrene insulation in building applications. *Build. Environ.* **2017**, *123*, 50–65. [CrossRef]
4. Ciulla, G.; Galatioto, A.; Ricciu, R. Energy and economic analysis and feasibility of retrofit actions in Italian residential historical buildings. *Energy Build.* **2016**, *128*, 649–659. [CrossRef]
5. Orlik-Kożdoń, B.; Nowoświat, A. Modelling and testing of a granular insulating material. *J. Build. Phys.* **2018**, *42*, 6–15. [CrossRef]
6. Hansen, T.K.; Bjarløv, S.P.; Peuhkuri, R. The effects of wind-driven rain on the hygrothermal conditions behind wooden beam ends and at the interfaces between internal insulation and existing solid masonry. *Energy Build.* **2019**, *196*, 255–268. [CrossRef]
7. Manzan, M.; De Zorzi, E.Z.; Lorenzi, W. Numerical simulation and sensitivity analysis of a steel framed internal insulation system. *Energy Build.* **2018**, *158*, 1703–1710. [CrossRef]
8. Orlik-Kożdoń, B.; Steidl, T. Impact of internal insulation on the hygrothermal performance of brick wall. *J. Build. Phys.* **2017**, *41*, 120–134. [CrossRef]
9. Nowoświat, A.; Pokorska-Silva, I. The influence of thermal mass on the cooling off process of buildings. *Period. Polytech. Civ. Eng.* **2018**, *62*, 173–179. [CrossRef]
10. Krause, P. The numeric calculation of selected thermal bridges in the walls of AAC. *Cement Wapno Beton* **2017**, *22*, 371–380.
11. Pokorska-Silva, I.; Nowoświat, A.; Fedorowicz, L. Identification of thermal parameters of a building envelope based on the cooling process of a building object. *J. Build. Phys.* **2019**. in Press. [CrossRef]
12. Khoukhi, M. The combined effect of heat and moisture transfer dependent thermal conductivity of polystyrene insulation material: Impact on building Energy performance. *Energy Build.* **2018**, *169*, 228–235. [CrossRef]
13. Havinga, L.; Schellen, H. The impact of convective vapour transport on the hygrothermal risk of the internal insulation of post-war lightweight prefab housing. *Energy Build.* **2019**, *204*, 109418. [CrossRef]
14. Budaiwi, I.; Abdou, A. Effect of thermal conductivity change of moist fibrous insulation on Energy performance of buildings under hot-humid conditions. *Energy Build.* **2013**, *60*, 388–399. [CrossRef]
15. Kontoleon, K.J.; Giarma, G. Dynamic thermal response of building material layers in aspect of their moisture content. *Appl. Energy* **2016**, *170*, 76–91. [CrossRef]

16. Orlik-Kozdoń, B.; Nowoswiat, A.; Krause, P.; Ponikiewski, T. A numerical and experimental investigation of temperature field in place of anchors in ETICS System. *Constr. Build. Mater.* **2018**, *167*, 553–565. [CrossRef]
17. Nowoświat, A.; Skrzypczyk, J.; Krause, P.; Steidl, T.; Winkler-Skalna, A. Estimation of thermal transmittance based on temperature measurements with the application of perturbation numbers. *Heat Mass Transf.* **2018**, *54*, 1477–1489. [CrossRef]
18. Dissanayake, D.M.K.W.; Jayasinghe, C.; Jayasinghe, M.T.R. A comparative embodied Energy analysis of a house with recycled expanded polystyrene (EPS) based foam concrete wall panels. *Energy Build.* **2017**, *135*, 85–94. [CrossRef]
19. Firlag, S.; Piasecki, M. NZEB renovation definition in a heating dominated climate: Case study of Poland. *Appl. Sci.* **2018**, *8*, 1605. [CrossRef]
20. Lakatos, A.; Deak, I.; Berardi, U. Thermal Characterization of different graphite polystyrene. *Int. Rev. Appl. Sci. Eng.* **2018**, *9*, 163–168. [CrossRef]
21. Ye, R.; Lin, W.; Yuan, K.; Fang, X.; Zhang, Z. Experimental and numerical investigations on the thermal performance of building plane containing CaCl2 6H2O/expanded graphite composite phase material. *Appl. Energy* **2017**, *193*, 325–335. [CrossRef]
22. Faravelli, T.; Pinciroli, M.; Pisano, F.; Bozzano, G.; Dente, M.; Ranzi, E. Thermal degradation of polystyrene. *J. Anal. Appl. Pyrolysis* **2001**, *60*, 103–121. [CrossRef]
23. Kussauer, R.; Ruprecht, M. *Die häufigsten Mängel bei Beschichtungen und Wärmedämm-Verbundsystemen: Erkennen, Vermeiden, Beheben*; Rudolf Müller: Köln, Germany, 2014.
24. Mehta, S.; Biederman, S.; Shivkumar, S. Thermal degradation of foamed polystyrene. *J. Mater. Sci.* **1995**, *30*, 2944–2949. [CrossRef]
25. Shu-Kai, Y.; Chien-Hsiung, H.; Kuo-Chung, C.; Tsu-Huang, C.; Weng-Jeng, G.; Sea-Fues, W. Effect of dispersion method and process variables on the properties of supercritical CO2 foamed polystyrene/graphite nanocomposite foam. *Polym. Eng. Sci.* **2013**, *53*, 2061–2072.
26. Tran, M.P.; Gong, P.; Detrembleur, C.; Thomassin, J.M.; Buahom, P.; Saniei, M.; Kenig, S.; Parka, C.B.; Lee, S.E. Reducing Thermal Conductivity of Polymeric Foams with High Volume Expansion Made From Polystyrene/Expanded Graphite. *PIERS Online* **2016**, *4*, 1870–1882.
27. Vlcek, J. A field method for determination of emissivity with imaging radiometers. *Photogramm. Eng. Remote Sens.* **1982**, *48*, 609–614.

Article

Integrated Measuring and Control System for Thermal Analysis of Buildings Components in Hot Box Experiments

Tullio de Rubeis [1,*], Mirco Muttillo [1], Iole Nardi [2], Leonardo Pantoli [1], Vincenzo Stornelli [1] and Dario Ambrosini [1]

[1] Department of Industrial and Information Engineering and Economics (DIIIE), University of L'Aquila, Piazzale Pontieri 1, Monteluco di Roio, I 67100 L'Aquila, Italy; mirco.muttillo@univaq.it (M.M.); leonardo.pantoli@univaq.it (L.P.); vincenzo.stornelli@univaq.it (V.S.); dario.ambrosini@univaq.it (D.A.)

[2] Energy Efficiency Unit Department (DUEE-SPS-ESU), ENEA Casaccia, S.M. Di Galeria, 00123 Rome, Italy; iole.nardi@enea.it

* Correspondence: tullio.derubeis@univaq.it; Tel.: +39-0862-434342

Received: 7 May 2019; Accepted: 25 May 2019; Published: 29 May 2019

Abstract: In this paper, a novel integrated measuring and control system for hot box experiments is presented. The system, based on a general-purpose microcontroller and on a wireless sensors network, is able to fully control the thermal phenomena inside the chambers, as well as the heat flux that involves the specimen wall. Thanks to the continuous measurements of air and surfaces temperatures and energy input into the hot chamber, the thermal behavior of each hot box component is analyzed. A specific algorithm allows the post-process of the measured data for evaluating the specimen wall thermal quantities and for creating 2D and 3D thermal models of each component. The system reliability is tested on a real case represented by a double insulating X-lam wall. The results of the 72 h experiment show the system's capability to maintain stable temperature set points inside the chambers and to log the temperatures measured by the 135 probes, allowing to know both the U-value of the sample (equal to 0.216 ± 0.01 W/m²K) and the thermal models of all the hot box components. The U-value obtained via hot box method has been compared with the values gathered through theoretical calculation and heat flow meter measurements, showing differences of less than 20%. Finally, thanks to the data postprocessing, the 2D and 3D thermal models of the specimen wall and of the chambers have been recreated.

Keywords: hot box; thermal models; measuring and control system; digital temperature probes; data post-processing

---

1. Introduction

The building sector, responsible for more than one third of the total energy use and associated greenhouse gas emissions [1], needs continuous and constant energy efficiency improvement. The future of this sector is strongly related to the reduction of such alarming quota, that can be pursued by improving the thermal performance of structural elements, responsible for the heat losses, and by increasing the efficiency of buildings' technical plants.

The dispersing elements have a relevant impact on the buildings 'energy consumption [2] and their thermal characteristics can be experimentally studied by means of hot boxes able to recreate real and repeatable operating conditions [3]. A hot box allows to analyze real size structural components subject to a known thermal forcing (steady and dynamic) imposed as boundary conditions.

A hot box is essentially constituted in terms of two main chambers (hot and cold), while the building component under investigation is interposed between them. Known thermal conditions are created, in order to reproduce a typical thermal stress that characterizes the actual use of the sample: for instance, 20 °C are imposed in hot chamber and 0 °C in cold chamber, with the aim of having enough temperature difference for giving rise to an appreciable thermal flow. The considerable dimensions of common hot boxes and the actual dimensions of the analyzed object determine a burdensome control of the thermal phenomena, especially when the requirements of technical standards are met. In fact, hot box experiments require measuring and control systems able to save temperature values from several (order of hundreds) probes and for long-lasting campaigns, and to maintain imposed boundary conditions by controlling the heating and cooling systems that equip the hot box itself. Add to this that the sizes and features of the hot box components directly influence the number of installed probes and, therefore, the control system complexity. In literature, there are many examples of laboratory tests carried out using hot boxes, morphologically different from each other. Caruana et al. [4] employed a hot box to investigate the thermal properties of a new building block (specimen dimensions equal to 165×190 cm), to improve its U-value without changing compressive strength, physical dimensions or manufacturing process. Gullbrekken et al. [5] discussed how natural convection in air-permeable glass wool insulation affects the thermal transmittance in walls, roofs and floors. The study was carried out by means of a rotatable guarded hot-box (with a metering area equal to 245×245 cm). Prata et al. [6], studied the dynamic thermal behavior of a Linear Thermal Bridge (LTB) in a wooden building corner by means of a calibrated hot box. The specimen was made up of two cross laminated timber (CLT) panels bolted together to simulate the dynamic thermal behavior of a wooden building corner (each panel had dimensions equal to 100×215 cm). The work of Lechowska et al. [7] presented experiments in a guarded hot box for improving the PVC window frame thermal transmittance without frame geometrical dimension and material variations (the external dimensions of the window frame were 150×150 cm).

The sizes of the specimen (that can be up to 3 m), the maintenance of stable thermal conditions, and the knowledge of the thermal phenomena that happen between the layers of the sample, and between sample and chambers require complex measuring and control systems.

Therefore, despite its basic working principle, a hot box requires the use of a manifold equipment, often difficult to manage. Indeed, if the hot box equipment is not adequate or the number of temperature probes is not sufficient, a partial knowledge of thermal phenomena and not suitable thermal conditions inside the chambers are obtained. Some authors claim that, given the state of the art, climate chambers can be precisely controlled and programmed with temperature cycles [8]. Being missing an analysis of the measuring and control systems commonly employed in hot box experiments, a literature insight on such systems is presented in the following (Section 2), highlighting the frequent employment of commercial (and sometimes expensive) devices.

Given that: (i) a hot box requires a burdensome management of the temperature probes and of the devices installed in it; (ii) commonly, separated measuring and control systems equip hot boxes, causing possible rough temperature regulation; (iii) by now, there is no possibility for commercial devices to real-time monitor the thermal distributions inside the chambers (i.e., it is impossible to assess potential thermal stratifications); there is the need for a novel, cheap, integrated and reliable measuring and control system for thermal analyses in hot box experiments, presented in this paper. Moreover, to test the capabilities of the system and the data post-processing of the measured values, a real application on a X-lam sample wall with double insulating layer is performed.

The paper is structured as follows. Section 2 proposes a review of the measuring and control systems commonly employed in hot boxes experiments. The new system and its properties are presented in Section 3. Section 4 shows the results obtained from the application of the system to a real case. The conclusions are reported in Section 5.

2. Common Systems and Literature Background

The measuring and control system, presented in this work, equips a Guarded Hot Box (GHB), constituted by three boxes: a guard box, a metering box (inside the hot one) and a cold box. A detailed description of GHB and its operation is presented in previous works [9–11]. The knowledge of energy balance inside the hot box allows to understand how complex the thermal phenomena (that characterize the operation of a GHB) and the management of the systems (that control the thermal steady-state conditions) are. The heat flows distribution in the GHB is shown in Figure 1, where Φ_1 represents the heat flow rate through the specimen, Φ_p is the total power input into the hot chamber, Φ_2 is an imbalance, i.e., the heat flow rate parallel to the specimen, Φ_3 is the heat flow rate through metering box walls. Φ_5 is the heat flow rate parallel to the specimen surface at the edges of the specimen, called peripheral heat loss.

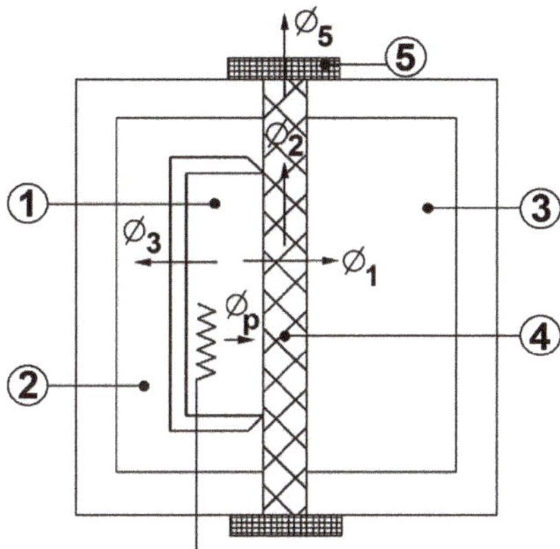

Figure 1. Energy flows in the GHB: (1) metering box; (2) guard box; (3) cold box; (4) specimen wall; (5) tempering ring.

Known the thermal energy input into the hot chamber and the internal flanking losses inside guard and metering boxes, it is possible to determine the amount of heat that passes through the sample, that represents the main goal of hot box experiments. What retrieved by theoretical analysis should be confirmed by measurements.

Hot box experiments shall be conducted imposing temperature differences between hot and cold chambers usually chosen for the end-use application: 20 °C is a common temperature difference for building applications. These conditions are generally guaranteed using electric resistances, for the energy input into the hot chamber, and refrigerating unit for the energy input into the cold chamber.

The temperature sensors are installed on both the surfaces of the sample (hot and cold sides) and on the walls of metering and guard boxes to quantify the effects of flanking losses and the heat flow that passes through the specimen. Further sensors are placed to monitor the air temperature values inside chambers and laboratory. A current and voltage meter allows to know the thermal energy input into the hot chamber by the Joule effect, being the heating supplied by electric resistances. The minimum number of temperature probes to be installed is defined by the standard UNI EN ISO 8990 [12], and it shall be of at least two probes for square meter.

Before presenting the integrated measuring and control system proposed in this work, a literature review of the common measuring and control systems employed for hot box experiments is proposed. The works were selected based on the following research questions:

- what types of hot boxes are commonly used?
- what types of probes are commonly employed to measure temperatures on the specimen and inside the chambers?
- what types of measuring and control systems are commonly used?

Based on the parameters mentioned above, the works selected from literature are summarized in Table 1.

Table 1. Literature review on common logging and control systems installed in hot boxes.

Authors	Year	Hot Box Typology	Temperature Probe Typology (and Number)	Measuring System	Control System
Koo et al. [13]	2018	C	T-type TC (5 [a])	n.a.	n.a.
Biswas et al. [14,15]	2018	G	TC (25 [a])	n.a.	n.a.
Woltman et al. [3]	2017	C	T-type TC	OM-320 Data Acquisition System	OMRON E5CK digital controller
Caruana et al. [4]	2017	C	TC (18 [a] + 20 [b])	n.a.	n.a.
Gullbrekken et al. [5]	2017	G	T-type TC (39 [a])	n.a.	n.a.
Prata et al. [6]	2017	C	T-type TC (48 [a])	GL820 Midi DataLogger	n.a.
Lechowska et al. [7]	2017	G	n.a.	Ahlborn Almemo (Ahlborn Mess—und Regelungstechnik GmbH, Deutschland)	AMR Ahlborn WinControl system
Sassine et al. [16]	2017	S	TC (6 [a] + 2 [b])	Digital multimeter	Microcomputer
Peters et al. [17]	2017	C	TR	Arduino data logger + HOBO loggers	n.a.
Douzane et al. [18]	2016	G	T-type TC	n.a.	n.a.
Buratti et al. [19,20]	2016	C	TR (8 [a] + 9 [b])	PC	PID control system
Basak et al. [21]	2016	G	K-type TC	Agilent Data Acquisition Systems 34970A	MATLAB fuzzy scripts
Grynning et al. [22]	2015	G	TC	n.a.	n.a.
Seitz et al. [23]	2015	C	T-type TC (64 [a, b])	Four 16-channel data acquisition cards (two MCCDAQ USB-2416 with AI-EXP32	Omega 2110J
Pernigotto et al. [24,25]	2015	C	T-type TC (24 [a])	Digital multimeter	PID control system
Faye et al. [26]	2015	C	K-type TC	n.a.	n.a.
Meng et al. [27]	2015	S	T-type TC (10 [a])	JTRG-II automatic tester	Voltage regulator
Manzan et al. [28]	2015	C	TR (4 [a] + 2 [b])	Babuc A by LSI	On/off thermostat
Seitz et al. [29]	2014	G	TC	n.a.	n.a.
Vereecken et al. [30]	2014	C	n.a.	n.a.	n.a.
Ghosh et al. [31]	2014	G	K-type TC	Agilent Data Acquisition Systems 34970A	PID logic P logic P logic with duty cycle control
Sousa et al. [32]	2014	C	n.a.	n.a.	n.a.
Kus et al. [33]	2013	C	K-type YC (34 [a, b] Test 1 - 52 [a, b] Test 2)	Ahlborn Almemo	AMR Ahlborn WinControl system
Chen et al. [34]	2012	G	T-type TC (372 [b]) and TR (33 [b])	n.a.	PID control system
Martin et al. [35]	2012	G	T-type TC (172 [a, b])	n.a.	Indirect system
Grynning et al. [36]	2011	C	TC (38 [a])	n.a.	n.a.
Asdrubali et al. [37]	2011	C	T-type TC (142 [b])	PC	PID control system
Qin et al. [38]	2009	S	TC (6 [a])	n.a.	n.a.

[a] installed on the specimen. [b] installed in the hot box. LEGEND: TC (thermocouple), TR (thermoresistance), G (guarded hot box); S (simple hot box); C (calibrated hot box); (definitions provided by [27,39]).

Therefore, the literature review shows that the calibrated hot box is the most used [3,4,6,13,17,19,20,23–26,28,30,32,33,36,37] followed by the guarded hot box [5,7,14,15,18,21,22,29,31,34,35] and the simple hot box [16,27,38]. The measuring systems always employ analog probes and the thermocouples are more widely used [3–6,13–16,18,21–27,29,31,33–38] with respect to the thermoresistances [17,19,20,28,34]. The number of probes installed that, as mentioned, depends on the size of the hot box, is very variable: some cases have less than 20 probes [13,16,19,20,27,28,38], most cases have a number of probes between 20 and 100 [4–6,14,15,23–25,33,36], while in three cases there are more than 100 probes [34,35,37]. The measuring and control systems are mainly commercial [3,6,7,16,19–21,23–25,27,28,31,33,37] except for one customized system [17].

Based on the literature review, the following observations can be made: (1) all the analyzed cases employ analog temperature sensors; (2) except for one case, all the analyzed hot box facilities use commercial and separated measuring and control systems; (3) none of the cases analyzed would appear to show the possibility of real-time monitoring the thermal distributions.

The first observation allows to underline how the management of many analog sensors is complex because of several disadvantages due to losses and signal noise along the connection cables [40] that require compensation circuits for each probe [41–43].

The second observation highlights that, besides being commercial, the measuring and control systems are also separated and independent: the measuring system permits to monitor and measure the thermal properties of the sample inside the chambers, while the control systems permit to turn on/off the heating and cooling systems of the hot box to reach the wished thermal conditions. However, the choice of employing separated systems can cause a rough temperature regulation inside the chambers. In fact, the control systems regulate the temperatures by using their own sensing probes that differ from those employed to measure the thermal properties of the sample. Therefore, if the regulation of the wished thermal conditions is difficult or approximate, the fluctuations between the two chambers become considerable and the experiments may be far from the steady-state condition.

The last observation points out that the lack of real-time monitoring determines the inability of displaying anomalous thermal stratifications inside the chambers and their incidence.

Therefore, the main problems that can be encountered for hot box experiments can be summarized as follows: (1) burdensome management of the temperature probes installed; (2) possible considerable thermal fluctuations between the two chambers; (3) difficult control of thermal stratifications inside the chambers.

The hot box employed in this work has quite large dimensions (300 × 300 × 320 cm) and summing up the number of probes installed in the chambers, the total amount of sensors needed is extremely high and, therefore, their management becomes complex. Moreover, the experimental tests need steady-state conditions guaranteed by small temperature fluctuations between the two chambers as indicated by the standard UNI EN ISO 8990 [12]). Therefore, the management of controllers represents a further difficulty.

Based on this analysis, the measuring and control system presented in this work tries to overcome these difficulties, in order to simplify the carrying out and accuracy of the experiments. In particular, the main novelties of the proposed systems are:

1. measuring, control and post-processing phases integrated into one system;
2. scalability of the number of temperature probes usable, without the need of compensation circuit and additional devices;
3. arbitrary choice of number and typology (ambient or surface) of temperature probes usable for regulating temperature setups inside the chambers;
4. management of temperature fluctuations between hot and cold chambers (through a specific PI regulator);
5. instantaneous visualization of the thermal phenomena that characterize the experiment and control of the correct evolution of the experimental tests;

6.   low-cost of the system.

3. Proposal of a New Integrated System

Given the literature background on hot boxes and their weak points, an integrated measuring and control system, based on several sensors and actuators, for laboratory experiments in hot box is presented. The system is made up of a master control unit and three slave units respectively dedicated to cold chamber, hot chamber and specimen wall under analysis (Figure 2). Through a wireless "ad hoc" protocol, the master unit communicates with the slaves for the parameters setting and for receiving the values measured inside the chambers. The UART (Universal Asynchronous Receiver-Transmitter) protocol is employed to realize a master-PC communication allowing the data post-processing, which are later sent to a web server. Although the data post-processing is obtained through a dedicated designed software, data can be elaborated by any custom object-oriented software. The master control unit can log all the measured values and manage the devices and equipment (slave units) that control thermal conditions inside the chambers. In detail, it allows: (1) to measure ambient and surface temperatures needed to evaluate the thermal models inside the chambers and the thermal properties of the specimen; (2) to measure the energy input for heat up the hot chamber needed to know the heat flux that passes through the sample; (3) to control cooling and thermal-power systems.

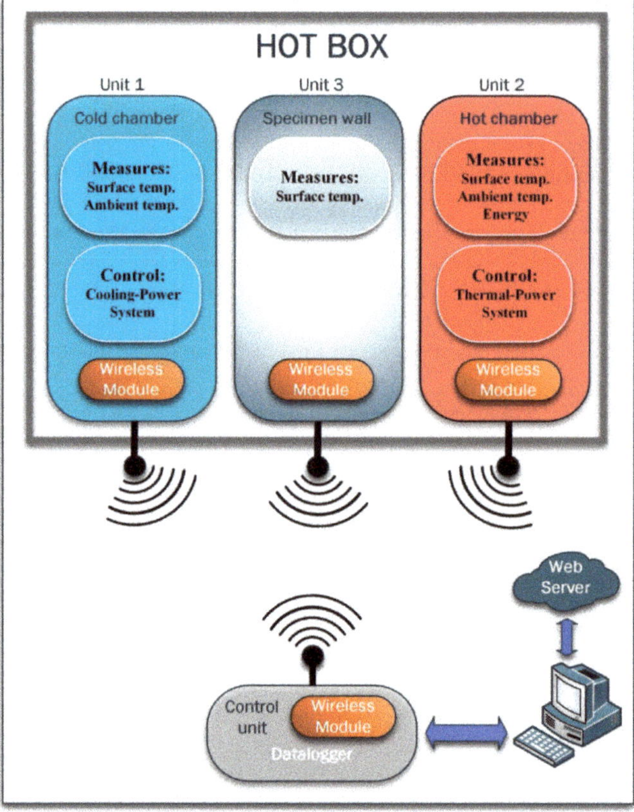

Figure 2. Architecture of the proposed system.

The working principle of the system consists of three distinct phases. The first phase includes the system start up and the configuration of the parameters of interest (temperatures set points, data acquisition and measurement rates) through a Human-Machine Interface (HMI). In the second phase, the system maintains the desired conditions controlling the machines that equip the hot box and, at the same time, acquiring the measured data. In the third and last phase, the post-processing of the data takes place thanks to a semi-automatic interaction between the system algorithm and a dedicated MATLAB® code. An overall schematic of the system operating phases is shown in Figure 3.

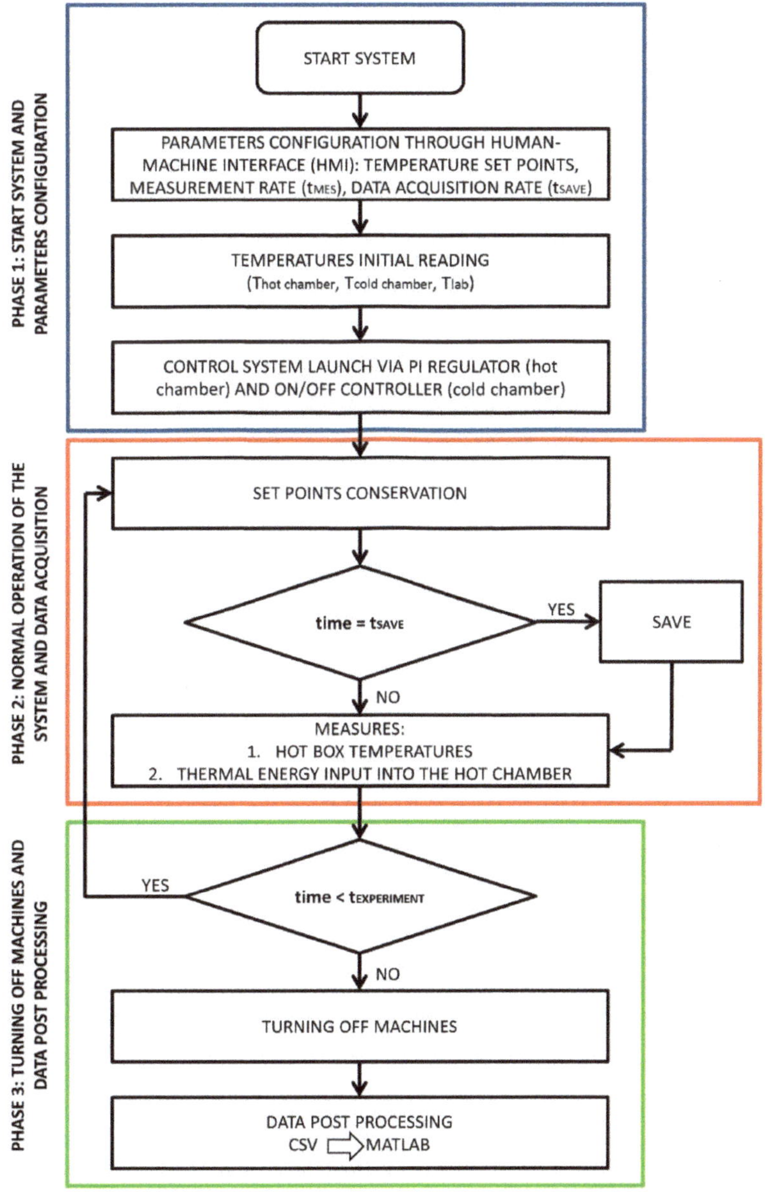

Figure 3. Flow chart of the integrated measuring and control system operation.

To give a universal feature to the system, each unit has the same PCB (printed circuit board), based on a general-purpose microcontroller ATmega2560 for the input/output (I/O) control management. Therefore, each control unit has enabled only the specific I/Os to its functionality. The PCB is programmable via USB port, it has an external DAC (digital analog converter) device for the hot chamber regulation and a digital output for the cold chamber control. Moreover, the system is equipped with an SD card reader for the data backup, 16 input channels for digital probes, a built-in wireless module, a real time clock (RTC) with a backup battery to work when there is no power supply, and additional connectable I/Os usable for any system upgrade. All the units are supplied by external power sources.

A 3D view of the board employed for the system is shown in Figure 4.

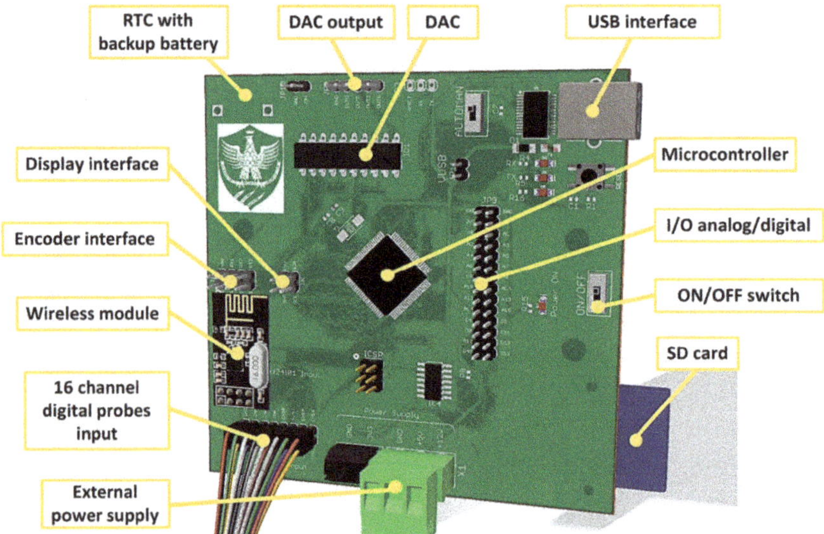

Figure 4. A 3D view of the basic PCB used for the system.

In detail, the unit dedicated to the specimen wall (Figure 5) allows to measure the surface temperature values of both sides of the sample. The probes distribution has been divided into two channels having 25 probes each to minimize the number of cables and simplify their installation. The real-time visualization of the surfaces thermal models of the sample is carried out through an HMI, consisting of a 3.5″ touch display, with a designed graphical interface. The touch display interrogation permits to show the surface temperature value corresponding to a specific probe, both with a numerical value and by means of a false-color scale. The spatial discretization reproduced on the touch panel is the same of that chosen for the probes' installation, which in our case is equal to 50 × 50 cm.

The PCB of the hot chamber control unit (Figure 6) is the most complex, due to the many measurement and control operations it must carry out. Indeed, unlike the unit of the specimen wall, which performs only measuring actions, the hot chamber unit allows to control the machines that equip the chamber and measure both temperatures (surface and ambient) and thermal energy input. All these operations take place simultaneously. The thermal energy is input into the hot chamber by means of electric resistances (Joule effect), to achieve and maintain the pursued set point temperature. The electric resistances control happens through a PI (proportional-integral) regulator, whose algorithm has been implemented on the microcontroller. The PI parameters were determined by means of the Ziegler-Nichols method [44]. The control phase is performed by voltage regulation of the electric resistances thanks to a full AC wave control circuit, managed by a trigger module. Specifically, the trigger module controls the gates of the full AC wave control circuit components modulating the output voltage. The input control voltage of the trigger module ranges from 0 to 5 V and it is generated by an external 8-bit parallel DAC managed by the microcontroller via eight digital outputs. The measurement of the energy fed into the hot chamber occurs through a current-voltage method by means of two Agilent 34401A instruments connected to a computer via GPIB interface.

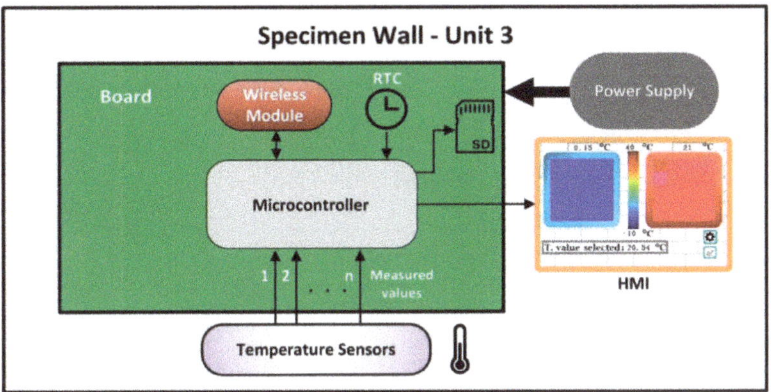

Figure 5. Specimen wall unit architecture.

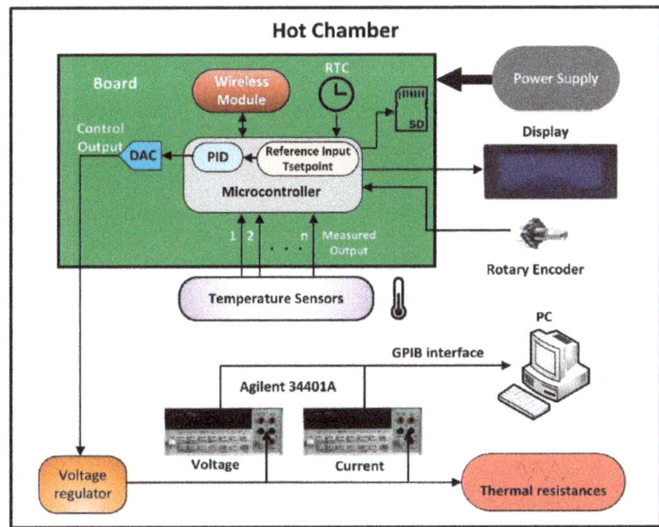

Figure 6. Hot chamber control unit architecture.

Analogously to what described for the hot chamber, the cold chamber control unit (Figure 7) permits to perform simultaneously temperature measurements (surface and ambient) and the management of the refrigerating unit that provides the necessary energy for maintaining the wished temperature set point. The microcontroller regulates the cold chamber temperature by means of an on/off control of the refrigerating unit operating on a relay.

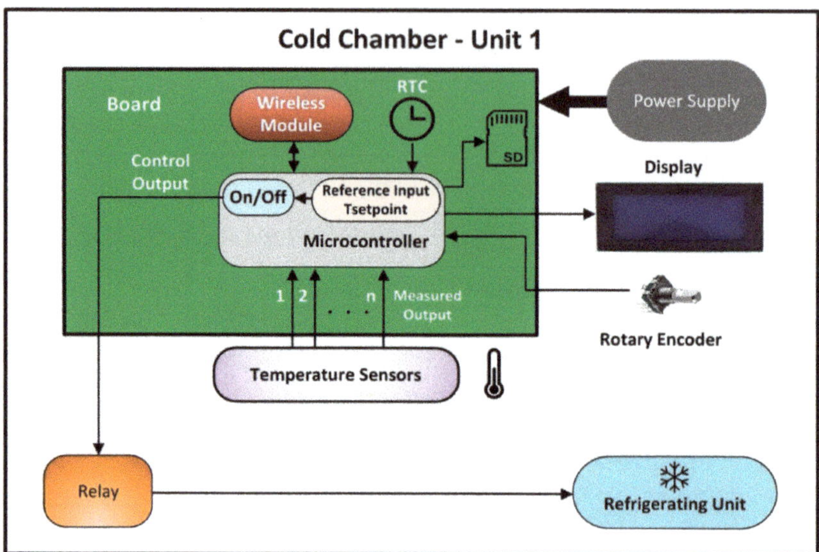

Figure 7. Cold chamber control unit architecture.

Unlike what usually happens with traditional hot box controllers, the architecture of the proposed control units of both the chambers allows to choose the reference temperature probes (surface or ambient) to be used for the wished set point temperature control. Moreover, to guarantee high measurements precision and high flexibility of the system employment, the number of probes can be arbitrarily increased (without the need for further hardware components) and the procedures for parameters configuration (set-point temperatures, acquisition time rate, etc.) can be performed run-time, without interrupting the normal system operation. The control unit parameters configuration takes place via rotary encoder and their visualization on 20 × 4 LCD displays.

The temperature sensors chosen for the proposed system are the DS18B20 digital thermometers of Maxim Integrated [45]. These probes, frequently employed for thermal experiments [46], communicate with a proprietary 1-Wire protocol that allow to connect many probes on the same data line simplifying the installation phase. Moreover, being digital sensors, they are not affected by disturbances on the transmission lines. This choice allows to use long connection cables (order of 100 m) avoiding the employment of specific matching circuits, as it happens for thermocouples and RTDs (Resistance Temperature Detectors) sensors. The digital thermometers operate in the range 3–5.5 V and their temperature interval ranges between −55 °C and 125 °C with a settable resolution. With a 12-bit resolution the analog-digital conversion time is equal to 750 ms. Each sensor univocally has a 64-bit address identified by the microcontroller. This feature is appropriate when the system has many sensors because they can be connected on the same bus allowing theoretically infinite devices. The digital thermometers DS18B20 allow to receive a broadcast command by the microcontroller in order to obtain the same start conversion time for the probes connected along the bus.

The current consumption of the digital sensors is equal to 750 nA, in idle mode, and 1 mA, in conversion mode. The worst case happens when all the probes are in conversion mode. During the normal operation, the system, without considering probes, has a current absorption equal to 100 mA and, when the wireless transmission is activated, the electric consumption grows up to 500 mA.

4. Application on a Real Case

In this section, the sample wall, the probes arrangement, and the results of the experimental test obtained through the application of the proposed measuring and control system are presented.

4.1. Sample Wall Description

The specimen wall (Figure 8) is constituted by a double insulation X-lam bearing member, also known as CLT (Cross Laminated Timber). This type of specimen was chosen to test the energy performance of rather innovative walls that, thanks to their energy and seismic properties, can represent an alternative to the classic masonry walls. Indeed, the employment of new materials for the building envelope is experiencing a growing spread [47]. The specimen has real dimensions equal to 300×300 cm. The X-lam panel is made up of solid wood with crossed layers (several superimposed layers glued one on the other), so that the grain of each layer is rotated in the plane of 90° with respect to the adjacent layers. Each layer is made up of dried spruce boards. The side of the wall facing the hot chamber is insulated with mineral wool, while the other one has expanded polystyrene (EPS) mixed with graphite. The X-lam panel and the insulation layers are sandwiched between double plasterboard layers.

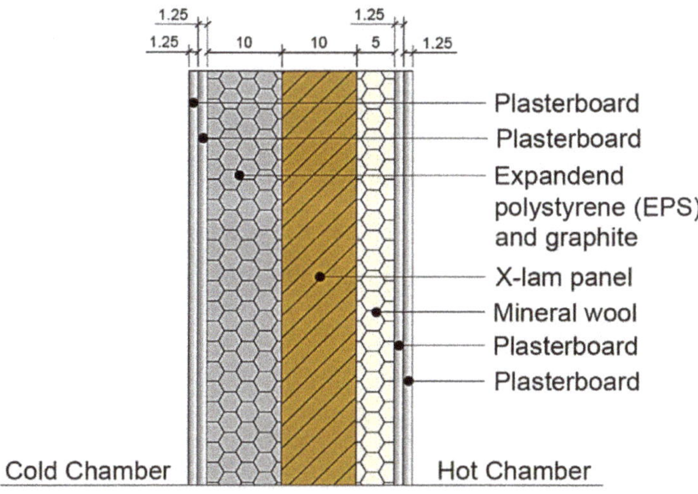

Figure 8. The specimen wall (all the dimensions are expressed in centimeters).

The thermal properties of the wall's layers and the calculated thermal resistance (according to ISO 6946 [48]) of the wall are shown in Table 2.

Table 2. Thermal properties of the wall's layers and calculated thermal resistance of the wall.

	Description	Thickness (cm)	Density (kg/m³)	Conductivity (W/mK)	Resistance (m²K/W)
	External surface resistance	-	-	-	0.04
1	Plasterboard	1.25	900	0.210	0.06
2	Plasterboard	1.25	900	0.210	0.06
3	EPS and graphite	10.00	32	0.031	3.23
4	X-lam panel	10.00	470	0.130	0.77
5	Mineral wool	5.00	135	0.039	1.28
6	Plasterboard	1.25	900	0.210	0.06
7	Plasterboard	1.25	900	0.210	0.06
	Internal surface resistance				0.13
	Overall thermal resistance				5.69

4.2. Sensors Arrangement

The hot box was equipped with 135 sensors to measure surface and ambient temperatures including the temperature inside the laboratory. An overall view of the hot box with the sample wall and the installed temperature probes is shown in Figure 9a. Each side of the sample wall was equipped with 25 probes (symmetrically positioned on the two sides), in a grid 5 by 5 probes (Figure 9b). The internal surfaces of the cold chamber had a total of 42 temperature sensors, while the internal walls of the metering box had 24 probes. To calculate the mean radiative temperature, both in hot and cold chambers, the temperatures of the apparatus surfaces "seen from the specimen" (baffles) must be known. Therefore, 9 surface temperature sensors were installed in the cold chamber and 6 probes in the hot chamber; although the probes number differs, requirements of UNI EN ISO 8990 [12], already explained in Section 2, have been fulfilled. Both chambers and laboratory have been equipped with ambient temperature probes.

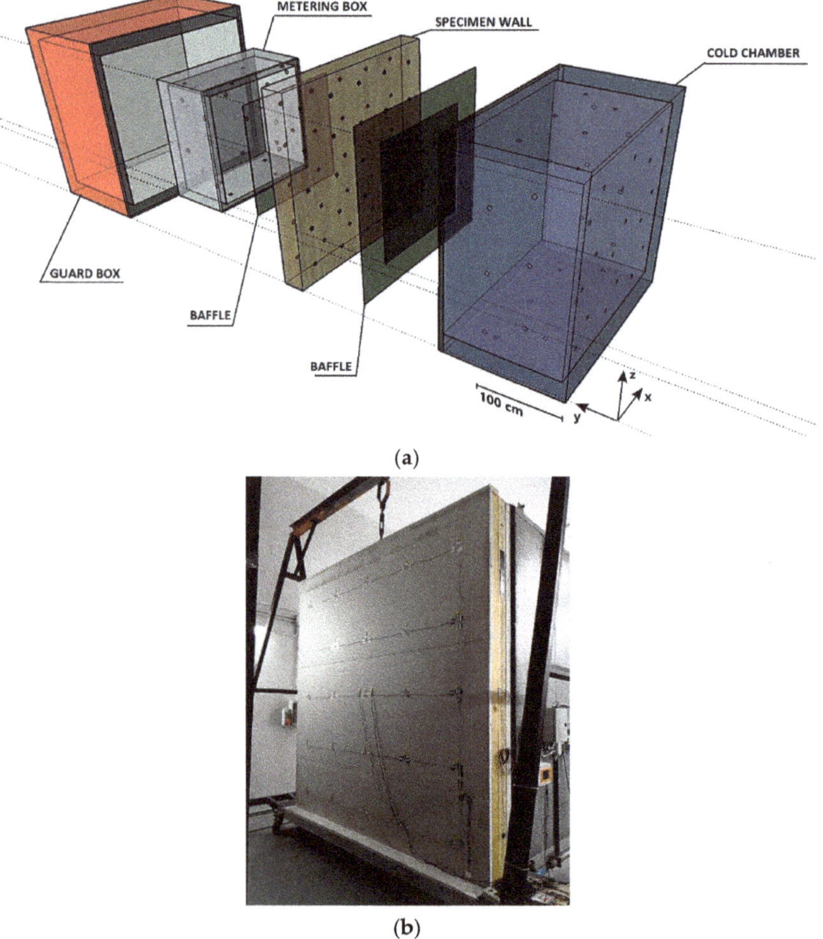

(a)

(b)

Figure 9. Guarded hot box apparatus: (**a**) 3D view of the hot box; (**b**) Probes arrangement on the specimen wall surface.

4.3. Results of the Measuring and Control System Application

The testing phase of the system had a total duration of 88 h, from 1st March at 10:20 am to 5th March at 2:20 am, 2019. If the initial achievement of the steady-state condition and the thermal equilibrium of the hot box after the machines turning off are not considered, the actual duration of the test was 72 h. The turning on of the electric resistances and the introduction of the thermal energy into the hot chamber have led to a temperature increase, until the set point value, equal to 20 °C, was reached. The thermal conditions of the cold chamber were regulated by the refrigeration unit, also activated by the measuring and control system, until reaching the set point value equal to 0 °C. The parameters configuration of the test is summarized in Table 3. In this experimental phase, the control of the machines was carried out by means of the ambient temperature probes of metering and cold boxes, although this choice is arbitrary and non-binding.

Table 3. Parameters configuration.

Parameter	Value	U.M.
Set point temperature—hot chamber	20.0	°C
Set point temperature—cold chamber	0.0	°C
Data acquisition rate	10	min

Since the system allowed to measure and log temperature values along time of each hot box component and the thermal energy input into the hot chamber via electric resistances, it was possible to identify three different phases of the experiment through which the proposed system and the hot box response were analyzed.

The first phase (phase "A" in Figure 10) includes the system activation and the set points achievement in both the chambers and it shows how quickly the set point values were reached. The second phase (phase "B" in Figure 10) is characterized by the maintenance of the steady-state conditions between the two chambers minimizing the temperature fluctuations. At the end of the second phase, the machines (electric resistances and refrigerating unit) were turned off to return to the initial conditions (phase "C" in Figure 10). The total value of the energy input into the hot chamber during the phase "B" was equal to 1.06 kWh. It is worth noting that the measure of energy fed takes place only when the phase "A" is concluded, namely when the steady-state conditions are verified inside the chambers. During the phase "B", the temperature values in the metering box were subject to minimum fluctuations, always lower than ±0.3 °C. Furthermore, Figure 10a highlights that, although the guard box was rather influenced by thermal fluctuations of the laboratory, it allowed to maintain very stable thermal conditions inside the metering box. Small temperature fluctuations, due to the on/off setting of the refrigeration unit, can also be observed in the cold chamber, where the fluctuations were always lower than ±1.4 °C.

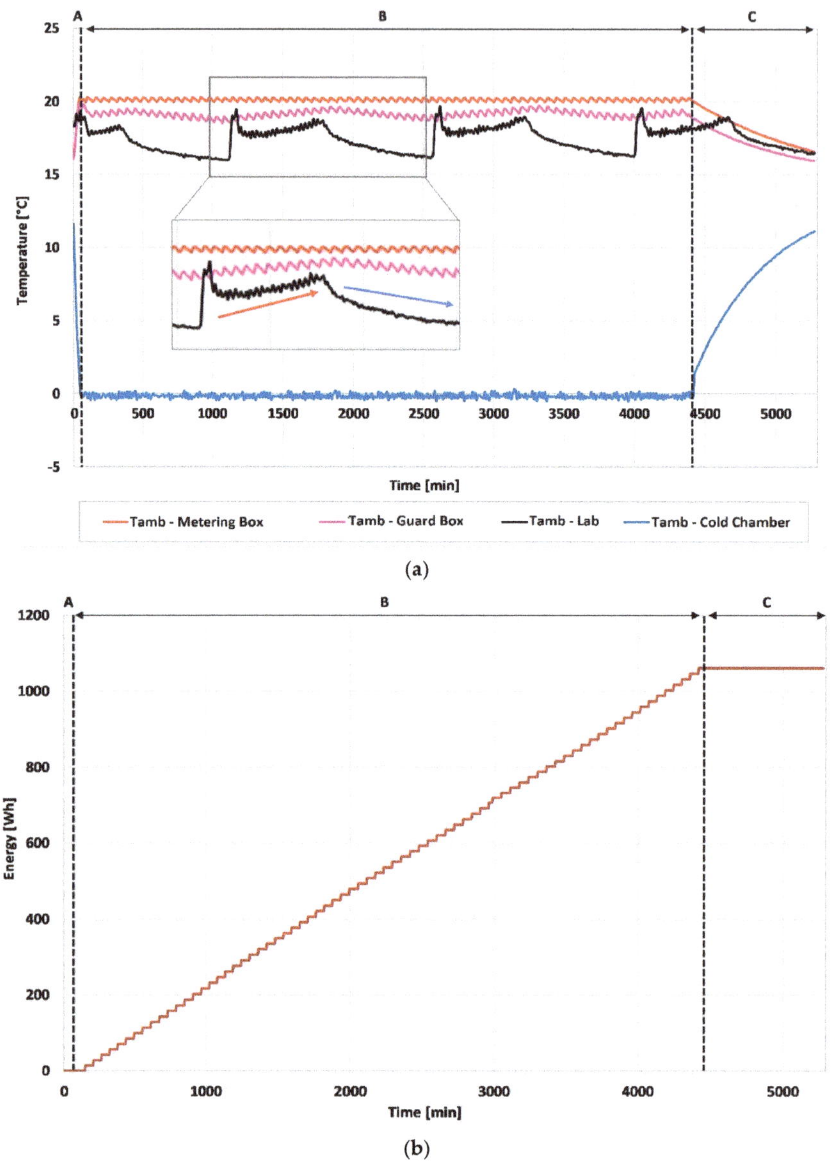

Figure 10. Hot box experiment: (**a**) Temperature trends inside chambers and laboratory, and highlights of temperatures interplay; (**b**) Thermal energy input into the hot chamber.

4.4. Experimental Analysis of the Specimen Wall Thermal Properties

During the experimental phase, heat flow meter and guarded hot box approaches were performed to carry out the performance of the specimen wall and to test the proposed system.

A comparison between these two approaches allows to underline some useful observations. The heat flow meter (HFM) method is widely employed for in-situ measurements of thermal transmittance of building components thanks to its ease of use, given the simplicity of the involved probes (a thermopile for the flux, and thermoresistances or thermopiles for the temperatures) and the availability of proprietary software that permits to process the measured data and to retrieve the U-value.

The guarded hot box approach allows to set and control the temperatures inside the chambers and, therefore, it permits repeatable conditions and the assessment of the influence of boundary conditions on the thermal behavior of the analyzed building component. Despite these pros, the con is mainly due to the sizes of such apparatus and its cost. Indeed, many probes are needed, besides an accurate system for temperature setting and control.

Therefore, if on one hand the heat flow meter approach lets to know punctual spatial information about thermal transmittance of building components with ease of use in non-repeatable conditions, on the other hand the GHB permits to evaluate the U-value of building elements on a wide spatial scale and in repeatable conditions, but with a more complex measuring and control system and higher costs.

In this work, a heat flow meter was installed on the specimen surfaces following the recommendations provided by ISO 9869 [49]. The analysis had a duration of 72 h and it was carried out through a Hukseflux HFP01 heat flow meter, whose characteristics are summarized in Table 4.

Table 4. Technical specifications of the heat flow meter.

Instrument	Type	Measuring Range	Resolution
Fluxmeter	Hukseflux HFP01	−2000 to 2000 W/m^2	50 μV/W/m^{-2}
Temperature probes	LSI Lastem EST124-Pt100	−40 to 80 °C	0.01 °C
Datalogger	LSI Lastem M-Log ELO008	−300 to 1200 mV	±100 μV

The results of the HFM campaign (Figure 11), obtained via progressive average method, showed a mean heat flux equal to 3.39 W/m^2, and a U-value equal to 0.177 ± 0.01 W/m^{2}K, determined considering internal and external surface resistances (R_{si} and R_{se}), equal to 0.13 and 0.04 m^{2}K/W respectively, as provided by the standard UNI EN ISO 6946 [48].

It is worth noting that the U-value determined through the HFM campaign is very close to the value obtained with theoretical calculation equal to 0.176 ± 0.03 W/m^{2}K (Table 2).

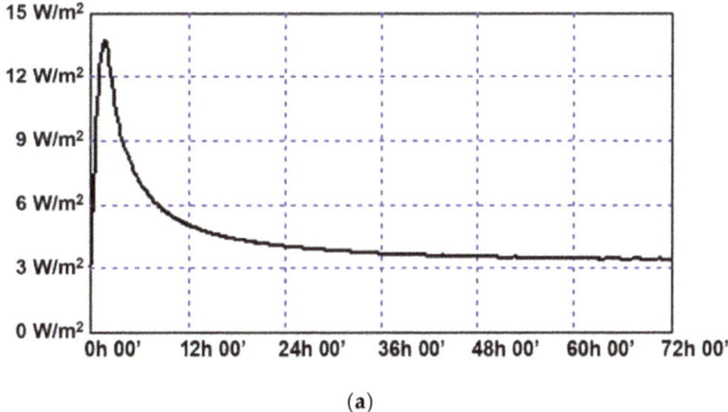

(a)

Figure 11. *Cont.*

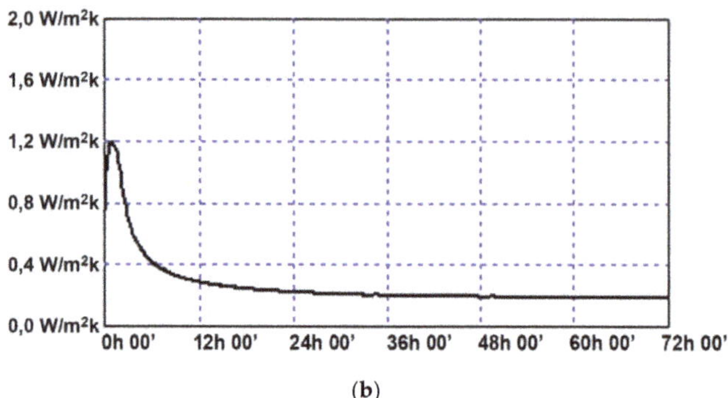

(**b**)

Figure 11. Results of the heat flow meter campaign: (**a**) Specific heat flux; (**b**) Thermal conductance.

The performance of the specimen wall was experimentally evaluated in guarded hot box by means of the measuring and control system proposed in this work. The data measured by the proposed system were processed by using a designed MATLAB® GUI (Graphical User Interface) which, after importing the data, allowed to determine the U-value of the sample. The post-processing algorithm was realized following the standard UNI EN ISO 8990 [12], according to which the value of the thermal transmittance (U) is calculated by Equation (1).

$$U = \frac{\varnothing}{A(T_{n1} - T_{n2})}\left[\mathrm{W/m^2K}\right] \tag{1}$$

where $\varnothing$ is the power supplied to the metering box [W], A is the metering area [m²], T_{n1} and T_{n2} are the environmental temperatures inside the chambers, hot and cold side respectively [°C], calculated by Equation (2).

$$T_n = \frac{T_a \frac{\varnothing}{A} + Eh_r(T_a - T_r')\,T_s}{\frac{\varnothing}{A} + Eh_r(T_a - T_r')} \quad [°\mathrm{C}] \tag{2}$$

where T_a is the measured mean air temperature [°C], T_r' is the measured mean baffle temperature [°C], T_s is the measured mean surface temperature [°C], E is the emissivity factor (assumed equal to 0.9 as provided by the UNI 8990), h_r is the calculated radiation coefficient [W/m²K], provided by Equation (3) [12].

$$h_r = 4\sigma T_m^3 \quad \left[\mathrm{W/m^2K}\right] \tag{3}$$

where σ is the Stefan's constant and T_m is the calculated appropriate mean radiant absolute temperature provided by Equation (4).

$$T_m^3 = \frac{\left(T_r'^2 + T_S^2\right)(T_r' + T_S)}{4} \quad \left[\mathrm{W/m^2K}\right] \tag{4}$$

Therefore, the proposed system and the probes that equip the hot box allowed to determine the experimental U-value of the specimen wall that resulted equal to 0.216 ± 0.01 W/m²K. A comparison between the U-values obtained with the different approaches is showed in Table 5.

Table 5. U-values of the sample with the different approaches.

Approach	U-Value [W/m²K]	Percentage Variation [%] [a]
Theoretical calculation	0.176 ± 0.03	-
Heat flow meter	0.177 ± 0.01	0.57
Guarded hot box	0.216 ± 0.01	18.37

[a] with respect to the theoretical value.

4.5. Uncertainty Analysis

Uncertainty analysis of the thermal transmittance values was carried out by using the Holman's method [50,51], according to which, if a set of measurements is supposed, the calculated result uncertainty is estimated on the basis of the uncertainties in the primary measurements. The result R is a given function of the independent variables x_1, x_2, ..., x_n; w_r is the result's uncertainty and w_1, w_2, ..., w_n are the uncertainties in the independent variables. Therefore, the uncertainty in the result is determined by Equation (5) [50,51].

$$w_R = \left[\left(\frac{\delta R}{\delta x_1} w_1 \right)^2 + \left(\frac{\delta R}{\delta x_2} w_2 \right)^2 + \dots + \left(\frac{\delta R}{\delta x_n} w_n \right)^2 \right]^{1/2} \tag{5}$$

The uncertainties w_n of the data measured (e.g., temperature probes, heat flux, etc.) have been evaluated from the manufacturers' datasheets.

Based on the uncertainties obtained, it is worth noting that the digital nature of the probes and the system architecture allow to easily increase the number of sensors installed to improve the measurement precision. Indeed, this flexibility of the number of sensors would be more complex for systems equipped with analog probes due to losses and noise along the connection cables that require the use of compensation circuits for each probe.

4.6. 2D and 3D Thermal Model Visualization

In addition to the standard analysis of temperatures and energy trends (Figure 10), thanks to the considerable number of probes installed, the system allowed to perform a post-processing analysis of the measured data through which 2D and 3D thermal models of the specimen wall and hot box surfaces were created. Indeed, the designed MATLAB® GUI, besides allowing the U-value evaluation, allowed to carry out 2D and 3D thermal distributions at any desired time. An example of 2D view of the sample wall surfaces is shown in Figure 12. This display mode is useful to visualize the thermal evolution on the wall surfaces during the experiments and to check that there are no thermal anomalies such that measurements results can be compromised. Moreover, a video containing the 2D thermal evolution of the sample wall surfaces is shown in Supplementary Materials Video S1.

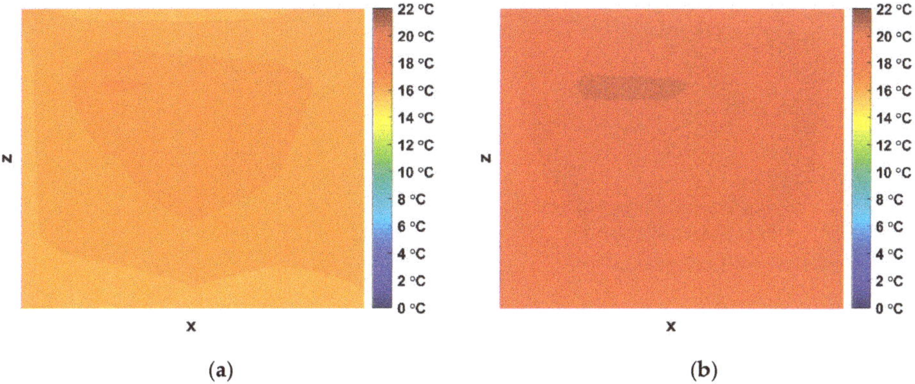

Figure 12. *Cont.*

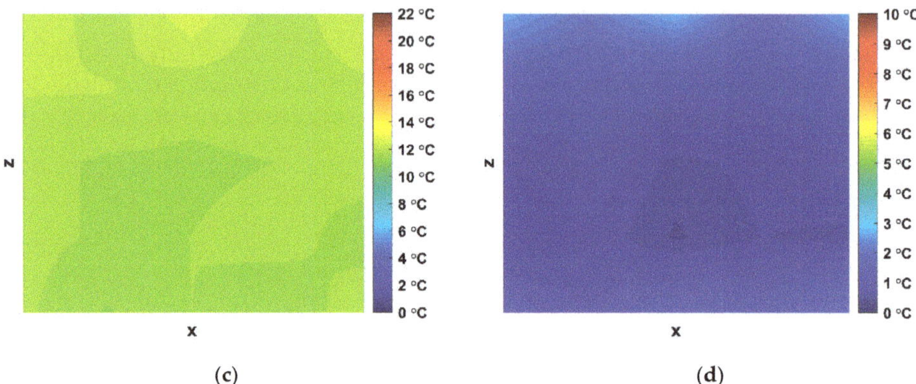

Figure 12. 2D Thermal models of the sample wall surfaces: (**a**) Hot facing side at the beginning; (**b**) Hot facing side at steady condition (after 24 h); (**c**) Cold facing side at the beginning; (**d**) Cold facing side at steady condition (after 24 h).

Furthermore, the analysis of thermal models allowed to create a three-dimensional representation of the temperatures' distribution inside the chambers, by using a 3D CAD software. An example of the three-dimensional representation is shown in Figure 13. It is worth noting that the surface temperature distributions of the two chambers is rather uniform with very small thermal variations. The cold chamber is characterized by a vertical thermal stratification, while the metering box thermal distribution is less uniform, due to the positioning of the electric resistances inside the hot chamber.

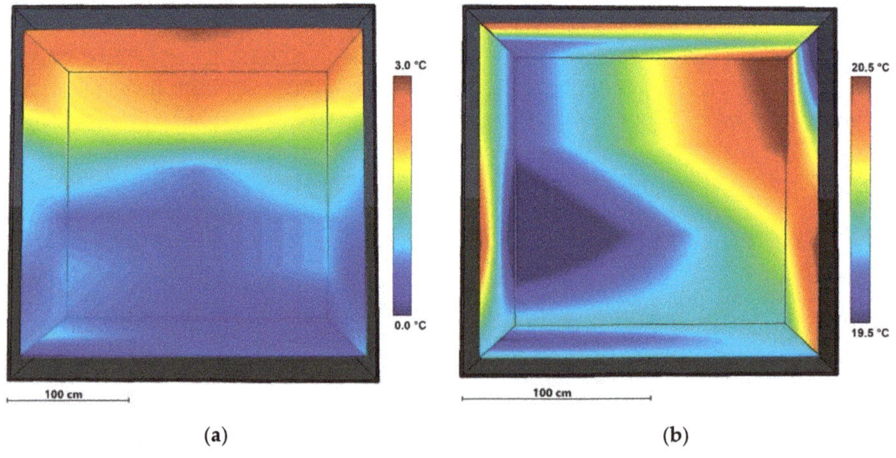

Figure 13. 3D thermal models: (**a**) Cold box; (**b**) Metering box.

4.7. Pros and Cons of the System

Based on the experience carried out in this work and after the testing phase of the sample wall by means of the hot box approach, pros and cons of the proposed measuring and control system can be highlighted. In this sense, Figure 14 summarizes the main advantages and disadvantages deriving from the use of the proposed system.

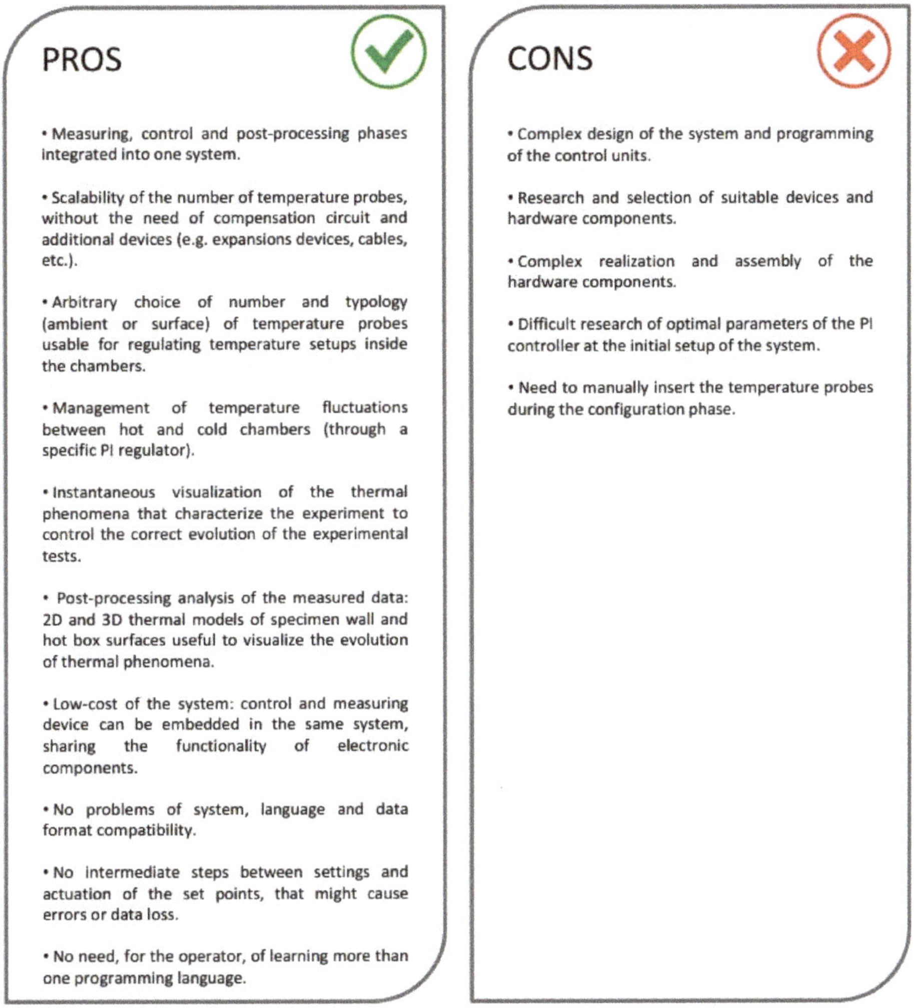

PROS

• Measuring, control and post-processing phases integrated into one system.

• Scalability of the number of temperature probes, without the need of compensation circuit and additional devices (e.g. expansions devices, cables, etc.).

• Arbitrary choice of number and typology (ambient or surface) of temperature probes usable for regulating temperature setups inside the chambers.

• Management of temperature fluctuations between hot and cold chambers (through a specific PI regulator).

• Instantaneous visualization of the thermal phenomena that characterize the experiment to control the correct evolution of the experimental tests.

• Post-processing analysis of the measured data: 2D and 3D thermal models of specimen wall and hot box surfaces useful to visualize the evolution of thermal phenomena.

• Low-cost of the system: control and measuring device can be embedded in the same system, sharing the functionality of electronic components.

• No problems of system, language and data format compatibility.

• No intermediate steps between settings and actuation of the set points, that might cause errors or data loss.

• No need, for the operator, of learning more than one programming language.

CONS

• Complex design of the system and programming of the control units.

• Research and selection of suitable devices and hardware components.

• Complex realization and assembly of the hardware components.

• Difficult research of optimal parameters of the PI controller at the initial setup of the system.

• Need to manually insert the temperature probes during the configuration phase.

Figure 14. Pros and Cons of the proposed measuring and control system.

5. Conclusions

In this paper, after a detailed literature review, an integrated measuring and control system for hot box experiments is presented and its capabilities are tested through a real application on a X-lam sample wall with double insulating layer.

The system, based on a general-purpose microcontroller, digital thermometers, and on the use of an "ad hoc" wireless sensors network, is described both at hardware and firmware levels. The novelties of the proposed system and its main properties are presented.

The system's capability has been tested on a double insulation X-lam wall. The results of the 72 h experiment have shown the system's capability to maintain the wished thermal conditions with small fluctuations (maximum temperature fluctuations in hot and cold chambers equal to ±0.3 °C and ±1.4 °C, respectively) and to measure temperatures and energy input into the hot chamber that resulted equal to 1.06 kWh. The U-value of the wall, equal to 0.216 ± 0.01 W/m^{2}K, was determined by means of the data post-processing of the measured data and it has been compared with the transmittance values obtained through theoretical calculation (equal to 0.176 ± 0.03 W/m^{2}K) and heat flow meter measurements (equal to 0.177 ± 0.01 W/m^{2}K). Moreover, the data post-processing allowed to create 2D and 3D thermal models of specimen wall and chambers.

Finally, the proposed system can represent a convincing improvement with respect to the traditional approaches used in hot box experiments, and, therefore, its employment can be an alternative for those who carry out this kind of analysis. The ease of installation and management of the system, and the low costs, could also favor a more widespread of hot boxes, increasing the research of materials with high energy performance.

Supplementary Materials: The following are available online at http://www.mdpi.com/1996-1073/12/11/2053/s1, Video S1: 2D thermal evolution of the sample wall surfaces.

Author Contributions: Conceptualization, T.d.R. and M.M.; Data curation, T.d.R. and M.M.; Investigation, T.d.R. and M.M.; Methodology, T.d.R.; Resources, T.d.R. and I.N.; Software, M.M. and L.P.; Supervision, V.S. and D.A.; Validation, V.S. and D.A.; Visualization, T.d.R.; Writing—original draft, T.d.R.; Writing—review & editing, I.N., L.P., V.S. and D.A.

Funding: This research did not receive any specific grant from funding agencies in the public, commercial, or not-for-profit sectors.

Acknowledgments: The authors wish to thank gratefully G. Pasqualoni and N. Zaccagnini of the University of L'Aquila—DIIIE Dept., for the fundamental support during the system construction.

Conflicts of Interest: The authors declare no conflict of interest.

References

1. International Energy Agency (IEA). *World Energy Outlook*; IEA: Paris, France, 2016.
2. Pargana, N.; Pinheiro, M.D.; Silvestre, J.D.; de Brito, J. Comparative environmental life cycle assessment of thermal insulation materials of buildings. *Energy Build.* **2014**, *82*, 466–481. [CrossRef]
3. Woltman, G.; Noel, M.; Fam, A. Experimental and numerical investigations of thermal properties of insulated concrete sandwich panels with fiberglass shear connectors. *Energy Build.* **2017**, *145*, 22–31. [CrossRef]
4. Caruana, C.; Yousif, C.; Bacher, P.; Buhagiar, S.; Grima, C. Determination of thermal characteristics of standard and improved hollow concrete blocks using different measurement techniques. *J. Build. Eng.* **2017**, *13*, 336–346. [CrossRef]
5. Gullbrekken, L.; Uvsløkk, S.; Kvande, T.; Time, B. Hot-box measurements of highly insulated wall, roof and floor structures. *J. Build. Phys.* **2017**, *41*, 58–77. [CrossRef]
6. Prata, J.; Simões, N.; Tadeu, A. Heat transfer measurements of a Linear Thermal Bridge in a wooden building corner. *Energy Build.* **2017**, *158*, 194–208. [CrossRef]
7. Lechowska, A.A.; Schnotale, J.A.; Baldinelli, G. Window frame thermal transmittance improvements without frame geometry variations: An experimentally validated CFD analysis. *Energy Build.* **2017**, *145*, 188–199. [CrossRef]
8. Trgala, K.; Pavelek, M.; Wimmer, R. Energy performance of five different building envelope structures using a modified Guarded Hot Box apparatus—Comparative analysis. *Energy Build.* **2019**, *195*, 116–125. [CrossRef]
9. De Rubeis, T.; Nardi, I.; Muttillo, M. Development of a low-cost temperature data monitoring. An upgrade for hot box apparatus. *J. Phys. Conf. Ser.* **2017**, *923*, 012039. [CrossRef]
10. Nardi, I.; Paoletti, D.; Ambrosini, D.; de Rubeis, T.; Sfarra, S. U-value assessment by infrared thermography: A comparison of different calculation methods in a Guarded Hot Box. *Energy Build.* **2016**, *122*, 211–221. [CrossRef]
11. Nardi, I.; Paoletti, D.; Ambrosini, D.; de Rubeis, T.; Sfarra, S. Validation of quantitative IR thermography for estimating the U-value by a hot box apparatus. *J. Phys. Conf. Ser.* **2015**, *655*, 012006. [CrossRef]
12. UNI EN ISO 8990. *Thermal Insulation—Determination of Steady-State Thermal Transmission Properties—Calibrated and Guarded Hot Box*; International Standard Organization: Geneva, Switzerland, 1999.
13. Koo, S.Y.; Park, S.; Song, J.-H.; Song, S.-Y. Effect of Surface Thermal Resistance on the Simulation Accuracy of the Condensation Risk Assessment for a High-Performance Window. *Energies* **2018**, *11*, 382. [CrossRef]
14. Biswas, K. Development and Validation of Numerical Models for Evaluation of Foam-Vacuum Insulation Panel Composite Boards, Including Edge Effects. *Energies* **2018**, *11*, 2228–2244. [CrossRef]

15. Biswas, K.; Desjarlais, A.; Smith, D.; Letts, J.; Yao, J.; Jiang, T. Development and thermal performance verification of composite insulation boards containing foam-encapsulated vacuum insulation panels. *Appl. Energy* **2018**, *228*, 1159–1172. [CrossRef]

16. Sassine, E.; Younsi, Z.; Cherif, Y.; Chauchois, A.; Antczak, E. Experimental determination of thermal properties of brick wall for existing construction in the north of France. *J. Build. Eng.* **2017**, *14*, 15–23. [CrossRef]

17. Peters, B.; Sharag-Eldin, A.; Callaghan, B. Development of a simple Hot Box to determine the thermal characteristics of a three-dimensional printed bricks. In Proceedings of the ARCC 2017 Conference—Architecture of Complexity, Salt Lake City, UT, USA, 14–17 June 2017.

18. Douzane, O.; Promis, G.; Roucoult, J.M.; Tran Le, A.D.; Langlet, T. Hygrothermal performance of a straw bale building: In situ and laboratory investigations. *J. Build. Eng.* **2016**, *8*, 91–98. [CrossRef]

19. Buratti, C.; Belloni, E.; Lunghi, L.; Borri, A.; Castori, G.; Corradi, M. Mechanical characterization and thermal conductivity measurements using of a new "small hot-box" apparatus: Innovative insulating reinforced coatings analysis. *J. Build. Eng.* **2016**, *7*, 63–70. [CrossRef]

20. Buratti, C.; Belloni, E.; Lunghi, L.; Barbanera, M. Thermal Conductivity Measurements by Means of a New "Small Hot-Box" Apparatus: Manufacturing, Calibration and Preliminary Experimental Tests on Different Materials. *Int. J. Thermophys.* **2016**, *37*, 47. [CrossRef]

21. Basak, C.K.; Mitra, D.; Ghosh, A.; Sarkar, G.; Neogi, S. Performance Evaluation of a Guarded Hot Box Test Facility Using Fuzzy Logic Controller for Different Building Material Samples. *Energy Procedia* **2016**, *90*, 185–190. [CrossRef]

22. Grynning, S.; Misiopecki, C.; Uvsløkk, S.; Time, B.; Gustavsen, A. Thermal performance of in-between shading systems in multilayer glazing units: Hot-box measurements and numerical simulations. *J. Build. Phys.* **2015**, *39*, 147–169. [CrossRef]

23. Seitz, S.; MacDougall, C. Design of an Affordable Hot Box Testing Apparatus. In Proceedings of the 16th NOCMAT Non-Conventional Materials and Technologies, Winnipeg, MB, Canada, 10–13 August 2015.

24. Pernigotto, G.; Prada, A.; Patuzzi, F.; Baratieri, M.; Gasparella, A. Characterization of the dynamic thermal properties of the opaque elements through experimental and numerical tests. *Energy Procedia* **2015**, *78*, 3234–3239. [CrossRef]

25. Pernigotto, G.; Prada, A.; Patuzzi, F.; Baratieri, M.; Gasparella, A. Experimental characterization of the dynamic thermal properties of opaque elements under dynamic periodic solicitation. In Proceedings of the Building Simulation Applications BSA 2015—2nd IBPSA Italy Conference, Bozen, Italy, 4–6 February 2015; pp. 547–555.

26. Faye, M.; Lartigue, B.; Sambou, V. A new procedure for the experimental measurement of the effective heat capacity of wall elements. *Energy Build.* **2015**, *103*, 62–69. [CrossRef]

27. Meng, X.; Gao, Y.; Wang, Y.; Yan, B.; Zhang, W.; Long, E. Feasibility experiment on the simple hot box-heat flow meter method and the optimization based on simulation reproduction. *Appl. Eng.* **2015**, *83*, 48–56. [CrossRef]

28. Manzan, M.; Zandegiacomo De Zorzi, E.; Lorenzi, W. Experimental and numerical comparison of internal insulation systems for building refurbishment. *Energy Procedia* **2015**, *82*, 493–498. [CrossRef]

29. Seitz, A.; Biswas, K.; Childs, K.; Carbary, L.; Serino, R. High-Performance External Insulation and Finish System Incorporating Vacuum Insulation Panels—Foam Panel Composite and Hot Box Testing. *Next-Gener. Therm. Insul. Chall. Oppor.* **2014**, 81–100. [CrossRef]

30. Vereecken, E.; Roels, S. A comparison of the hygric performance of interior insulation systems: A hot box–cold box experiment. *Energy Build.* **2014**, *80*, 37–44. [CrossRef]

31. Ghosh, A.; Ghosh, S.; Neogi, S. Performance evaluation of a guarded hot box U-value measurement facility under different software based temperature control strategies. *Energy Procedia* **2014**, *54*, 448–454. [CrossRef]

32. Sousa, H.; de Sousa, R.M. Sensibility analysis of the thermal resistance of masonry through numerical simulations of laboratory tests. In Proceedings of the 9th International Masonry Conference, Guimaraes, Portugal, 7–9 July 2014.

33. Kus, H.; Özkan, E.; Göcer, Ö.; Edis, E. Hot box measurements of pumice aggregate concrete hollow block walls. *Constr. Build. Mater.* **2013**, *38*, 837–845. [CrossRef]

34. Chen, F.; Wittkopf, S.K. Summer condition thermal transmittance measurement of fenestration systems using calorimetric hot box. *Energy Build.* **2012**, *53*, 47–56. [CrossRef]

35. Martin, K.; Campos-Celador, A.; Escudero, C.; Gómez, I.; Sala, J.M. Analysis of a thermal bridge in a guarded hot box testing facility. *Energy Build.* **2012**, *50*, 139–149. [CrossRef]

36. Grynning, S.; Jelle, B.P.; Uvsløkk, S.; Gustavsen, A.; Baetens, R.; Caps, R.; Meløysund, V. Hot box investigations and theoretical assessments of miscellaneous vacuum insulation panel configurations in building envelopes. *J. Build. Phys.* **2011**, *34*, 297–324. [CrossRef]

37. Asdrubali, F.; Baldinelli, G. Thermal transmittance measurements with the hot box method: Calibration, experimental procedures, and uncertainty analyses of three different approaches. *Energy Build.* **2011**, *43*, 1618–1626. [CrossRef]

38. Qin, M.; Belarbi, R.; Aït-Mokhtar, A.; Nilsson, L.O. Coupled heat and moisture transfer in multi-layer building materials. *Constr. Build. Mater.* **2009**, *23*, 967–975. [CrossRef]

39. ASTM C1363-11. *Standard Test Method for Thermal Performance of Building Materials and Envelope Assemblies by Means of a Hot Box Apparatus*; ASTM: West Conshohocken, PA, USA, 2011.

40. Smith, I.B. Applications and limitations of thermocouples for measuring temperatures. *J. Am. Inst. Electr. Eng.* **1923**, *42*, 171–178. [CrossRef]

41. Depari, A.; Sisinni, E.; Flammini, A.; Ferri, G.; Stornelli, V.; Barile, G.; Parente, F.R. Autobalancing Analog Front End for Full-Range Differential Capacitive Sensing. *IEEE Trans. Instrum. Meas.* **2018**, *67*, 885–893. [CrossRef]

42. Barile, G.; Ferri, G.; Parente, F.R.; Stornelli, V.; Sisinni, E.; Depari, A.; Flammini, A. A CMOS full-range linear integrated interface for differential capacitive sensor readout. *Sens. Actuators A-Phys.* **2018**, *281*, 130–140. [CrossRef]

43. Ferri, G.; Stornelli, V.; Parente, F.R.; Barile, G. Full range analog Wheatstone bridge-based automatic circuit for differential capacitance sensor evaluation. *Int. J. Circ. Appl.* **2016**, *45*, 2149–2156. [CrossRef]

44. Hambali, N.; Rahim Ang, A.A.; Ishak, A.A.; Janin, Z. Various PID Controller Tuning for Air Temperature Oven System. In Proceedings of the IEEE International Conference on Smart Instrumentation, Measurement and Applications (ICSIMA), Kuala Lumpur, Malaysia, 26–27 November 2013. [CrossRef]

45. DS18B20, Digital Thermometer Datasheet. Available online: https://datasheets.maximintegrated.com (accessed on 12 February 2019).

46. Cid, N.; Ogando, A.; Gomez, M.A. Acquisition system verification for energy efficiency analysis of building materials. *Energies* **2017**, *10*, 1254–1266. [CrossRef]

47. Nardi, I.; de Rubeis, T.; Buzzi, E.; Sfarra, S.; Ambrosini, D.; Paoletti, D. Modeling and optimization of the thermal performance of a wood-cement block in a low-energy house construction. *Energies* **2016**, *9*, 677. [CrossRef]

48. UNI EN ISO 6946. *Building Components and Building Elements—Thermal Resistance and Thermal Transmittance—Calculation Method*; International Standard Organization: Geneva, Switzerland, 2018.

49. UNI ISO 9869. *Thermal Insulation. Building Elements. In-Situ Measurement of Thermal Resistance and Thermal Transmittance*; International Standard Organization: Geneva, Switzerland, 2015.

50. Holman, J.P. *Experimental Methods for Engineers*, 8th ed.; McGraw-Hill series in mechanical engineering; McGraw-Hill: New York, NY, USA, 2001; ISBN-13 978-0-07-352930-1.

51. De Rubeis, T.; Nardi, I.; Ambrosini, D.; Paoletti, D. Is a self-sufficient building energy efficient? Lesson learned from a case study in Mediterranean climate. *Appl. Energy* **2018**, *218*, 131–145. [CrossRef]

Article

Effects of the Aspect Ratio of a Rectangular Thermosyphon on Its Thermal Performance

Chia-Wang Yu [1], C. S. Huang [2], C. T. Tzeng [1] and Chi-Ming Lai [3,*]

[1]   Department of Architecture, National Cheng Kung University, Tainan 701, Taiwan;
      jiawang_yu@hotmail.com (C.-W.Y.); ctmt@mail.ncku.edu.tw (C.T.T.)
[2]   Department of Mechanical Engineering, Southern Taiwan University of Science and Technology,
      Tainan 710, Taiwan; huang.c.s1018@gmail.com
[3]   Department of Civil Engineering, National Cheng Kung University, Tainan 701, Taiwan
*   Correspondence: cmlai@mail.ncku.edu.tw; Tel.: +886-6-2757575 (ext. 63136)

Received: 13 September 2019; Accepted: 20 October 2019; Published: 22 October 2019

Abstract: The natural convection behaviors of rectangular thermosyphons with different aspect ratios were experimentally analyzed in this study. The experimental model consisted of a loop body, a heating section, a cooling section, and adiabatic sections. The heating and cooling sections were located in the vertical portions of the rectangular loop. The length of the vertical cooling section and the lengths of the upper and lower adiabatic sections were fixed at 300 mm and 200 mm, respectively. The inner diameter of the loop was fixed at 11 mm, and the cooling end temperature was 30 °C. The relevant parameters and their ranges were as follows: The aspect ratios were 6, 4.5, and 3.5 (with potential differences of 41, 27, and 18, respectively, between the cold and hot ends), and the input thermal power ranged from 30 to 60 W (with a heat flux of 600 to 3800 W/m^2). The results show that it is feasible to obtain solar heat gain by installing a rectangular thermosyphon inside the metal curtain wall and that increasing the height of the opaque part of the metal curtain wall can increase the aspect ratio of the rectangular thermosyphon installed inside the wall and thus improve the heat transfer efficiency.

Keywords: heat transfer; energy; thermosyphon; natural circulation loop

---

1. Introduction

A thermosyphon is a type of natural circulation loop. It relies on the proper arrangement of the heating zone and cooling zone to cause a change in the density of the fluid within the loop, and the resulting thermal buoyancy drives the working fluid to transfer thermal energy. The cooling section and the heating section are usually placed either above and below the loop or on the left and right sides of the loop. When the working fluid is heated in the heating section, its density decreases, generating thermal buoyancy and driving it to flow upward and dissipate heat in the cooling section. Gravity resists the upward flow of the working fluid and helps its downward flow (when the flow is in the same direction as gravity). Because no external driving force is required, thermosyphons have considerable operational reliability. This self-adjusting mechanism and stability have led thermosyphons to be used in a wide range of applications, such as solar heating and cooling systems, coolers for nuclear power plant reactors, geothermal energy systems, waste heat recovery systems, and electronic cooling systems. There are many considerations in the design of a thermosyphon with good thermal efficiency, including the choices of the working fluid and wall material, the locations of the heating and cooling sections, and the geometric parameters of the loop. Studies of the performance of natural convection loops and the effects of various parameters can be found in the existing literature [1–3].

Ismail and Abogderah [4] compared the theoretical predictions with experimental results of a flat-plate solar collector with heat pipes as energy transport. The experimental results of the proposed solar collector were also compared with a conventional commercial solar collector. The results showed that the instantaneous efficiencies of the proposed collector are lower than the conventional one in the morning and higher when the heat pipes reach their operating temperatures. Mathioulakis and Belessiotis [5] theoretically and experimentally studied the thermal performance of a solar water system with an integrated wickless gravity assisted loop heat pipe. The results validated a theoretical model for the collector that can be used for the optimization of the system design. Misale et al. [6] analyzed the influence of thermal boundary conditions on the flow regimes inside the pipes and the stability of the thermosyphon. Their results showed that the higher the heating power is, the greater the flow rate is. However, after the flow becomes a two-phase flow, the flow rate decreases due to the formation of bubbles, and an increase in the height difference of the loop also increases the flow rate and heat transfer gain.

Lai et al. [7] explored the thermal performance of a rectangular natural circulation loop. The results showed that the average velocity of the fluid increases with an increase in the heating power and aspect ratio or a decrease in the length of the cooling section. Desrayaud et al. [8] numerically investigated the thermal behavior of a rectangular natural circulation loop with horizontal heat exchangers. It was shown that the vortices result in the occurrence of the oscillations and cause the growth of temperature gradients. Huminic and Huminic [9] numerically investigated the nanofluid heat transfer in thermosyphon heat pipes. The results showed that using the nanofluid has better heat transfer characteristics than using water and the volume concentration of nanoparticles has a significant effect in reducing the temperature difference between the evaporator and condenser. Martinopoulos et al. [10] experimentally evaluated the thermal performance and the optimization parameters of a phase-change flat plate solar collector. The collector with a 50% volume filament at a 40° inclination showed a better system efficiency that increased proportionally with the increased mass flow rates.

Ho et al. [11] obtained the relationship between the Rayleigh number (Ra) and the Reynolds number (Re), as well as the Nusselt number (Nu) of a single-phase thermosyphon, using experiments and numerical simulations. Vijayan et al. [12] studied the effects of the heater and cooler orientations in a single-phase thermosyphon. Three oscillatory modes and instabilities were observed in the experiments. Misale et al. [13] tested thermosyphons at different inclination angles and found that the inclination angle affected the heat transfer performance and that the best performance can be achieved at an inclination angle of 0° (i.e., vertical with respect to the ground). Swapnalee and Vijayan [14] obtained the relation between Re and the Grashof number (Gr) from experiments and simulations of a single-phase thermosyphon and used the geometric parameter Ng to modify Gr to make the prediction model of the relation applicable to four different types of heating and cooling. Thomas and Sobhan [15] experimentally studied the stability and transient performance of a vertical heater-vertical cooler natural circulation loop with metal oxide nanoparticles. The results indicated that nanofluids containing aluminum oxide and copper oxide have better heat transfer performance as the working fluid than pure water. Huang et al. [16] experimentally analyzed the natural convection of a rectangular thermosyphon with an aspect ratio (AR) of 3.5. They found that the value of the dimensionless heat transfer coefficient, Nu, is generally between 5 and 10, and that the power of the heating section and the height difference between the cooling and heating sections are the main factors affecting the natural convection intensity of the thermosyphon.

The thermal performance and applications of a rectangular thermosiphon with the heating and cooling sections on the opposite vertical legs are worthy of in-depth investigation, but have rarely been studied in the literature. To the best knowledge of the authors, a study responding to the design diversity of balcony exterior walls or curtain walls via the loop aspect ratios, as shown in Figure 1, has not yet been conducted. Therefore, this study, which is a follow-up to [16], experimentally investigated the effect of the AR of a rectangular thermosyphon loop on its natural convection performance using boundary conditions of a constant heat flux and a fixed wall temperature for the heating and cooling sections, respectively, of the loop.

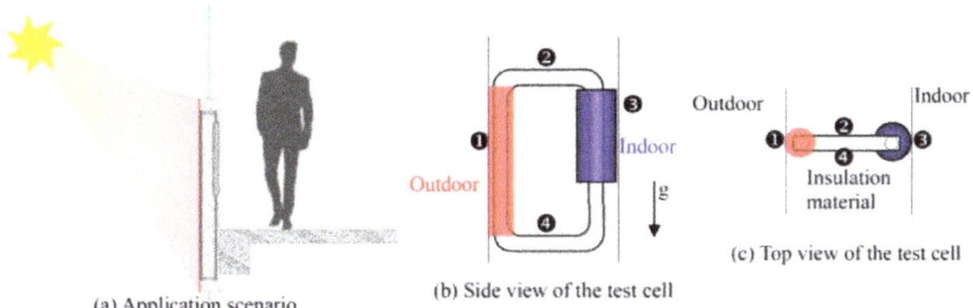

Figure 1. Scenario and test cell illustration.

2. Materials and Methods

2.1. Scenarios and Test Cell Development

Our goal was to develop a structure within the metal curtain wall to capture solar heat energy (Figure 1a). The exterior wall of the structure that absorbs solar heat was simplified to a single vertical tube ❶ with a constant heat flux boundary. The structure was designed to be a rectangular thermosyphon to investigate its thermal performance in detail, as shown in Figure 1b,c. The heat insulation materials filled the gap between the exterior and interior walls of the structure. When hot water is required by the occupants, cold water can be fed into the heat exchanger ❸ at the indoor side. The side view in Figure 1b clearly illustrates the components of the circulation loop by removing all the heat insulation materials between the outdoor and indoor ends.

During the day, the working fluid in the circulation loop becomes a high-temperature liquid after it absorbs the solar heat from an outdoor heat source ❶. The fluid is driven by thermal buoyancy to create natural convection and then flows along the horizontal circulation branch ❷ to the indoor heat sink ❸, where heat is dissipated, and then flows under gravity along another horizontal circulation branch ❹ back to the outdoor heat source. Thus, a naturally flowing rectangular thermosyphon is formed, in which the outdoor heat source ❶ is represented by a vertical tube heated by a constant heat flux, and the indoor heat sink ❹ is the vertical section with an isothermal temperature boundary.

2.2. Experimental Test Cell

The test cell can be divided into the rectangular loop, the heating section, the cooling section, and the adiabatic sections. The heating section and the cooling section are both located in the vertical portions of the rectangular loop, as shown in Figure 2, and are introduced below.

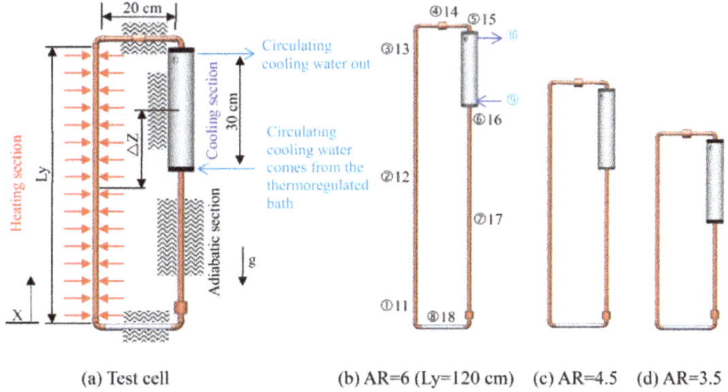

Figure 2. Experimental test cell and geometric parameters.

2.2.1. Rectangular Loop

The loop in the experiment was made of red copper tube with a high thermal conductivity, and the entire loop had the same cross-sectional area. The round copper tube had an outer diameter of 12.7 mm and an inner diameter (Di) of 11 mm. The experimental test cell was installed inside the curtain wall. The lengths of the upper and lower adiabatic sections, *Lx*, of the experimental test cell were similar to the thickness of the curtain wall. In building practice, the thickness of the wall was approximately 200 mm with little variation. Therefore, *Lx* was fixed at 200 mm in this study. The length of the vertical heating section, Ly, was similar to the height of the curtain wall, and this height varied with the elevation design of the building. Therefore, *Ly* was set to 1200, 900, and 700 mm in this study. As a result, the ARs (AR = *Ly/Lx*) of the experimental test cell were 6, 4.5, and 3.5, respectively, as shown in Figure 2b–d. To investigate the influence of the geometry on the heat transfer capability of the experimental test cell, the length of the cooling end *Lc* was fixed at 300 mm. Therefore, the potential differences of the experimental test cell $\Delta Z = \frac{1}{2}\left(\frac{Ly-Lc}{Di}\right)$ were 41, 27 and 18, respectively.

2.2.2. Heating Section

The heating section is a constant heat flux boundary for the simulation of actual applications. Therefore, the heating wire was used to generate heating at a fixed thermal power. To increase the uniformity of the heating section, the heating wire was directly processed in the heating section and was mainly composed of mica paper, heating wire, and heat insulating material. Because the tubes of the experimental test cell were made of conductive red copper, the surfaces of the copper tubes in the heating section had to be insulated before winding the heating wire around the heating section. Therefore, soft muscovite paper, which is an excellent insulator and is thermally resistant to 500 and 550 °C, was wound around the heating section for insulation. Based on the input electric power of 60 W and the maximum power supply voltage of 60 V, the electric resistance was determined to be 60 Ω. Thus, the heating wire should have a similar resistance. A 2.5 m-long heating wire with a diameter of 0.25 mm and a resistance of 25.82 Ω/m was selected for the experiment.

2.2.3. Cooling Section

The cooling section is a simulated isothermal boundary condition. The cooling water sleeve was 300 mm long and was mainly composed of a water sleeve body and two water sleeve caps. The water sleeve body was a cylinder with an outer diameter of 61 mm and an inner diameter of 40 mm. Each end of the sleeve was covered by a water sleeve cap, at the center of which was an opening with a diameter of approximately 11 mm. The thermosyphon could pass through the cooling water sleeve via these openings, and the joints were fitted with O-ring grooves to prevent leakage of the cooling water sleeve.

2.2.4. Adiabatic Sections

Other than the heating and cooling sections, the remainder of the cell was composed of the adiabatic sections. To eliminate the effects of heat loss and ambient environmental factors during the experiment, a 3 mm-thick insulating tape and a 20 mm-thick insulating pipe were wrapped around the body to effectively simulate the adiabatic boundary conditions.

2.3. Experimental Apparatus

The device and data acquisition system included a data acquisition unit (Yokogawa MX-100, Yokogawa Electric, Tokyo, Japan), a PC, a flow meter (Fluidwell F110-P, Fluidwell BV, Veghel, The Netherlands), a DC power supply unit (Gwinstek SPD-3606, GW Instek, New Taipei City, Taiwan) and a thermoregulated bath (RCB-412).

2.3.1. Power Supply System

The heating wire of the heating section was directly connected to the power supply. The output voltage and current were adjusted to provide the input electric power q_h of 30–60 W that was required in the experiment. If we take 300 W/m^2 as the nominal solar heat gain on the wall, then the service areas of each test cell are 0.1 m^2 (= 30/300; q_h = 30 W) and 0.2 m^2 (= 60/300; q_h = 60 W).

2.3.2. Thermoregulated Bath

The isothermal boundary of the cooling section was established by circulating water from the thermoregulated water bath to establish a cooling section temperature T_c of 30 °C required for the experiment.

2.3.3. Liquid Flow Meter

The flow meter was used to monitor the cooling water output by the thermoregulated bath in each set of experiments, with an average flow rate of approximately 35 mL/s.

2.3.4. Thermocouples

Type-K thermocouples were used in the experiment to measure the temperature distribution at different points along the loop, including the wall temperature and the water temperature inside the tube. A total of 18-point thermocouples was embedded in each experimental test cell at the lower, middle, and upper positions of the heating section, at the inlet and outlet of the water sleeve of the cooling section, and along the adiabatic sections. The numbering and measurement locations of the thermocouples are shown in Figure 2b, where the thermocouples numbered from 1 to 8 measure the temperatures at the center of the fluid in the tube, 9 and 10 measure the temperatures at the inlet and outlet of the water sleeve, and 11 to 18 measure the temperatures at the outer tube wall.

2.4. Experimental Procedures

At the beginning of the experiment, to prevent the initial temperature of the cooling section from starting the flow of the fluid in the tube, the power supply device needed to provide a high electric power to the heating wire. The output power was then reduced after the temperature of the heating section became higher than that of the cooling section to achieve a steady-state heating boundary. The experimental steps are as follows:

1. Set the temperature of the thermoregulated bath to the cooling section temperature of 30 °C required for the experiment and keep the inlet and outlet valves of the thermo-regulated bath closed.
2. Set the required power output (30, 40, 50, and 60 W) on the power supply.
3. When the temperature of the heating section is higher than the set temperature of the cooling section wall, open the inlet and outlet valves of the thermo-regulated bath.
4. After the temperature of the thermosyphon loop reaches a steady state, increase the input power and carry out the next steady-state experiment.
5. Repeat step 4 until the input power reaches 60 W, when the experiment is completed.

2.5. Parameters

The relevant heat transfer parameters were calculated from the temperature measured by each thermocouple, as well as the voltage and current supplied to the heating wire in the experimental test cell.

1. The input power $q_h = VI$ (W) was obtained from the voltage V and the current I provided by the power supply. During the experiment, the heating from the power supply was not completely transferred to the fluid in the heating section. A small amount of heat entered the cooling section

via axial heat conduction q_a (W) of the thermosyphon or escape into the environment. Therefore, this axial heat transfer was first subtracted from the input thermal power, that is, the corrected actual input thermal power was $q_{in} = q_h - q_a$ (W). Then, the heat flux q_h'' was calculated as q_{in}/A (W/m^2), where A is the area of the heating section.

2.   Modified Rayleigh number, Ra^*

$$Ra^* = \frac{g\beta q_h'' L_y^4}{k\alpha v} \tag{1}$$

where g is the acceleration due to gravity (m/s), β is the thermal expansion coefficient of water (1/K), v is the kinematic viscosity (m^2/s), and α is the thermal diffusivity $\frac{k}{\rho C_p}$ (m^2/s), ρ is the density (kg/m^3), C_p is the specific heat (J/kg K), and k is the thermal conductivity (W/m). The values of these physical properties are based on the average water temperature of the heating section (35 °C).

3.   Reynolds number, Re

$$Re = \frac{VD_i}{v} \tag{2}$$

where the fluid flow velocity V (m/s) in the tube is observed through the transparent section of the loop in the experiment.

4.   Nusselt number

(1)   The average thermal convection coefficient h at the hot end was:

$$h = \frac{q_h''}{\overline{T}_w - \overline{T}_m} \tag{3}$$

where $\overline{T}_w$ (K) and $\overline{T}_m$ (K) are the average temperature of the heating wall and the average water temperature of the heating section, respectively.

(2)   The Nusselt number Nu was calculated as follows:

$$Nu = \frac{hD_i}{k} \tag{4}$$

5.   Thermal resistance of the working fluid flow, R_{flow}

$$R_{flow} = \frac{1}{\dot{m}C_p} = \frac{1}{\rho V A_f C_p} \tag{5}$$

2.6. Experimental Uncertainty

The heat input was measured by an electronic wattmeter with an accuracy of 0.01 W. The type-K thermocouples used for the temperature measurements were accurate to 0.1 °C. These errors are believed to be inconsequential to the results of the experiment.

3. Results and Discussion

This experiment mainly explored the effects of the AR on the heat transfer of a rectangular thermosyphon. The main parameters were: AR = 6, 4.5, and 3.5 (the potential differences between the cold and hot ends ΔZ are 41, 27, 18, respectively), the input thermal power ranged from 30 to 60 W, the heat flux was in the range of 600 to 3800 W/m^2, the cooling section temperature Tc is 30 °C, and the cold end length Lc is 30 m.

Due to space limitations of this article, only the experimental results for AR = 6 were discussed. The heat flow phenomena for ARs of 4.5 and 3.5 were similar to those for AR = 6 and were not discussed in this paper. Figure 3 shows the distribution of the steady-state wall temperature of the thermosyphon loop at different heating powers and with a cold wall temperature of 30 °C and a circulating water flow rate of 35 mL/s. The dotted lines partition the figure into four blocks, and the point with a dimensionless distance X_{loop} of 0 was set at the lower-left corner of the loop system (Figure 2a). The X_{loop} range of 0–0.43 corresponds to the heating section, the range of 0.43–0.5 corresponds to the adiabatic section between the outlet of the heating section and the inlet of the cooling section, the range of 0.5–0.61 corresponds to the cooling section, the range of 0.61–1 corresponds to the adiabatic section after the cooling section, and X_{loop} of 1 indicates the return to the starting point of the loop (i.e., the starting point of the heating section).

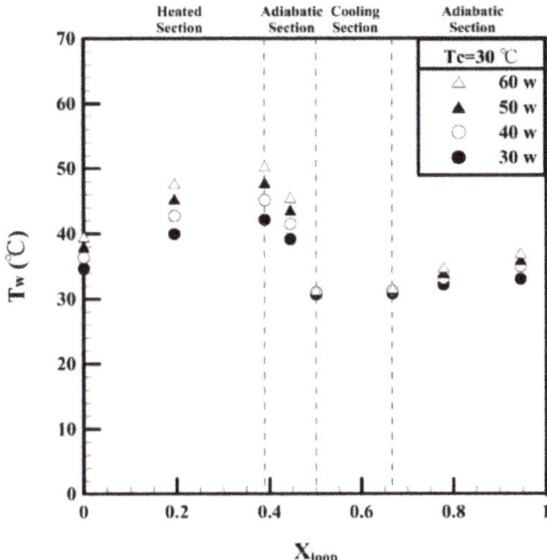

Figure 3. Steady-state temperature distribution of the wall of the loop with different heating powers.

The variation of the wall temperature T_w of the thermosyphon shows that after entering the heating section, the wall temperature of the loop began to increase linearly and reached the highest temperature at the end of the flow passage. After entering the adiabatic section, the temperature decreased considerably. After entering the cooling section, the wall temperature quickly decreased to the set cold wall temperature. The results confirm that the cold wall temperature met the set isothermal state and that the temperature increased after exiting the cooling section.

Next, the influence of different heating wattages on the temperature distribution was analyzed. The results show that the higher the wattage was, the higher the heating wall temperature iswas. In addition, the slope of the temperature variation increased, which reflects the boundary condition of the constant heat flux. In the adiabatic zone between the outlet of the heating section and the inlet of the cooling section, the wall temperature was affected by the low temperature of the cooling section due to the axial heat transfer of the tube wall. Therefore, the temperature decreased considerably after entering the adiabatic section. In the cooling section, the cooling wall temperature was not significantly affected by the wattage, indicating the isothermal condition of the cooling section. In the adiabatic zone between the outlet of the cooling section and the inlet of the heating section, the wall temperature was affected again by the axial heat transfer of the tube wall. The temperature of the cooling section was affected by the high temperature of the heating section, and the temperature increased.

Figure 4 shows the distribution of the maximum wall temperature $T_{w,max}$ under the specified hot water intake mode (i.e., the cold end circulating water flow rate of 35 mL/s). The results show that at AR = 3.5, as the heating power increased, the maximum wall temperature increased linearly from 42.1 °C (30 W) to 50.3 °C (60 W). The maximum wall temperature decreased nonlinearly with increasing AR. With AR = 6, the minimum temperatures were 39.6 °C (30 W) and 46.2 °C (60 W).

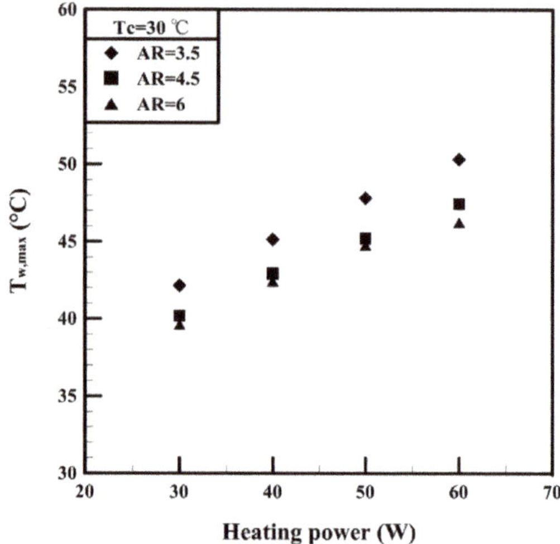

Figure 4. Distribution of the maximum wall temperature.

Figure 5 shows the relationship between *Re* and *Ra**. With the same heating power, *Ra** increased with the increase of aspect ratio. Within the range of parameter values in this study, *Re* ranged from 129 to 213, indicating a laminar flow pattern. When AR = 6, Re increased with increasing *Ra**, but the upward trend became moderate in the high *Ra** range. When AR = 3.5, Re also increased with increasing *Ra**, but both the amplitude and the slope of the increase were larger than those with AR = 6.

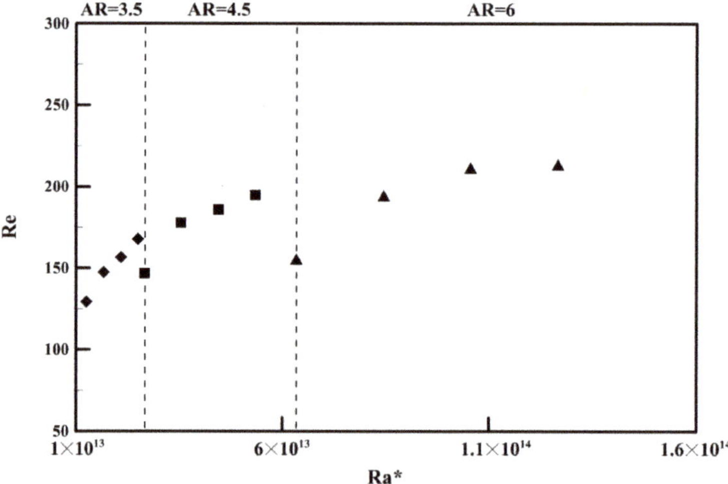

Figure 5. Relationship between the Reynolds number (*Re*) and the modified Rayleigh number (*Ra**).

Figure 6 shows the relationship between the average Nu and *Ra** in the heating section. Within the range of parameter values in this study, Nu was between 4.3 and 8.4. With a fixed AR, as *Ra** increased, the variation in the amplitude of Nu was small. When AR = 3.5, as *Ra** increased, Nu increased slightly. Nu increased with increasing AR, indicating that the AR affected the heat transfer performance of the rectangular thermosyphon, that is, with the same heating power, a higher AR corresponded to a better heat transfer capability. Figures 5 and 6 show that as the AR increased, the potential difference ΔZ

of the heating section and cooling section also increased, thereby increasing the fluid flow rate and the heat transfer efficiency in the tube. This can be used as a reference for incorporating the studied thermosyphons into building design. When the rectangular thermosyphon inside the metal curtain wall is used to obtain the solar heated water, the opaque part of the metal curtain wall can be raised to make the thermosyphon have a high AR to enhance the heat transfer efficiency.

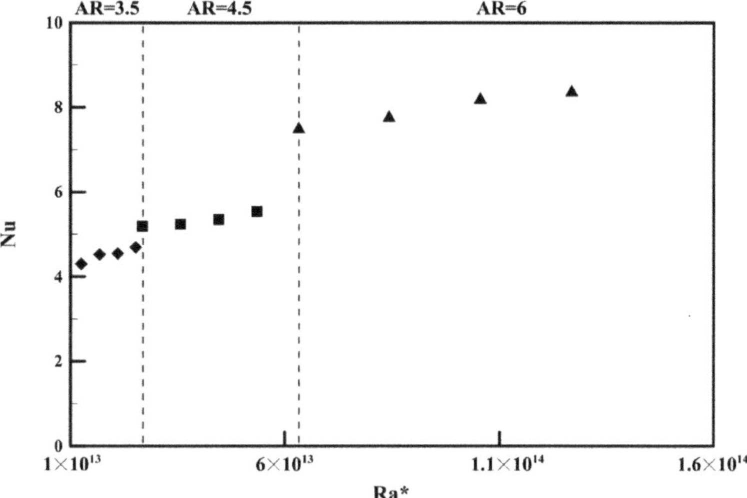

Figure 6. Relationship between the average Nusselt number (*Nu*) and the modified Rayleigh number (*Ra**) of the heating section.

Figure 7 shows the relationship between the thermal resistance of the working fluid flow (R_{flow}) and *Ra** of the heating section. The higher *Ra** was, the smaller R_{flow} was, that is, a higher heating power resulted in greater thermal buoyancy, thereby increasing the flow rate and decreasing R_{flow}. Within the range of parameter values in this study, R_{flow} was between 0.18 and 0.29 K/W.

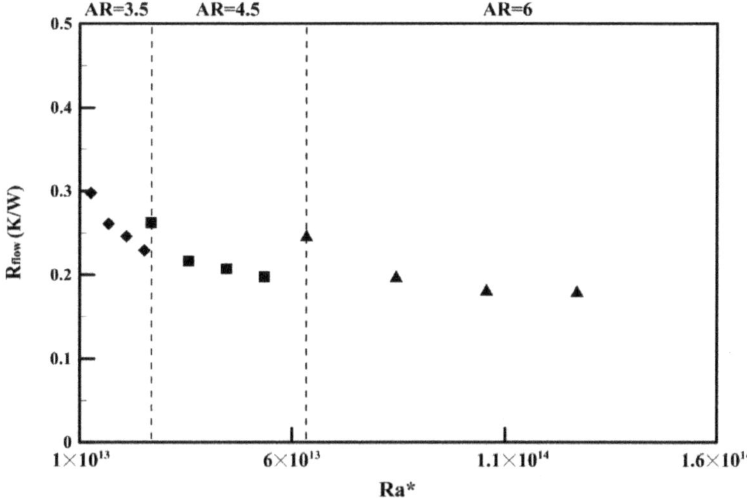

Figure 7. Relationship between the thermal resistance of the working fluid flow (R_{flow}) and *Ra**.

4. Conclusions

The main purpose of this study was to investigate the effects of different geometric aspect ratios and the heating wattage on the thermal capacity of a rectangular thermosyphon loop with an inner diameter of 11 mm. Based on the experimental results, the following conclusions can be summarized:

1. It is feasible to install a rectangular thermosyphon inside a metal curtain wall to obtain solar heated water.
2. When the AR increased, the maximum wall temperature decreased nonlinearly. The lowest temperature can be reduced to 39.6 °C and 46.2 °C at 30 W and 60 W, respectively, with AR = 6.
3. Within the range of parameter values in this study, Nu was between 4.3 and 8.4. The higher the AR was, the higher Nu was, indicating that the AR can affect the heat transfer efficiency of the rectangular thermosyphon.
4. When the rectangular thermosyphon is used inside the metal curtain wall to obtain a solar heat gain, the opaque part of the metal curtain wall can be raised to give the thermosyphon a higher AR to enhance the heat transfer efficiency.
5. The larger Ra* was, the lower the thermal resistance of the working fluid flow was, that is, a greater heating power or larger AR resulted in greater thermal buoyancy, thereby causing the flow to increase and the thermal resistance of the working fluid flow to decrease.

The results are limited to the chosen specific aspect ratios (6, 4.5, and 3.5). To respond to the design diversity of balcony exterior walls or curtain walls via the loop geometry configurations, thermosyphons with different aspect ratios could be tested. Although a computational modeling was not the focus of this study, it is worthy of future consideration.

Author Contributions: C.-M.L. and C.S.H. designed the study; C.-W.Y. and C.S.H. performed the experiments; C.-W.Y., and C.S.H. analyzed the data; C.T.T. and C.-M.L. wrote the manuscript.

Funding: This research received no external funding

Conflicts of Interest: The authors declare no conflict of interest.

References

1. Esen, M.; Esen, H. Experimental investigation of a two-phase closed thermosyphon solar water heater. *Sol. Energy* **2005**, *79*, 459–468. [CrossRef]
2. Buschmann, M.H. Nanofluids in thermosyphons and heat pipes: Overview of recent experiments and modelling approaches. *Int. J. Therm. Sci.* **2013**, *72*, 1–17. [CrossRef]
3. Ho, C.J.; Chen, Y.Z.; Tu, F.-J.; Lai, C.-M. Thermal performance of water-based suspensions of phase change nanocapsules in a natural circulation loop with a mini-channel heat sink and heat source. *Appl. Therm. Eng.* **2014**, *64*, 376–384. [CrossRef]
4. Ismail, K.A.R.; Abogderah, M.M. Performance of a Heat Pipe Solar Collector. *J. Sol. Energy Eng.* **1998**, *120*, 51–59. [CrossRef]
5. Mathioulakis, E.; Belessiotis, V. A new heat-pipe type solar domestic hot water system. *Sol. Energy* **2002**, *72*, 13–20. [CrossRef]
6. Misale, M.; Garibaldi, P.; Tarozzi, L.; Barozzi, G.S. Influence of thermal boundary conditions on the dynamic behaviour of a rectangular single-phase natural circulation loop. *Int. J. Heat Fluid Flow* **2011**, *32*, 413–423. [CrossRef]
7. Lai, C.-M.; Chen, R.-H.; Huang, C.S. Development and thermal performance of a wall heat collection prototype. *Build. Environ.* **2012**, *57*, 156–164. [CrossRef]
8. Desrayaud, G.; Fichera, A.; Lauriat, G. Two-dimensional numerical analysis of a rectangular closed-loop thermosiphon. *Appl. Therm. Eng.* **2013**, *50*, 187–196. [CrossRef]
9. Huminic, G.; Huminic, A. Numerical study on heat transfer characteristics of thermosyphon heat pipes using nanofluids. *Energy Convers. Manag.* **2013**, *76*, 393–399. [CrossRef]
10. Martinopoulos, G.; Ikonomopoulos, A.; Tsilingiridis, G. Initial evaluation of a phase change solar collector for desalination applications. *Desalination* **2016**, *399*, 165–170. [CrossRef]

11. Ho, C.J.; Chiou, S.P.; Hu, C.S. Heat transfer characteristics of a rectangular natural circulation loop containing water near its density extreme. *Int. J. Heat Mass Transf.* **1997**, *40*, 3553–3558. [CrossRef]

12. Vijayan, P.K.; Sharma, M.; Saha, D. Steady state and stability characteristics of single-phase natural circulation in a rectangular loop with different heater and cooler orientations. *Exp. Therm. Fluid Sci.* **2007**, *31*, 925–945. [CrossRef]

13. Misale, M.; Garibaldi, P.; Passos, J.C.; de Bitencourt, G.G. Experiments in a single-phase natural circulation mini-loop. *Exp. Therm. Fluid Sci.* **2007**, *31*, 1111–1120. [CrossRef]

14. Swapnalee, B.T.; Vijayan, P.K. A generalized flow equation for single phase natural circulation loops obeying multiple friction laws. *Int. J. Heat Mass Transf.* **2011**, *54*, 2618–2629. [CrossRef]

15. Thomas, S.; Sobhan, C.B. Stability and transient performance of vertical heater vertical cooler natural circulation loops with metal oxide nanoparticle suspensions. *Heat Transf. Eng.* **2018**, *39*, 861–873. [CrossRef]

16. Huang, C.S.; Yu, C.-W.; Chen, R.H.; Tzeng, C.-T.; Lai, C.-M. Experimental observation of natural convection heat transfer performance of a rectangular thermosyphon. *Energies* **2019**, *12*, 1702. [CrossRef]

Article

Experimental Observation of Natural Convection Heat Transfer Performance of a Rectangular Thermosyphon

C. S. Huang [1], Chia-Wang Yu [2], R. H. Chen [3], Chun-Ta Tzeng [2] and Chi-Ming Lai [4,*]

[1] Department of Mechanical Engineering, Southern Taiwan University of Science and Technology, Tainan 710, Taiwan; huang.c.s1018@gmail.com

[2] Department of Architecture, National Cheng Kung University, Tainan 701, Taiwan; jiawang_yu@hotmail.com (C.-W.Y.); ctmt@mail.ncku.edu.tw (C.-T.T.)

[3] Department of Mechanical and Energy Engineering, National Chiayi University, Chiayi 600, Taiwan; chenrh@mail.ncyu.edu.tw

[4] Department of Civil Engineering, National Cheng Kung University, Tainan 701, Taiwan

*   Correspondence: cmlai@mail.ncku.edu.tw; Tel.: +886-6-2757575

Received: 8 March 2019; Accepted: 3 May 2019; Published: 6 May 2019

Abstract: This study experimentally investigates the natural convection heat transfer performance of a rectangular thermosyphon with an aspect ratio of 3.5. The experimental model is divided into a loop body, a heating section, a cooling section, and two adiabatic sections. The heating section and the cooling section are located in the vertical legs of the rectangular loop. The length of the vertical heating section and the length of the upper and lower horizontal insulation sections are 700 mm and 200 mm, respectively, and the inner diameter of the loop is 11 mm. The relevant parameters and their ranges are as follows: the input thermal power is 30–60 W (with a heat flux in the range of 60–3800 W/m^2); the temperature in the cooling section is 30, 40, or 50 °C; and the potential difference between the hot and cold sections is 5, 11, or 18 for the cooling section lengths of 60, 45, and 30 cm, respectively. The results indicate that the value of the dimensionless heat transfer coefficient, the Nusselt number, is generally between 5 and 10. The heating power is the main factor affecting the natural convection intensity of the thermosyphon.

Keywords: heat transfer; energy; thermosyphon; natural circulation loop

---

1. Introduction

A thermosyphon is a type of natural convection loop. By proper arrangement of the heating zone and the cooling zone, the fluid in the loop changes in density, and the resulting thermal buoyancy drives the heat transfer of the working fluid. The cooling section and the heating section are usually placed on the upper and lower sides or on the left and right sides of the loop, respectively. Heating of the working fluid produces a lower density, which generates thermal buoyancy and upward flow, and the heat is radiated toward the cooling section. Gravity works against the working fluid as it flows upwards and assists it as it flows downward (i.e., in the same direction as the gravity). Because there is no need for an external driving force, a thermosyphon has considerable operational reliability. The self-adjusting mechanism and stability enable thermosyphons to be applied in an extensive range of applications, such as solar heating and cooling systems, coolers for reactors in nuclear power plants, geothermal energy systems, waste heat recovery systems, and electronic cooling systems. There are many considerations when designing a thermosyphon with good thermal efficiency, such as the choice of working fluid, the choice of wall material, the location of the heating and cooling sections, and the geometry of the loop [1–10].

Garrity et al. [1] studied the instability of a two-phase thermosiphon with a microchannel evaporator and a condenser. Vijayan et al. [2] analyzed the steady state and stability of single-phase, two-phase, and supercritical natural convection in a rectangular loop using a 1-D theoretical model. Misale et al. [3] analyzed the influence of the thermal boundary conditions on the flow regimes inside the pipes and the stability of the thermosiphon. The results indicated that the larger the heating power is, the larger the flow rate; however, after becoming a two-phase flow, the flow rate is reduced due to the generation of bubbles. Lai et al. [4] explored the influences of the aspect ratio, potential difference, heating power, and cooling temperature on the thermal performance of a rectangular natural circulation loop. Delgado et al. [5] reviewed the information and application of two latent working fluids: phase change material (PCM) emulsions and microencapsulated PCM slurries.

Desrayaud et al. [6] numerically investigated the instability of a rectangular natural circulation loop with horizontal heat exchanging sections for various Rayleigh number (Ra) values. The results show that vortices induce the occurrence of oscillations and the growth of temperature gradients. Buschmann [7] analyzed the research on thermosyphons, heat pipes, and oscillating heat pipes operated with nanofluids based on 38 experimental studies and 4 modeling approaches. The results show that the effects related to the filling ratio, inclination angle, and operation temperature seem to be similar to those for classical working fluids. Huminic [8] investigated the effects of volume concentrations of nanoparticles and the operating temperature on the heat transfer performance of a thermosyphon heat pipe with nanofluids using a three-dimensional simulation. The results show that the volume concentration of nanoparticles significantly reduced the temperature difference between the evaporator and the condenser. Sureshkumar et al. [9] reviewed and summarized an improvement in the thermal efficiency and the thermal resistance of heat pipes with nanofluids. Gupta et al. [10] presented an overview of the heat transfer mechanisms of heat pipes in terms of thermal performance.

Ho et al. [11] obtained the relation of the Ra with the Reynolds number (Re) and Nusselt number (Nu) of a single-phase thermosyphon through experiments and numerical simulations. Vijayan et al. [12] studied the effect of the heater and cooler orientations on a single-phase thermosyphon. Three oscillatory modes and instabilities were observed in the experiments. Misale et al. [13] tested a thermosyphon with different tilting angles and found that the tilt angle affects the heat transfer effect, and the best effect was obtained with a tilt angle of 0° (vertical to the ground). Swapnalee and Vijayan [14] obtained the relationship formula between the Re and Gr through experiments and simulations of single-phase flow in a thermosyphon and used the geometric parameter N_g to modify the Gr, crafting a prediction model that was applicable in four different heating and cooling modes. Thomas and Sobhan [15] experimentally studied the stability and transient performance of a vertical heater–cooler natural circulation loop with metal oxide nanoparticles. Their results indicate that nanofluids containing aluminum oxide and copper oxide have superior heat transfer performances than pure water as the working fluid.

Because the literature on heating and cooling sections positioned in the two vertical sections of a rectangular thermosyphon is quite limited, despite the high potential of this application, this paper explores the natural convection heat transfer phenomenon of a rectangular thermosyphon with a geometric aspect ratio of 3.5. The boundary conditions of the loop heating section and the cooling section feature a fixed heat flux and a fixed wall temperature, respectively.

2. Materials and Methods

2.1. Scenarios and Test Cell Development

We aim to develop a structure that can harvest solar heat while attached to a metal wall (Figure 1a). In addition to extracting heat energy, this building façade prototype can also effectively buffer the heat of the sun. Therefore, an energy-harvesting façade, as part of a solar thermal system and integrated with the building envelope structure, has been developed. To simplify the boundary conditions, we simplified the prototype into a rectangular natural circulation loop to further explore its basic heat flow mode and thermal performance.

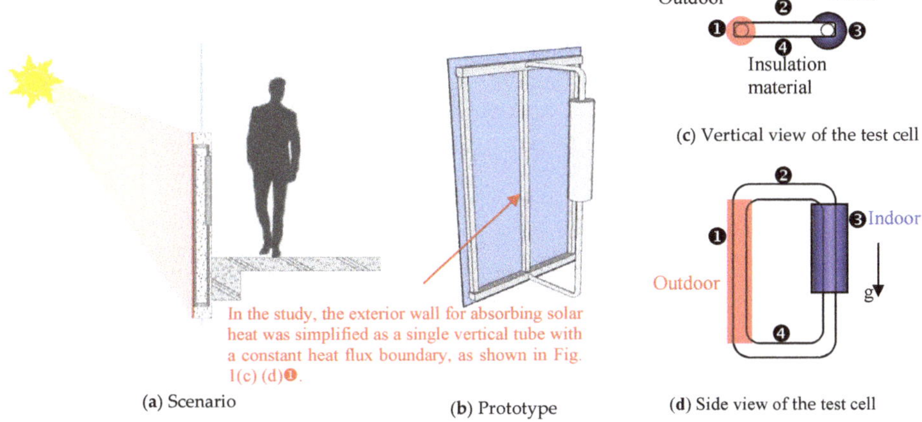

Figure 1. Scenario, prototype development, and test cell illustration.

The prototype design is shown in Figure 1b. To simplify the boundary conditions and observe the basic heat flow performance, we simplified the prototype to the experimental test cell shown in Figure 1c,d. The heat absorbing wall panel shown in Figure 1b is first simplified to a boundary with a constant heat flux, as shown by ❶ in Figure 1c,d. The vertical view in Figure 1c shows the components without the insulation above the device for easy viewing. The side view in Figure 1d shows the components without the heat insulation materials filling the space between the exterior wall and the interior wall to clearly illustrate the components of the circulation loop.

During the day, the working fluid in the circulation loop becomes a high-temperature fluid after absorbing heat from the outdoor heat source ❶. Driven by thermal buoyancy, the fluid forms natural convection, circulates along the horizontal circulation branch ❷ to the indoor heat sink, dissipates heat in the indoor heat sink ❸, and flows back to the outdoor heat source by another horizontal circulation branch ❹ via gravity; thus, a naturally flowing natural circulation loop is formed. The outdoor heat source ❶ is represented by a heated vertical tube; the indoor heat sink ❸ is a vertical section of an isothermal boundary.

2.2. Experimental Test Cell

The experimental test cell can be divided into a rectangular loop comprising a heating section, a cooling section, and two adiabatic sections. The heating section and the cooling section are both located in the vertical portions of the loop (Figure 2) and are described below.

2.2.1. The Rectangular Loop

The loop components in this test cell are all red brass tubes with high thermal conductivity, and the entire loop in the test cell has the same cross-sectional area, resulting in a constant tube flow value. The red brass tubes have an outer diameter of 12.7 mm and an inner diameter (D_i) of 11 mm. The length of the vertical heating section (Ly) is 700 mm, and the lengths (Lx) of the upper and lower adiabatic sections of the experimental test cell are both 200 mm; thus, the aspect ratio AR (=Ly/Lx) of the test cell is 3.5. To investigate the influence of the geometry of the experimental test cell on the heat transfer capability of the system, we set the cooling end length Lc to 30, 45, and 60 cm, as shown in Figure 2a–c. The resulting potential differences ($\Delta Z = \frac{1}{2}\left(\frac{L_y - L_c}{D_i}\right)$) are 18, 11, and 5, respectively.

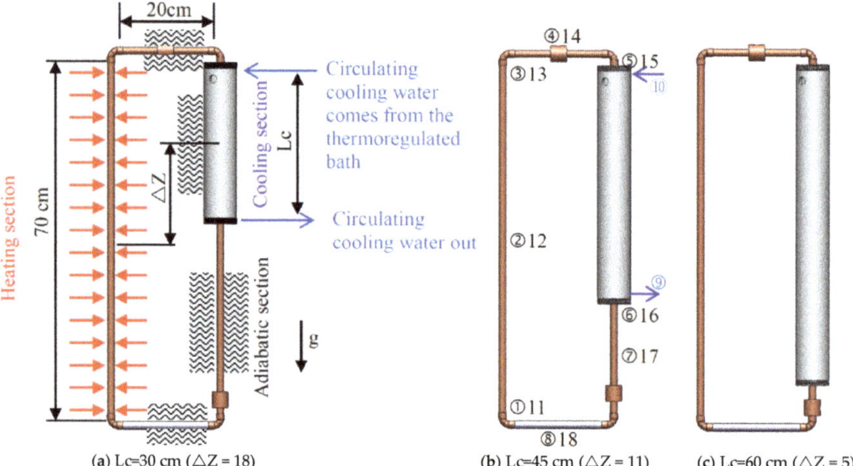

(a) Lc=30 cm (△Z = 18)   (b) Lc=45 cm (△Z = 11)   (c) Lc=60 cm (△Z = 5)

Figure 2. Experimental test cell and geometric parameters.

2.2.2. Heating Section

The heating section is a boundary with a constant heat flux that simulates the practical application, and therefore, heating wire is used to generate heating at a fixed power. To increase the uniformity of the heating section, the electric wire is directly processed into the heating section, which is mainly composed of muscovite paper, electric heating wire and heat insulation materials. Because the tube material of the experimental test cell is conductive red brass, we must insulate the surface of the heating section tube before winding the heating wire around the heating section. Therefore, the soft muscovite paper, which is an excellent insulator and heat-resistant to 500–550 °C, is wound on the heating section for insulation purposes. The input electric power and the maximum voltage of the power supply are 60 W and 60 V, respectively, so the resistance is 60 Ω. Therefore, this experiment selects a heating wire with a resistance close to 60 Ω. The heating wire is 2.5 m in length, 0.25 mm in diameter, and 25.82 Ω/m in resistance per unit length.

2.2.3. Cooling Section

The cooling section is a simulated isothermal boundary condition. To test the effects of different height differences between the hot and cold ends, the water sleeve is set to lengths of 300, 450, and 600 mm. The cooling water sleeve is mainly composed of a water sleeve body and a water sleeve cover. The water sleeve body is a cylinder with an outer diameter of 61 mm and an inner diameter of 40 mm. Both ends of the water sleeve are water sleeve covers, and the center of each water sleeve cover has a hole that is approximately 11 mm in diameter. The thermosyphon passes through the holes in the cooling water sleeve, and the joints are fitted with O-ring grooves to prevent water leakage from the cooling water sleeve.

2.2.4. Insulation Materials

Except for the heating section and the cooling section, the remaining sections are adiabatic. To eliminate the effects of the external environment and heat loss during the experiment, 3-mm-thick insulating tape and a 40-mm-thick insulation pipe are wrapped around the body to effectively simulate adiabatic boundaries.

2.3. Experimental Apparatus

The device and data acquisition system includes a data acquisition unit (Yokogawa MX-100, Tokyo, Japan), PC, a flow meter (Fluidwell F110-P, Veghel, The Netherlands), a DC power supply unit (Gwinstek SPD-3606, New Taipei City, Taiwan), and a thermoregulated bath (RCB-412, New Taipei City, Taiwan).

2.3.1. Power Supply System

The heating wire of the heating section is connected directly to the power supply (Gwinstek SPD-3606); the output voltage and current are adjusted to provide an input power q_h of 30–60 W, as required in the experiment.

2.3.2. Thermoregulated Bath

The isothermal boundary of the cooling section is established by circulating water from a thermoregulated water bath (RCB412) to establish the cooling section temperature Tc (30, 40, or 50 °C) required for the experiment.

2.3.3. Liquid Flow Meter

A flow meter (F110-P; Fluidwell) is used to monitor the amount of cooling water sent by the thermostatic water tanks in each test, with an average flow rate of approximately 35 mL/s.

2.3.4. Thermocouples

This study uses type-K thermocouples to measure the wall temperature and the fluid temperature along the loop. There are 17 thermocouple points embedded in each experimental test cell at the upper, middle, and lower positions of the heating section, at the inlet and outlet of the water sleeve in the cooling section, and along the adiabatic sections. The thermocouple points are shown in Figure 2b. The thermocouple points numbered ①–⑧ measure the temperatures at the center point of the fluid in the tube; the thermocouple points numbered ⑨ and ⑩ measure the inlet and outlet water temperatures of the cooling sleeve; and the thermocouple points numbered 11–17 measure the temperatures of the outer tube wall.

2.4. Parameters

The relevant heat transfer parameters were calculated from the temperature measured by each thermocouple in the experimental test cell and the voltage and current supplied to the electric heating wire.

1. The input power $q_h = VI$ can be obtained from the voltage V and the current I supplied by the power supply. Thus, the heat flux q_{flux} is q_h/A. During the experiment, the heating from the power supply is not completely transferred to the fluid in the heating section. A small amount of heat enters the cooling section via axial heat conduction q_a in the loop body or escapes into the environment, so the axial heat transfer q_a must be deducted from the input thermal power first. Therefore, the corrected actual input thermal power is $q_{in} = q_h - q_a$.
2. Modified Rayleigh number, Ra*

$$Ra^* = \frac{g\beta\left(\frac{q_{in}}{A}\frac{R_i}{k}\right)L_yD_i^2}{\alpha v} \tag{1}$$

3. Parameters of the working fluid

    (1) Flow rate of the working fluid ($\dot{V}$) The temperature of the fluid in the rectangular loop increases in the heating section, causing the fluid to flow to the cooling section by convection. The flow rate of this flow varies depending on the amount of thermal power input to the heating section and the cooling conditions in the cooling section. In this

experiment, the input heat power (q_{in}) and temperature difference ($\Delta T = T_3 - T_1$) were used to determine the fluid flow rate in the tube.

$$\dot{V} = \frac{q_{in}}{\rho C_p A \Delta T} \tag{2}$$

(2)   Reynolds number, Re

$$\text{Re} = \frac{\dot{V}(2R_i)}{\nu} \tag{3}$$

4.   Nusselt number

(1)   The average heat transfer coefficient of the cooling end is:

$$\overline{h} = \frac{q_c/A}{\overline{T_w} - \overline{T_c}} \tag{4}$$

(2)   The calculation of the Nu is as follows:

$$\overline{\text{Nu}} = \frac{\overline{h}(2R_i)}{k} \tag{5}$$

2.5. Experimental Uncertainty

The uncertainties in the measured quantities of this study were estimated to be ±0.1 °C for temperature, ±2% for the volumetric flow rate of the cooling water, and ±2% for the heat flux measured by the heat flow meters. Following the uncertainty propagation analysis [16], the estimated uncertainties for the deduced experimental results were as follows: 5.1–26.2% for the heating flux, 3.1–25.8% for the cooling power, and 4.3–21.8% for the average Nu.

3. Results and Discussion

This experiment mainly discusses the effects of the heating power, cooling section temperature Tc, and height difference between the hot and cold ends ΔZ on the heat transfer phenomenon in a rectangular thermosyphon. The input thermal power is 30–60 W (the heating flux is in the range of 600–3800 W/m^2); Tc is 30, 40, or 50 °C, and ΔZ is 5, 11, or 18 (Lc = 60, 45, or 30 cm, respectively).

Due to space limitations, only the experimental result of Tc = 40 °C is introduced; the heat flow phenomena for the other Tc values of 30 °C and 50 °C are similar to that of Tc = 40 °C and are not discussed further. Figure 3 shows the change in working fluid (water) temperature in the loop at Tc = 40 °C as the thermal power increases. In the upper graphs, the x-axis represents thermocouple measurement points from Channel 1 to Channel 8, and the y-axis is the fluid temperature in the loop. In the lower graphs, the x-axis is experimental time.

Figure 3a shows the temperature variation of the working fluid (water) along the loop as a function of the heating power with a height difference $\Delta Z = 5$ (Lc = 60 cm). First, the fluid in the thermosyphon is close to the stationary state. When the thermal power is set to 30 W, the temperature of the fluid in the heating section increases, and the fluid in the heating section slowly flows to the cooling section; it takes approximately 20 min to reach a steady state. Because the initial input of thermal power already initiated natural convection in the fluid in the thermosyphon, it only takes 5 min to reach a steady state after increasing the thermal power to 40 W, 50 W, and 60 W. When the thermal power is increased to 40 W, the temperature of the fluid in the middle section of the heating section (Channel 2) oscillates. The possible reason may be that the temperature boundary layer growth in the heating section is accompanied by a temperature mixing phenomenon formed by the fluid turning at the bend (near the heating section inlet). This oscillating phenomenon becomes more pronounced as the thermal power increases for a height difference ΔZ of 5.

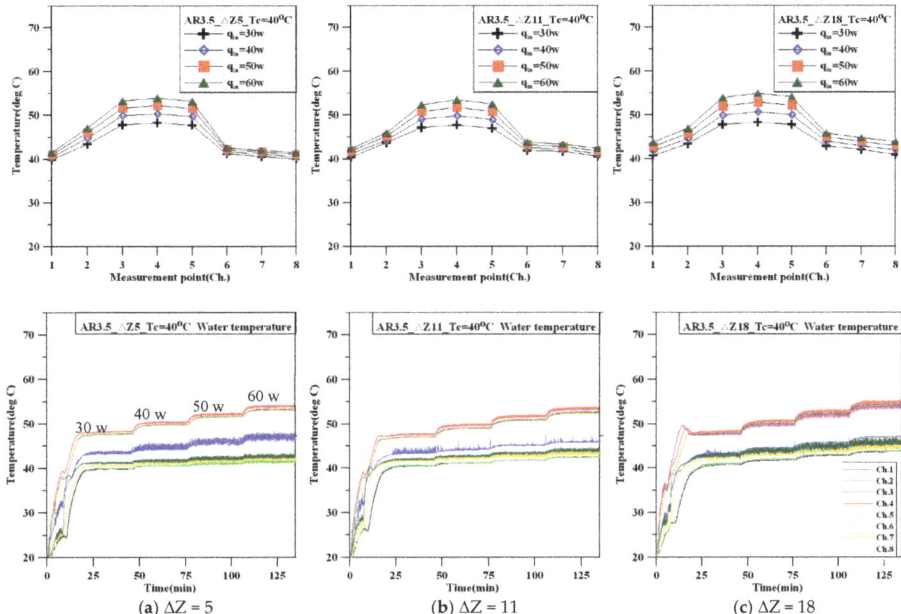

Figure 3. Fluid temperature variation along the loop as a function of the heating power (AR = 3.5, Tc = 40 °C, ΔZ = 5, 11, and 18).

Figure 3b shows the temperature variation of the working fluid along the loop as a function of the heating power with a height difference ΔZ of 11 (Lc = 45 cm). The temperature variation pattern of the fluid in the thermosyphon is similar to that of Figure 3a, and it takes approximately 20 min to initially reach a steady state. After that, with further increases in the thermal power, the steady state is reached in approximately 5 min. The steady-state average temperature of the fluid in the thermosyphon shown in Figure 3b is similar to that of Figure 3a. Additionally, after the input thermal power exceeds 40 W, the temperature oscillation in the middle section (Channel 2) of the heating section begins to decrease. With increasing thermal power, the oscillation at Channel 2 becomes less noticeable. Figure 3c shows the temperature variation with the height difference ΔZ of 18 (Lc = 30 cm). The temperature rise of the fluid in the thermosyphon is similar to the former two conditions, but the thermosyphon takes approximately 25 min to reach the initial steady state and approximately 5 min to reach new steady states when the heat power is increased.

Comparison of the three sets of data reveals that Figure 3c, with a larger ΔZ, has a higher average temperature (approximately 2 °C higher) after the steady state is reached. The temperature difference between the highest and lowest power in Figure 3c is slightly higher than in Figure 3a. Comparison of Figure 3a,b with Figure 3c shows that when ΔZ is larger (the length of the cooling section is shorter), fluid temperature oscillations were observed at the exits of both the heating section and the cooling section (Channel 3 and Channel 6). This oscillating phenomenon is more pronounced as the thermal power increases. However, as ΔZ increases, the temperature oscillation magnitude in the middle section of the heating section (Channel 2) begins to decrease.

Figure 4 shows the change in temperature at each point in the outer wall of the thermosyphon as the thermal power increases. The temperature changes in each graph are quite similar. Due to the increase in heating power and the temperature rise of the fluid in the tube, the temperature of the outer tube wall at the end of the heating section is the highest. After entering the adiabatic section, the wall temperature of the outer tube wall decreases slightly. This phenomenon is mainly caused by the axial heat transfer of the tube wall. After entering the cooling section, the temperature of the outer tube wall begins to drop to Tc after cooling. However, after the cooling section, the temperature of the outer wall of the adiabatic section rises slightly due to the axial heat transfer. This temperature rise phenomenon is more obvious for $\Delta Z = 18$ than for $\Delta Z = 5$ and 11.

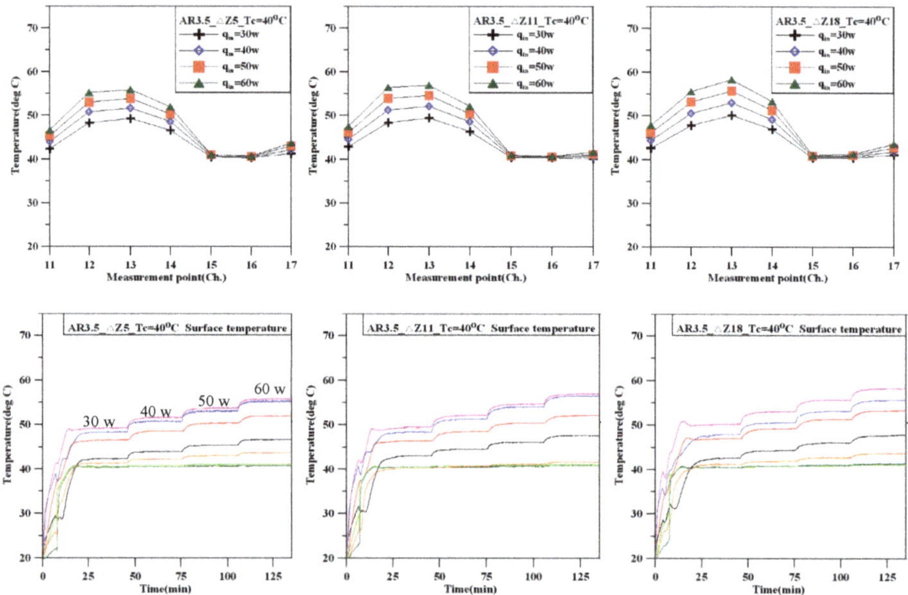

Figure 4. Exterior wall temperature variation along the loop as a function of the heating power.

Figure 5 shows the relationship between the Nu and the modified Rayleigh number (Ra*) of the loop. Overall, the Nu is approximately 5–10. As Ra* increases, the natural convection intensity of the working fluid increases, so Nu increases. Under the designated geometrical configuration, the height difference between the cold and hot ends of the loop also affects the natural convection strength of the loop fluid. When the Ra* is fixed, as the height difference ΔZ increases, the Nu also increases. The effect of the cooling end temperature Tc on Nu is less significant. Therefore, if it is desired to increase the natural convection effect of the fluid in the tube, in addition to increasing the power of the heating section, the height difference between the hot and cold ends is also one of the controllable factors.

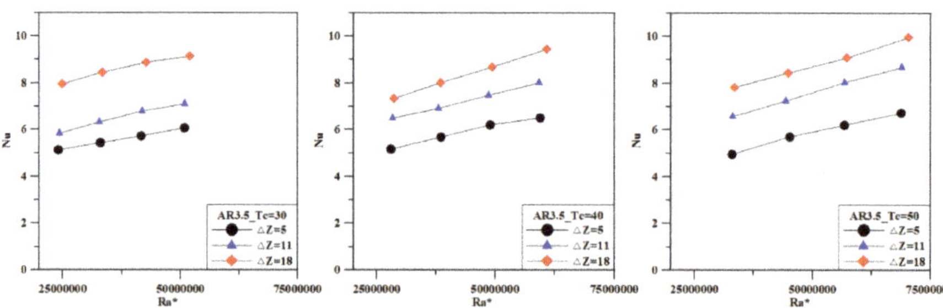

Figure 5. The relationship between the Nusselt number of loop flow and the modified Rayleigh number.

Figure 6 shows the relationship between the average flow velocity of the fluid and the cold end temperature. As the heating power increases, the average flow rate of the fluid increases significantly. With the increase in Tc, the average flow velocity of the fluid increases slightly in conjunction with a high heating power but decreases slightly in conjunction with a low heating power. The overall variation pattern indicates that the average flow velocity of the fluid is less affected by the cold end temperature than by the heating power. As the length of the cold end increases, the overall average flow rate decreases slightly (Figure 7).

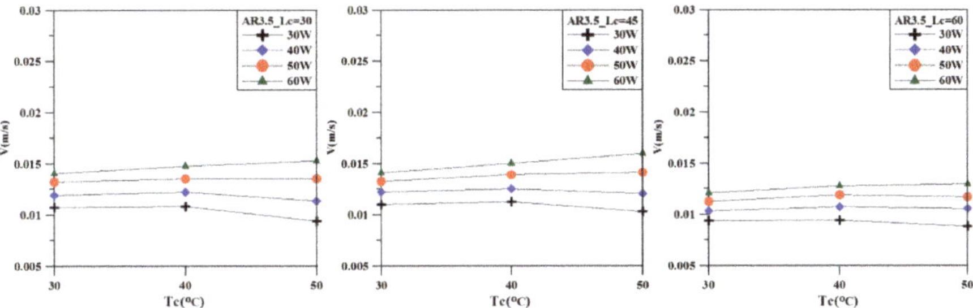

Figure 6. The relationship between average fluid velocity and temperatures of the cooling end Tc.

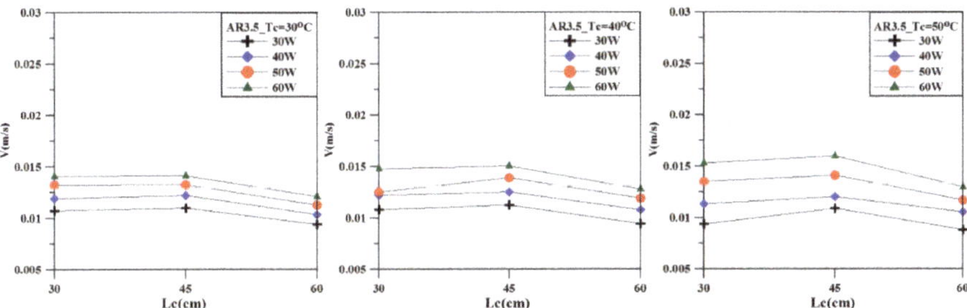

Figure 7. The relationship between the average fluid velocity and the length of the cooling end Lc.

Figure 8 shows the Re and heating power (q_{in}) of the fluid in the tube. When the input heating power is 30 W, the Re values of the tube flow for different ΔZ and Tc values are approximately 150–200. With an increase in the heating power or Tc, the Re increases, but the influence of heating power on the Re is more significant than that of Tc. Furthermore, the Re of $\Delta Z = 11$ is very close to that of $\Delta Z = 18$. A higher heating power, Tc, or ΔZ can result in higher fluid flow intensity. For 60 W, Tc = 50 °C, and $\Delta Z = 18$, the Re is approximately 345.

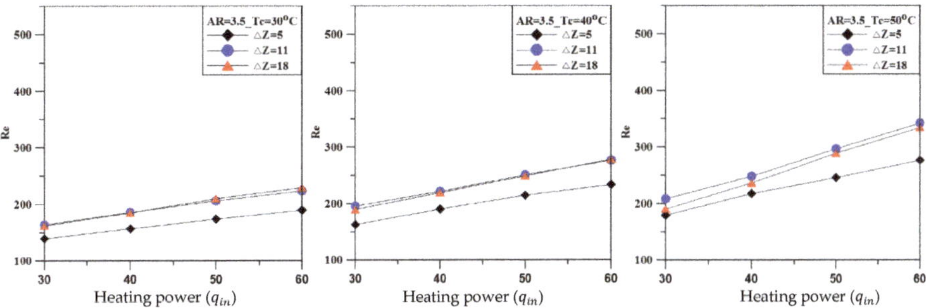

Figure 8. The relationship between the Reynolds number of pipe flow and heating power.

4. Conclusions

The main purpose of this study is to investigate the heat transfer phenomenon of a rectangular thermosyphon with an aspect (height-to-width) ratio of 3.5 and a tube inner diameter of 11 mm using different heating powers, height differences between the heating and cooling ends, and cold end temperatures. From the experimental results, the following conclusions can be summarized:

1. The outer wall of the end of the heating section has the highest wall temperature. Due to the influence of axial heat transfer through the thermosyphon wall, the wall temperature of the outer tube decreases slightly after exiting the heating section (entering the upper adiabatic section); similarly, after exiting the cooling section, the wall temperature of the adiabatic section increases slightly.
2. A higher heating power or a larger height difference between the hot and cold ends can increase the fluid flow in the loop, whereas the cooling temperature has little influence.
3. Overall, the Nu is approximately 5–10. If one wants to increase the natural convection effect of the fluid in the loop, in addition to increasing the heating power, the height difference between the hot and cold ends is also one of the controllable factors.
4. With a height difference of $\Delta Z = 5$ and a heating power of 40 W, the temperature of the fluid in the middle heating section oscillates, and an increase in the heating power also increases the oscillations. For $\Delta Z = 11$, the temperature oscillations in the middle heating section are reduced. For $\Delta Z = 18$, fluid temperature oscillations are observed at the exits of both the heating and cooling sections; however, the water temperature in the middle heating section does not oscillate. The oscillation phenomenon may result from the mixing of fluid with different temperatures, which is caused by the growth of temperature boundary layers and turning of the flow at the loop elbows after the fluid is heated or cooled.

Author Contributions: C.S.H., R.H.C., and C.-M.L. conceived of and designed the model. C.S.H. and C.-W.Y. performed the experimental work. C.S.H., C.-W.Y., and C.-M.L. analyzed the data. C.-T.T. and C.-M.L. wrote the paper.

Acknowledgments: Support from the National Science Council of the ROC through grant No. NSC 100-2221-E-006-240-MY2 is gratefully acknowledged.

Nomenclature

A	heating (or cooling) area (m^2)
AR	aspect ratio ($=L_y/L_x$)
Ch.	position of the thermocouple
C_p	specific heat capacity (kJ/kg K)
D_i	inner diameter of the loop tube (m)
g	acceleration due to gravity (m/s)
$\bar{h}$	average heat convection coefficient at the cooling end (W/m^2 °C)
I	electric current (A)
k	thermal conductivity of the working fluid (W/m)
L_c	length of the cooling end (m)
L_x	width of the test cell (200 mm)
L_y	height of the test cell (=length of the heating end) (700 mm)
Nu	Nusselt number
q_a	axial heat conduction along the loop wall (W)
q_c	heat transfer rate at the cooling section (W)
q_{flux}	heat flux at the heating section ($= q_h/A$)
q_h	heating power (W)

q_{in}	actual heating power (W)
Ra*	modified Rayleigh number
Re	Reynolds number
R_i	inner radius of the loop tube
T	temperature (°C)
T_c	temperature at the cooling section (°C)
$\overline{T}_c$	average fluid temperature at the cooling section (°C)
$\overline{T}_w$	average wall temperature at the cooling section (°C)
ΔT	temperature difference between the exit and inlet at the heating section (°C)
V	electric voltage (Volt)
$\dot{V}$	velocity of the loop working fluid (m/s)
ΔZ	potential difference $\left(=\frac{1}{2}\left(\frac{L_y - L_c}{D_i}\right)\right)$

Greek symbols

α	thermal diffusion coefficient (m^2/s)
β	thermal expansion coefficient (K^{-1})
ρ	density (kg/m^3)
ν	kinematic viscosity (m^2/s)

References

1. Garrity, P.T.; Klausner, J.F.; Mei, R. Instability phenomena in a two-phase microchannel thermosyphon. *Int. J. Heat Mass Transf.* **2009**, *52*, 1701–1708. [CrossRef]
2. Vijayan, P.K.; Sharma, M.; Pilkhwal, D.S.; Saha, D.; Sinha, R.K. A comparative study of single-phase, two-phase, and supercritical natural circulation in a rectangular loop. *J. Eng. Gas Turb. Power* **2010**, *132*, 102913. [CrossRef]
3. Misale, M.; Garibaldi, P.; Tarozzi, L.; Barozzi, G.S. Influence of thermal boundary conditions on the dynamic behaviour of a rectangular single-phase natural circulation loop. *Int. J. Heat Fluid Flow* **2011**, *32*, 413–423. [CrossRef]
4. Lai, C.-M.; Chen, R.-H.; Huang, C.S. Development and thermal performance of a wall heat collection prototype. *Build. Environ.* **2012**, *57*, 156–164. [CrossRef]
5. Delgado, M.; Lázaro, A.; Mazo, J.; Zalba, B. Review on phase change material emulsions and microencapsulated phase change material slurries: Materials, heat transfer studies and applications. *Renew. Sustain. Energy Rev.* **2012**, *16*, 253–273. [CrossRef]
6. Desrayaud, G.; Fichera, A.; Lauriat, G. Two-dimensional numerical analysis of a rectangular closed-loop thermosiphon. *Appl. Therm. Eng.* **2013**, *50*, 187–196. [CrossRef]
7. Buschmann, M.H. Nanofluids in thermosyphons and heat pipes: Overview of recent experiments and modelling approaches. *Int. J. Therm. Sci.* **2013**, *72*, 1–17. [CrossRef]
8. Huminic, G.; Huminic, A. Numerical study on heat transfer characteristics of thermosyphon heat pipes using nanofluids. *Energy Conversat. Manag.* **2013**, *76*, 393–399. [CrossRef]
9. Sureshkumar, R.; Mohideen, S.T.; Nethaji, N. Heat transfer characteristics of nanofluids in heat pipes: A review. *Renew. Sustain. Energy Rev.* **2013**, *20*, 397–410. [CrossRef]
10. Gupta, N.K.; Tiwari, A.K.; Ghosh, S.K. Heat transfer mechanisms in heat pipes using nanofluids—A review. *Exp. Therm. Fluid Sci.* **2018**, *90*, 84–100. [CrossRef]
11. Ho, C.J.; Chiou, S.P.; Hu, C.S. Heat transfer characteristics of a rectangular natural circulation loop containing water near its density extreme. *Int. J. Heat Mass Transf.* **1997**, *40*, 3553–3558. [CrossRef]
12. Vijayan, P.K.; Sharma, M.; Saha, D. Steady state and stability characteristics of single-phase natural circulation in a rectangular loop with different heater and cooler orientations. *Exp. Therm. Fluid Sci.* **2007**, *31*, 925–945. [CrossRef]
13. Misale, M.; Garibaldi, P.; Passos, J.C.; de Bitencourt, G.G. Experiments in a single-phase natural circulation mini-loop. *Exp. Therm. Fluid Sci.* **2007**, *31*, 1111–1120. [CrossRef]
14. Swapnalee, B.T.; Vijayan, P.K. A generalized flow equation for single phase natural circulation loops obeying multiple friction laws. *Int. J. Heat Mass Transf.* **2011**, *54*, 2618–2629. [CrossRef]

15. Thomas, S.; Sobhan, C.B. Stability and transient performance of vertical heater vertical cooler natural circulation loops with metal oxide nanoparticle suspensions. *J. Heat Transf. Eng.* **2018**, *39*, 861–873. [CrossRef]
16. Moffat, R.J. Using uncertainty analysis in the planning of an experiment. *J. Fluid Eng.* **1985**, *107*, 173–178. [CrossRef]

Article

Analysis of Convergence Characteristics of Average Method Regulated by ISO 9869-1 for Evaluating In Situ Thermal Resistance and Thermal Transmittance of Opaque Exterior Walls

Doo Sung Choi [1] and Myeong Jin Ko [2],*

[1]   Department of Architectural Engineering, Chungwoon University, Incheon 22100, Korea; trebelle@chungwoon.ac.kr
[2]   Department of Building System Technology, Daelim University College, Anyang 13916, Korea
*   Correspondence: whistlemj@nate.com; Tel.: +82-31-467-4825

Received: 17 March 2019; Accepted: 21 May 2019; Published: 24 May 2019

Abstract: In the last few decades, an average method which is regulated by ISO 9869-1 has been used to evaluate the in situ thermal transmittance (U-value) and thermal resistance (R-value) of building envelopes obtained from onsite measurements and to verify the validity of newly proposed methods. Nevertheless, only a few studies have investigated the test duration required to obtain reliable results using this method and the convergence characteristics of the results. This study aims to evaluate the convergence characteristics of the in situ values analyzed using the average method. The criteria for determining convergence (i.e., end of the test) using the average method are very strict, mainly because of the third condition, which compares the deviation of two values derived from the first and last periods of the same duration. To shorten the test duration, environmental variables should be kept constant throughout the test or an appropriate period should be selected. The convergence of the in situ U-value and R-value is affected more by the length of the test duration than by the temperature difference if the test environment meets literature-recommended conditions. Furthermore, there is no difference between the use of the U-value and R-value in determining the end of the test.

Keywords: thermal resistance; thermal transmittance; heat flow meter method; average method; convergence characteristics; opaque exterior wall

---

1. Introduction

To promote the spread of energy-efficient buildings, many countries around the world have enacted regulations and established policies. One of the most prominent and easy approaches by most countries is the imposition of increasingly strict specifications on the thermal performance of buildings. That is, the minimum required performance level of a building envelope has been tightened considerably over the past decade. For example, the mandatory thermal transmittance (U-value) for the exterior walls of residential buildings in Seoul, South Korea, has reduced from 0.48 W/m²·K in 2008 to 0.17 W/m²·K in 2018 [1].

The U-value is one of the most important properties used to evaluate the thermal performance of a building envelope. This property can be determined by theoretical or experimental methods. The theoretical U-value can be estimated using an approach regulated by the ISO 6946 standard [2] based on an electrical analogy and a steady-state condition. The theoretical U-value is used in the approval process for newly constructed or refurbished structures and in the certification process for energy-efficient buildings. However, these theoretical values do not accurately represent the in situ U-values because of various reasons associated with the design, construction, and operational stages. The discrepancies between the theoretical and in situ values provide misleading information on the energy performance of buildings, which not only prevents the owner from establishing a reasonable energy consumption plan but also may lead to economic losses resulting from missing the renovation period and selecting inappropriate retrofitting activities [3]. In particular, condensation has occurred even in recently constructed buildings with low theoretical U-values which have been certified as energy-efficient buildings; this indicates the limitations of the theoretical method and the importance of onsite measurement of U-values.

The heat flow meter method, which is regulated by the ISO 9869-1 standard [4], is a widely used method to measure the in situ U-value of building envelopes. This method estimates the in situ U-value by analyzing the measurement data of the heat flux through a test wall and the temperature difference between the inside and outside environments. According to the ISO standard, if the environmental condition is stable, the test should last at least 3 days; otherwise, the minimum test duration may be more than 7 days to obtain reliable results.

Many studies have previously evaluated the in situ U-value of walls using the standardized average method and compared the value with the theoretical value. For example, in a study by Adhikari et al. [5] on historical building walls, the differences between theoretical and measured U-values ranged from 2% to 58%. Cabeza et al. [6] measured the in situ U-values of experimental cubicles that used three typical insulation materials, namely polyurethane, polystyrene, and mineral wool. They found that the average differences between the experimental and theoretical U-values in two different weeks were 12% and 14%. Asdrubali et al. [7] conducted a study on some green buildings with low calculated U-values and found that the differences between the calculated and measured U-values ranged from 4% to 75%. Evangelisti et al. [8] evaluated the in situ U-values of three conventional exterior walls in the range of 0.504–1.897 W/m^2·K. They reported that the discrepancies between the theoretical U-value and measured U-value were in the range of 17–153%. Baker [9] evaluated the in situ U-values of traditional Scottish stone masonries with theoretical U-values ranging from 0.30 W/m^2·K to 2.65 W/m^2·K. The results showed that 44% of the total number of measurements were lower than the calculated U-value range, 42% were within the calculated range, and 14% were higher than the calculated range. Rye and Scott [10] reported that in 77% of the measurement cases, the software overestimated the U-values compared to onsite measurements. Other studies [11–14] have reported similar results that show discrepancies between theoretical and measured U-values, although the degree of discrepancy differs.

The above literature review indicates that many researchers have used the average method defined by the ISO 9869-1 standard [4] for data-processing. However, because the average method does not take into account the dynamic behavior of the walls, the test duration usually needs to be extended to improve the estimation accuracy of the in situ U-value. Therefore, the proper test duration and factors influencing the value are very interesting research topics. A study conducted by Rye and Scott [10] on traditional Scottish masonries showed that a period of at least a week is required before the U-value estimate stabilizes to within ±5% of the final value determined from data gathered over approximately 27 days. Asdrubali et al. [7] reported that when using the average method, the acquisition time can be 3 days if the indoor temperature is stable; otherwise, the time interval must be extended to 7 days. Gaspar et al. [12] showed that in the measurements of low U-value façades, temperature differences of above 19 °C require a test duration of 72 h; however, for lower temperature differences, the test duration must be extended to 144 h. Ahmad et al. [13] evaluated the in situ U-value and thermal resistance (R-value) of north- and east-facing walls made from reinforced precast concrete panels using the average method. The results showed that a test period of 6 days is sufficient to ascertain the in situ U-value and R-value of reinforced precast concrete walls. The results also indicated that, where the U-value depends on the wall orientation and outside weather conditions, the R-value is independent of the wall orientation. Ficco et al. [14] conducted in situ U-value measurements on existing buildings with theoretical U-values ranging from 0.37 W/m^2·K to 3.30 W/m^2·K. They estimated high relative uncertainties ranging from 8% at optimal operating conditions to approximately 50% at nonoptimal operating conditions. They also reported that temperature differences lower than 10

°C and low heat flow lead to unacceptable uncertainties. Deconinck and Roels [15] compared the performance of several semi-stationary and dynamic data analysis techniques used for evaluating the thermal property of building components using simulated datasets with different lengths and for different seasons. An analysis of the *R*-value using the average method showed that data periods of around 20 days or longer are required to obtain 5% accurate results in January. The simulation results also indicated that the *R*-values for the two summer scenarios in July showed the limited validity of the average method. Gaspar et al. [16] evaluated the minimum duration of in situ experimental campaigns to measure the *U*-value of the façades of existing buildings using the heat flow meter method. They determined the minimum test duration according to the criteria of data quality and variability of results proposed in ASTM C1155 [17] and the three convergence conditions described in ISO 9869-1 [4]. The results showed that the ISO criteria are more sensitive and provide more accurate results than the ASTM criteria but require a longer test duration.

The infrared thermography (IRT) method is widely employed in building diagnostics for qualitative evaluation to detect heat losses, air leakages, thermal bridges, sources of moisture, missing materials, and defects in insulation materials [18–22]. Furthermore, many studies [23–28] have recently proposed quantitative IRT methodologies for evaluating the in situ *U*-value of a building envelope. In addition, several researchers [29–31] have proposed the use of statistical approaches, in particular Bayesian inference, to infer the in situ thermal properties from heat flux and temperature measurements. It is noteworthy that the validity of these newly proposed methods is mainly verified using the average method, which is regulated by ISO 9869-1 [4].

The above literature review shows that many researchers have used the average method to obtain the in situ *U*-values and have reported the minimum measurement period and environmental conditions required when this method is used. The average method has also been used for the verification of newly proposed methods. Nevertheless, with regard to determining the in situ *U*-values using the average method, studies on the test duration required to obtain a reliable result and the causes that increase the test duration are still lacking. Furthermore, only a few works have investigated the convergence characteristics of the in situ *U*-value or *R*-value.

Therefore, this study aims to evaluate the convergence characteristics of the in situ *R*-value and *U*-value of an exterior wall analyzed using the average method as a data-processing technique. The convergence characteristics were analyzed according to the convergence conditions of the ISO 9869-1 standard [4] using datasets with different analysis periods in a measurement campaign of 21 consecutive days. In addition, the convergence characteristics of both the *R*-value and *U*-value were reviewed together to identify the difference between the use of the two values for determining the end of the test. A clearer understanding of the convergence characteristics will help researchers and diagnosticians to select an appropriate test duration and reduce the uncertainty of onsite measurements of the *R*-value and *U*-value.

The rest of this paper is organized as follows. Section 2 describes the case study and the method used in the research. Section 3 discusses the convergence characteristics of both the *R*-value and *U*-value. Finally, Section 4 presents the conclusions of the study and future research ideas.

2. Methods

2.1. Investigated Building

The building considered in this case study was a private house with a 52 m^2 floor area, which was constructed in 1990 and is located in the city of Gwangmyeong in the central region of Korea. The test object was the northwest-facing external wall to avoid direct solar radiation. Table 1 lists the materials of the test wall and their thermal conductivity, as well as their *R*-value and *U*-value calculated using the theoretical approach regulated by ISO 6946 [2]. Information on the building materials and surface resistances was obtained from the design documents. The calculated *U*-value can be determined as follows:

$$U_D = \frac{1}{R_{si} + \sum_i \frac{t_i}{\lambda_i} + R_{se}},$$

(1)

where U_D represents the U-value evaluated by the calculation method (W/m^2·K); t_i is the thickness of the i-th layer (m); λ_i is its thermal conductivity (W/m·K); and R_{si} and R_{se} are the interior and exterior surface resistances (m^2·K/W), respectively.

Table 1. Stratigraphies and thermophysical properties of the test wall.

Material Layer	t	λ	ρ	c	R	U_D (W/m^2·K)
Internal surface					0.110	
Mortar	10	1.400	2000	780	0.007	
Cement brick	90	0.600	1700	835	0.150	0.460
Glass wool	60	0.035	40	670	1.714	
Cement brick	90	0.600	1700	835	0.150	
External surface					0.043	

t: Thickness; λ: Thermal conductivity; ρ: Density; c: Specific heat capacity; R: Thermal resistance; U_D: Theoretical thermal transmittance.

2.2. In Situ Measurement

Onsite measurement was conducted from December 30, 2016 to January 19, 2017 in accordance with the ISO 9869-1 standard [4]. The standard states that surveys can last from a minimum of 3 days to more than 7 days. However, many studies [7,8,12–14,28,29] conducted measurements for approximately 1 week and sometimes even more than 2 weeks to obtain satisfactory results. In this study, the measurement was conducted for 21 days to identify the effects of increases and changes in the measurement period.

The measurement equipment consisted of devices for measuring the R-value and U-value, such as a heat flux sensor and temperature sensor, and devices for confirming the validity of the measurement conditions, such as a pyranometer and an infrared camera. The heat flux sensor (EKO MF-200, EKO Instruments, Tokyo, Japan) was installed on the inside surface of the test wall after identifying the best position using the infrared camera (FLIR T620, FLIR systems, Portland, OR, USA). Two thermocouples (Testo 0602 5792, Testo AG, Lenzkirch, Germany) were mounted on the inside surface near the heat flux sensor and on the opposite outside surface with adhesive tape. The inside air temperature and wind speed were measured in the vicinity of the test wall using a comfort probe (Testo 0628 0143, Testo AG, Lenzkirch, Germany). A hot-wire anemometer (Testo 0635 1543, Testo AG, Lenzkirch, Germany) was employed to measure the outside air temperature and local wind speed. The pyranometer (EKO MS 602, EKO Instruments, Tokyo, Japan) was installed perpendicular to the outside surface to identify the influence of direct solar radiation. The measurements were recorded using a data logger (Graphtec GL220, Graphtec Corporation, Yokohama, Japan) with a sampling period of 1 min. The main technical specifications of the measurement equipment are listed in Table 2, whereas Figure 1 shows the installation of the measurement equipment.

Table 2. Main technical specifications of measurement equipment.

Equipment (model)	Parameter	Range	Accuracy
Heat flux sensor (EKO MF-200)	Heat flux		±2%
Comfort probe (Testo 0628-0143)	Inside air temperature	0–50 °C	±0.5 °C
	Inside wind speed	0–5 m/s	± (0.03 m/s + 4%)
Hot-wire probe (Testo 0635-1543)	Outside air temperature	−20–70 °C	±0.5 °C
	Outside wind speed	0–20 m/s	±(0.03 m/s + 4%)
Thermocouple (Testo 0602-5792)	Surface temperature	−200–1000 °C	±(0.5 °C + 0.3%)
Pyranometer (EKO MS-602)	Solar radiation	0–2000 W/m^2	<25 W/m^2
Infrared camera (FLIR T-620)	Thermogram	7.5–14 µm	±2 °C

Figure 1. Photograph of the test wall and measurement equipment.

The measured data obtained from the 21 days of the experimental campaign are shown in Figure 2. The average air and surface temperature differences between the inside and outside environments throughout the monitoring period were 21.8 °C and 19.6 °C, respectively. These temperature differences were considerably higher than the recommended value of 10 °C. The maximum solar radiation incident on the outside surface of the northwest-facing test wall was 109.8 W/m^2; thus, the influence of direct solar radiation was considered negligible. The average indoor and outdoor wind speeds were approximately 0.07 m/s and 0.23 m/s, respectively. In particular, the average outdoor wind speed was significantly lower than the recommended value of 1 m/s to avoid excessive effects of convective phenomena. In addition, the influence of moisture content on the measurement results is considered negligible, as there was no rain during the measurement period.

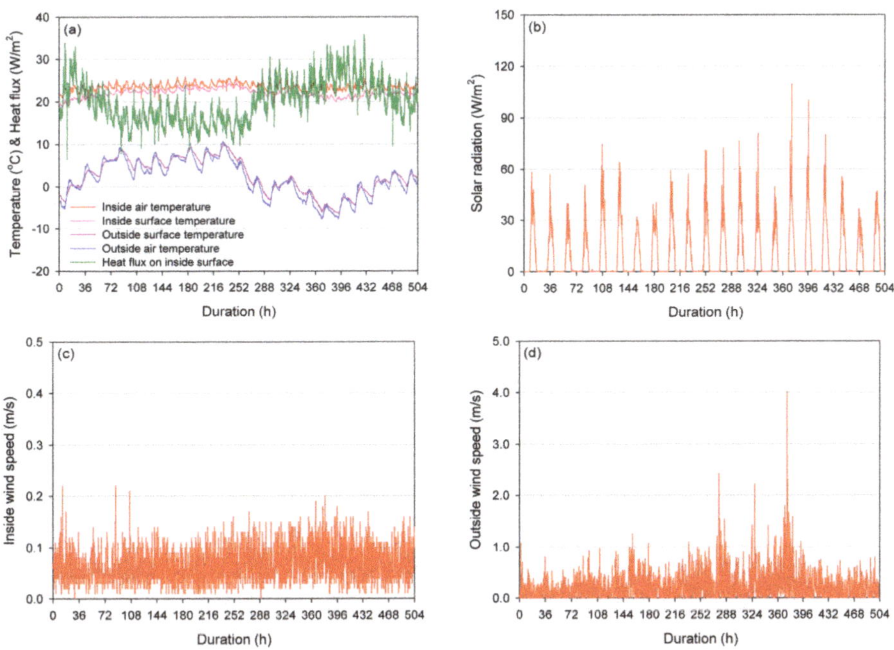

Figure 2. Data obtained from the monitoring process from December 30, 2016 to January 19, 2017: (**a**) Temperature and heat flux; (**b**) solar radiation; (**c**) inside wind speed; and (**d**) outside wind speed.

2.3. Data Analysis

The U-value was analyzed using the average method regulated by ISO 9869-1 [4]. The average method assumes that the U-value can be obtained by dividing the mean density of the heat flow rate by the mean temperature difference. If the average is taken over a sufficiently long period of time, a good estimation of the equivalent steady-state thermal behavior of the wall can be obtained. An estimate of the in situ U-value can be obtained as follows:

$$U_{AM} = \frac{\sum_{j=1}^{n} q_j}{\sum_{j=1}^{n}\left(T_{i,j} - T_{e,j}\right)},$$

(2)

where U_{AM} represents the U-value evaluated by the average method (W/m^2·K); q is the density of the heat flow rate (W/m^2); T_i and T_e are the interior and exterior air temperatures (K), respectively; and j represents the individual measurements.

According to the ISO 9869-1 standard [4], the end of the test should be determined using the R-value calculated from the surface temperature difference across the test wall. However, even though the difference between the use of the two values in determining when to terminate the test is not clearly known, several researchers [7,8,12,16] used the U-value instead of the R-value. An estimate of the R-value is obtained as follows:

$$R_{AM} = \frac{\sum_{j=1}^{n}\left(T_{si,j} - T_{se,j}\right)}{\sum_{j=1}^{n} q_j},$$

(3)

where R_{AM} represents the R-value evaluated by the average method (m^2·K/W), and T_{si} and T_{se} are the interior and exterior surface temperatures (K), respectively.

According to the ISO 9869-1 standard [4], when the estimate is computed after each measurement, an asymptotical value is obtained. If the following three convergence conditions are met simultaneously, this value can be considered to be the actual value, and the test should be terminated. The first convergence condition is that the test duration should exceed 72 h. The second convergence condition is that the R-value obtained at the end of the test does not deviate by more than ±5% from the value obtained 24 h prior to end of the test, as given in Equation (4). The third convergence condition is that the R-value obtained by analyzing data from the first time period during $\text{INT}(2 \times D_T/3)$ days does not deviate by more than ±5% from the values obtained by analyzing data from the last time period of the same duration, as given in Equation (5).

$$\left|\frac{R_{D_T} - R_{D_T - 24h}}{R_{D_T - 24h}}\right| \leq 5\%,$$

(4)

$$\left|\frac{R_{\text{INT}(2 \times D_T/3),\ first} - R_{\text{INT}(2 \times D_T/3),\ last}}{R_{\text{INT}(2 \times D_T/3),\ last}}\right| \leq 5\%,$$

(5)

where D_T is the duration of the test (days), and INT is the integer part.

In this study, the uncertainty associated with the U-value evaluation was estimated by the combined standard uncertainty determined according to the Guide to the Expression of Uncertainty in Measurement [32] while considering the accuracy of the measurement equipment as well as the operating conditions. The combined standard uncertainty for all the input quantities, which are independent, was obtained as

$$u_c(y) = \sqrt{\sum_{i=1}^{N}\left(\frac{\delta f}{\delta x_i}\right)^2 u^2(x_i)},$$

(6)

where $u_c(y)$ is the combined standard uncertainty, N is the number of input quantities X_i on which the measurand Y depends, f is the functional relationship between the measurand Y and input quantities X_i and between the output estimate y and input estimates x_i on which y depends, $u(x_i)$ is the standard uncertainty of the input estimate x_i, and $\delta f / \delta x_i$ is the sensitivity coefficient (c_i) or partial derivative with respect to the input quantity X_i of the functional relationship f.

As indicated by Equations (2) and (3), the R-value and U-value are associated with three variables; thus, each value has three sensitivity coefficients. Therefore, the combined standard uncertainties of the R-value and U-value were obtained by using the corresponding sensitivity coefficients—that is, the partial derivatives, as follows:

$$u_c(R) = \sqrt{\left(\frac{\delta R}{\delta q}\right)^2 u^2(q) + \left(\frac{\delta R}{\delta T_{si}}\right)^2 u^2(T_{si}) + \left(\frac{\delta R}{\delta T_{se}}\right)^2 u^2(T_{se})}, \tag{7}$$

$$u_c(U) = \sqrt{\left(\frac{\delta U}{\delta q}\right)^2 u^2(q) + \left(\frac{\delta U}{\delta T_i}\right)^2 u^2(T_i) + \left(\frac{\delta U}{\delta T_e}\right)^2 u^2(T_e)}. \tag{8}$$

Table 3 lists the uncertainty sources and their contributions toward calculating the standard uncertainty of input quantities. In this study, the uncertainty was expressed using the expanded uncertainty U, which was obtained by multiplying the combined standard uncertainty $u_c(y)$ by a coverage factor k. The value of the coverage factor was selected as 2, corresponding to a confidence level of 95.45%.

Table 3. Uncertainty sources and their contributions.

Type	Uncertainty Source	Systematic Uncertainty	Random Uncertainty
Instrument	Accuracy of thermocouples	±0.5 °C [1]	
	Accuracy of heat flux sensor	2% [1]	
	Accuracy of data logger	10% [2]	
	Thermocouple calibration	±2.2 °C [1]	
	Heat flux sensor calibration	3% [1]	
Operation	Poor contact between thermocouple and surface		5% [2]
	Poor contact between heat flux sensor and surface		5% [2]
	Modification of isotherms caused by heat flux sensor		2%–3% [2]
	Variation in temperatures and heat flux over time		±10% [2]

[1] Uncertainty value according to manufacturer's technical specifications. [2] Uncertainty value according to ISO 9869-1 [4].

3. Results and Discussion

3.1. Evolution of R-Value and U-Value

Figure 3 shows the evolution of the in situ R-value and U-value over the total test duration of 21 consecutive days with their corresponding expanded uncertainties analyzed in a cycle of 1 day using the average method. The in situ R-value and U-value tended to stabilize around the final values from the 12th day of the test. The R-value and U-value obtained at the end of the test were 0.988 ± 0.009 m²·K/W and 0.908 ± 0.007 W/m²·K, respectively, which are, respectively, 51.1% smaller and 113.0% larger than the theoretical values calculated according to the ISO 6946 standard. These results show that the thermal performance of the test wall deteriorated considerably for approximately 28 years after completion, and that it was necessary to measure the thermal performance of the wall onsite.

For the asymptotic values to be considered as the R-value and U-value, the three conditions mentioned in Section 2.3 must be satisfied. Tables 4 and 5 summarize the values used for determining the convergence of the R-value and U-value calculated by the average method, respectively. The deviation of the R-value and U-value according to the second and third convergence conditions is shown in Figure 4.

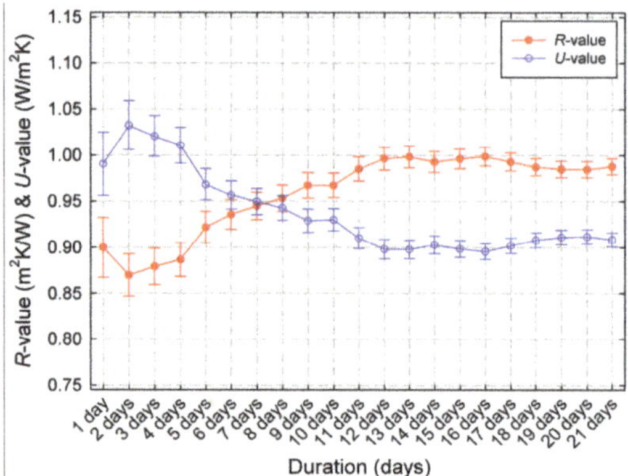

Figure 3. Evolution of in situ *R*-value and *U*-value with their expanded uncertainties.

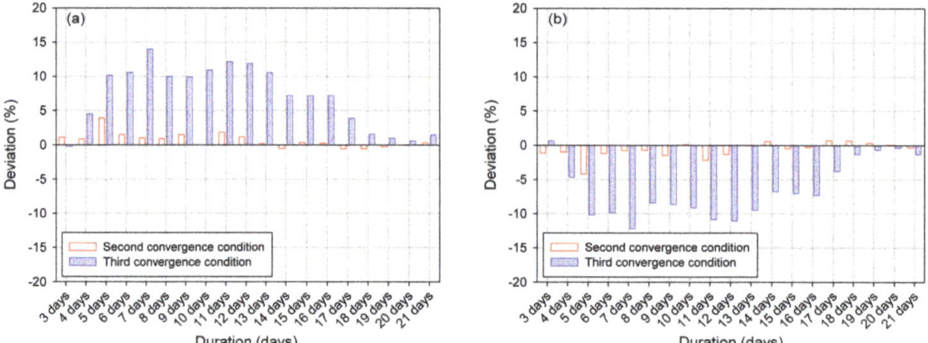

Figure 4. Deviation of (**a**) *R*-value and (**b**) *U*-value according to convergence conditions of the ISO 9869-1 standard.

As can be seen in Figure 4, the second condition that the deviation between the value obtained at the end of the test, and the value obtained 24 h before should be within ±5%, which is easily satisfied because there is no major difference in the data used for the calculation of the cumulative average. However, deviations for the third convergence condition according to Equation (5) are considerably larger than deviations for the second convergence condition. The decrease in the deviation with an increase in the analysis period is also not clear, and the third convergence condition was satisfied only 17 days after the start of measurement. These results were obtained because the analysis period was shortened from D_T to $INT(2 \times D_T/3)$, and the deviations of the *R*-value and *U*-value were calculated between the initial and latter periods of the same duration. The overlap period for the second convergence condition continued to increase proportionally as the measurement period became longer, but the overlap period for the third convergence condition did not exceed 50% of the comparison period. Therefore, to easily satisfy the third convergence condition, there should be slight changes in the environmental variables in the two periods.

As can be seen in Figure 2a, the temperature difference across the test wall during the entire test period tended to decrease at the beginning, then remained constant, and finally increased again. Though the test was conducted under a large temperature difference and stable environment as recommended in previous studies [7,12–14,33] and the ISO 9869-1 standard [4], the difficulty in satisfying the third convergence condition is attributed to these changes in the temperature differences during the entire test period. The results show that the third convergence condition is satisfied by the influence of cumulative averaging over a sufficiently long analysis period accompanied by an increase in the measurement period. For example, for the third convergence condition, the analysis period corresponding to the measurement period of 17 days in which the variation was stable within ±5% was 11 days, which is considerably longer than the test duration of previous studies [7,8,12,14,28].

When the measurement period is three and four days, both the convergence conditions are satisfied, but the deviations of the R-value and U-value obtained at the end of the test are more than 10%. The R-value and U-value whose deviations are within ±5% of their corresponding final values are derived from data obtained after the 7th day; however, both convergence conditions were satisfied only from the 17th day. Therefore, it is considered that the criteria for determining convergence by the average method based on ISO 9869-1 [4], particularly the third convergence condition, are very strict.

As can be observed in Tables 4 and 5, the duration of the test affects the measurement uncertainties. For U-values analyzed with their expanded uncertainties in Table 5, the measurement uncertainties decreased from ±0.022 W/m²·K on the third day to ±0.007 W/m²·K at the end of the test when the duration of the test was extended. This tendency is commonly confirmed in the R-values in Table 4. These results are in good agreement with the results in previous studies in [7,12,16,28], showing that an increase in the measurement period improves the measurement accuracy.

Table 4. R-values analyzed by average method and their expanded uncertainties.

D_T (days)	$INT(2 \times D_T/3)$ (days)	R-values with Their Expended Uncertainties (m²·K/W)			
		R_{D_T}	R_{D_T-24h}	$R_{INT(2 \times D_T/3), first}$	$R_{INT(2 \times D_T/3), last}$
3	2	0.879 ± 0.020	0.870 ± 0.023	0.870 ± 0.023	0.868 ± 0.025
4	2	0.887 ± 0.018	0.879 ± 0.020	0.870 ± 0.023	0.909 ± 0.029
5	3	0.922 ± 0.017	0.887 ± 0.018	0.879 ± 0.020	0.969 ± 0.025
6	4	0.936 ± 0.016	0.922 ± 0.017	0.887 ± 0.018	0.980 ± 0.022
7	4	0.945 ± 0.015	0.936 ± 0.016	0.887 ± 0.018	1.011 ± 0.023
8	5	0.954 ± 0.015	0.945 ± 0.015	0.922 ± 0.017	1.014 ± 0.021
9	6	0.967 ± 0.014	0.954 ± 0.015	0.936 ± 0.016	1.028 ± 0.020
10	6	0.967 ± 0.013	0.967 ± 0.014	0.936 ± 0.016	1.038 ± 0.020
11	7	0.985 ± 0.013	0.967 ± 0.013	0.945 ± 0.015	1.060 ± 0.019
12	8	0.996 ± 0.013	0.985 ± 0.013	0.954 ± 0.015	1.067 ± 0.017
13	8	0.998 ± 0.012	0.996 ± 0.013	0.954 ± 0.015	1.054 ± 0.016
14	9	0.993 ± 0.011	0.998 ± 0.012	0.967 ± 0.014	1.037 ± 0.015
15	10	0.996 ± 0.011	0.993 ± 0.011	0.967 ± 0.013	1.037 ± 0.014
16	10	0.999 ± 0.010	0.996 ± 0.011	0.967 ± 0.013	1.037 ± 0.013
17	11	0.993 ± 0.010	0.999 ± 0.010	0.985 ± 0.013	1.023 ± 0.012
18	12	0.987 ± 0.009	0.993 ± 0.010	0.996 ± 0.013	1.012 ± 0.011
19	12	0.985 ± 0.009	0.987 ± 0.009	0.996 ± 0.013	1.006 ± 0.011
20	13	0.985 ± 0.009	0.985 ± 0.009	0.998 ± 0.012	1.004 ± 0.011
21	14	0.988 ± 0.009	0.985 ± 0.009	0.993 ± 0.011	1.007 ± 0.010

Table 5. U-values analyzed by average method and their expanded uncertainties.

D_T (days)	$INT(2 \times D_T/3)$ (days)	U-values with Their Expended Uncertainties (W/ m²·K)			
		U_{D_T}	U_{D_T-24h}	$U_{INT(2 \times D_T/3), first}$	$U_{INT(2 \times D_T/3), last}$
3	2	1.021 ± 0.022	1.032 ± 0.026	1.032 ± 0.026	1.039 ± 0.029
4	2	1.011 ± 0.020	1.021 ± 0.022	1.032 ± 0.026	0.984 ± 0.030
5	3	0.968 ± 0.017	1.011 ± 0.020	1.021 ± 0.022	0.917 ± 0.022
6	4	0.957 ± 0.015	0.968 ± 0.017	1.011 ± 0.020	0.911 ± 0.019
7	4	0.950 ± 0.014	0.957 ± 0.015	1.011 ± 0.020	0.887 ± 0.019
8	5	0.942 ± 0.013	0.950 ± 0.014	0.968 ± 0.017	0.887 ± 0.017
9	6	0.929 ± 0.013	0.942 ± 0.013	0.957 ± 0.015	0.874 ± 0.015
10	6	0.930 ± 0.012	0.929 ± 0.013	0.957 ± 0.015	0.869 ± 0.015
11	7	0.910 ± 0.011	0.930 ± 0.012	0.950 ± 0.014	0.846 ± 0.014
12	8	0.898 ± 0.010	0.910 ± 0.011	0.942 ± 0.013	0.838 ± 0.012
13	8	0.898 ± 0.010	0.898 ± 0.010	0.942 ± 0.013	0.853 ± 0.012

Table 5. *Cont.*

D_T (days)	INT(2×D_T/3) (days)	U-values with Their Expended Uncertainties (W/ m²·K)			
		U_{D_T}	U_{D_T-24h}	$U_{INT(2×D_T/3),\,first}$	$U_{INT(2×D_T/3),\,last}$
14	9	0.903 ± 0.010	0.898 ± 0.010	0.929 ± 0.013	0.866 ± 0.011
15	10	0.898 ± 0.009	0.903 ± 0.010	0.930 ± 0.012	0.865 ± 0.011
16	10	0.896 ± 0.009	0.898 ± 0.009	0.930 ± 0.012	0.862 ± 0.010
17	11	0.902 ± 0.008	0.896 ± 0.009	0.910 ± 0.011	0.875 ± 0.010
18	12	0.908 ± 0.008	0.902 ± 0.008	0.898 ± 0.010	0.886 ± 0.009
19	12	0.911 ± 0.008	0.908 ± 0.008	0.898 ± 0.010	0.892 ± 0.009
20	13	0.912 ± 0.008	0.911 ± 0.008	0.898 ± 0.010	0.894 ± 0.009
21	14	0.908 ± 0.007	0.912 ± 0.008	0.903 ± 0.010	0.891 ± 0.009

3.2. Effect of Variation in Analysis Period

Reliable results can be obtained when the measurement is conducted using the heat flow meter method under stable environmental conditions such as a high temperature difference across the test wall, low wind speed, and an avoidance of direct solar radiation. However, the period in which the above stable environmental conditions are satisfied is actually not long. In addition, measurements conducted in buildings where residents are living cause inconvenience to their daily lives. Therefore, it is difficult to measure the in situ *R*-value and *U*-value of a wall for a long period of time for many buildings.

In this study, for a single measurement campaign of 21 days, we reviewed the convergence characteristics of the in situ *R*-value and *U*-value under the assumption that many measurements are conducted on the same test wall by shifting the measurement start date by 1 day and setting the analysis period to be different. For example, if the analysis period is set to 3 days and the period is continuously moved forward by 1 day from the start date of the test, 19 tests can be considered to have been conducted. In this example, out of the 19 cases, there are seven cases (37%) where no days are duplicated and 10 cases (53%) where 1 day is duplicated. As all days during the original measurement period of 21 days met the recommended environmental conditions, the approach proposed in this study can be considered reasonable. Therefore, this approach can be used to study the convergence characteristics of the in situ *R*-value and *U*-value according to variation in the analysis period under the aforementioned constraints. However, this approach has limitations in that the number of cases decreases as the analysis period becomes longer and the environmental variables in these cases become similar to each other because of overlap.

Figure 5 shows the *R*-values and *U*-values evaluated according to the approach described above for different analysis periods. The shorter the analysis period, the larger the dispersion of the *R*-values and *U*-values. As the analysis period becomes longer, these values tended to converge near the final values analyzed using the measurement data for 21 days. In addition, the percentages of cases deviating by more than ±5% from the final values were as high as approximately 40% when the analysis period was short; however, these percentages gradually decreased, and such cases were not found when the analysis period was longer than 13 days.

The convergence of the *R*-values and *U*-values obtained for different analysis periods was examined according to the two convergence conditions of the ISO 9869-1 standard, and the results are shown in Figures 6 and 7. The second convergence condition, which compares the deviation of the two values with a 24 h difference, is unsatisfactory only in three cases for the *R*-values and four cases for the *U*-values among the cases with analysis periods of 3 and 4 days. This is because when the measurement is conducted in a stable environment, as is the case in this study, a rapid change exceeding more than ±5% is unlikely to occur, considering a 24 h difference. On the other hand, the third convergence condition is not met in any of the cases until the analysis period is 16 days, and the deviation is also very large compared to that in the second condition. These results indicate that the fulfillment of the convergence conditions according to the ISO 9869-1 standard is largely dependent on the third condition. In other words, to obtain reliable results in a short period of time, it is necessary to keep the environmental variables constant throughout the measurement period, or an appropriate period must be selected.

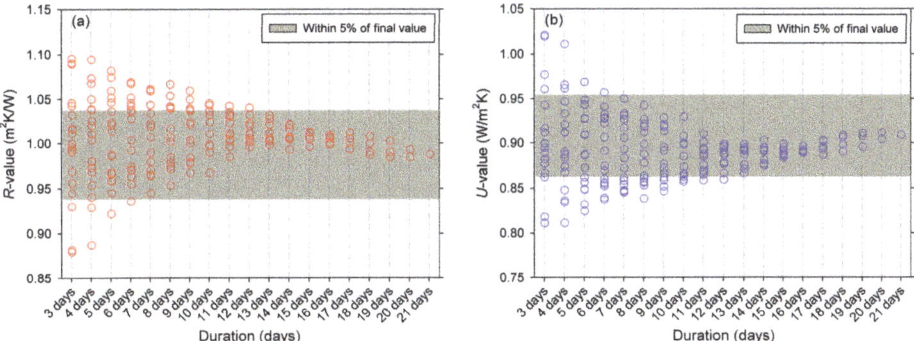

Figure 5. Distributions of (**a**) *R*-values and (**b**) *U*-values evaluated for different analysis periods.

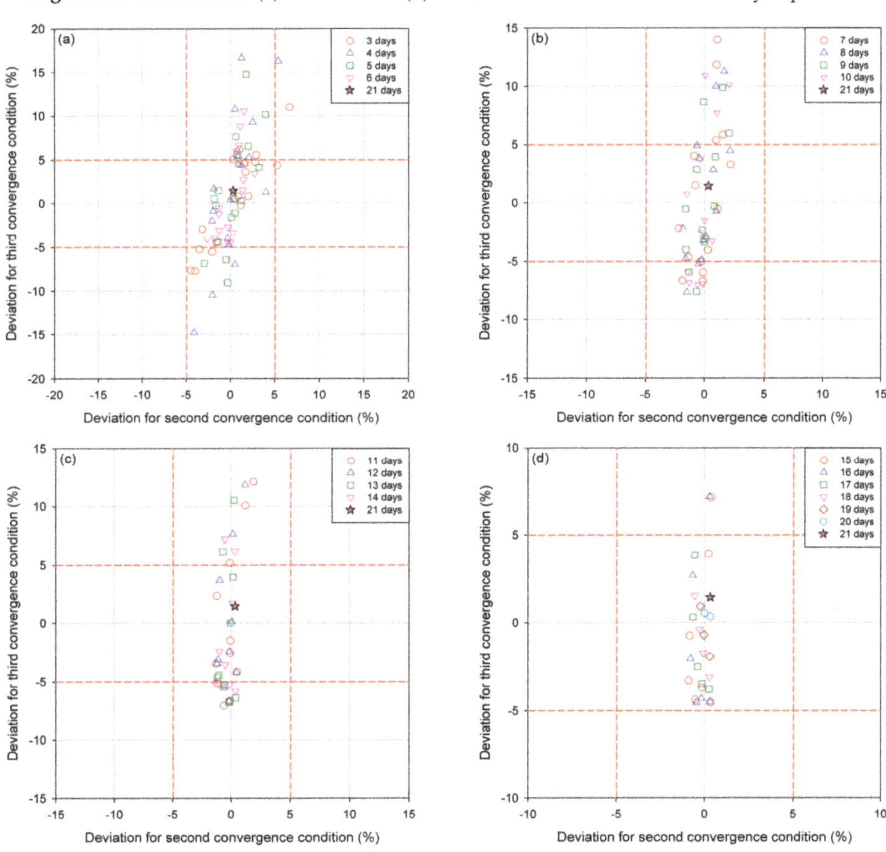

Figure 6. Convergence characteristics of *R*-values evaluated for different analysis periods according to conditions of the ISO 9869-1 standard. For ease of understanding, the plots are divided into four sub-figures based on the length of the analysis period: (**a**) 3–6 days; (**b**) 7–10 days; (**c**) 11–14 days; (**d**) 15–20 days.

As shown in Figure 5, the *R*-values and *U*-values in all the cases with an analysis period of 13 days or more were within ±5% of the respective final values. However, Figures 6 and 7 show that both convergence conditions begin to be satisfied in the cases where the analysis period is more than 17 days. Therefore, the findings show that a minimum test duration of more than 2 weeks is required, even if the daily air temperature difference is maintained at a minimum of 16.5 °C throughout the entire test duration and the environmental variables are not changed considerably.

According to the average method based on the ISO 9869-1 standard [4], the test may be ended only when the convergence conditions obtained using the *R*-value are fulfilled. In this study, two convergence conditions were analyzed using both the *R*-value and *U*-value, and very similar convergence characteristics were identified. As can be observed in Figures 6 and 7, the deviations for the two convergence conditions appear symmetrical because the *R*-value and *U*-value are essentially reciprocal. However, except for the deviations with symmetrical form, other aspects such as the fulfillment of the convergence conditions and the proportion of cases for which the convergence conditions are fulfilled were very similar for the two values. Therefore, the findings show that it is possible to determine the end of the test, that is its convergence, using the U-value instead of the R-value.

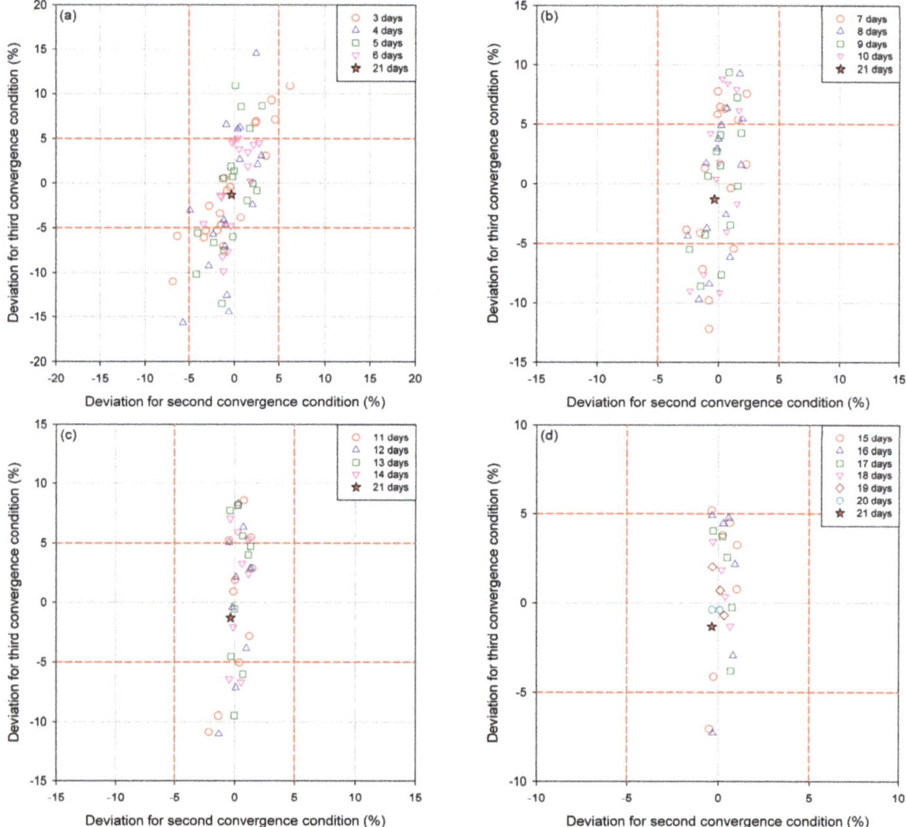

Figure 7. Convergence characteristics of *U*-values evaluated for different analysis periods according to conditions of the ISO 9869-1 standard. For ease of understanding, the plots are divided into four sub-figures based on the length of the analysis period: (**a**) 3–6 days; (**b**) 7–10 days; (**c**) 11–14 days; (**d**) 15–20 days.

3.3. Effect of Temperature Difference

A large temperature difference between the inside and outside environments has often been referred to as one of the key factors to obtain reliable results through the heat flow meter method. Thus, the influence of the temperature difference on the accuracy of the results derived from different analysis periods was analyzed. Figure 8 shows the average surface and air temperature differences between the inside and outside environments for different analysis periods described in Section 3.2. As the analysis period decreased, the average temperature differences showed dispersed distribution because of changes in the outside air temperature throughout the test duration. When the analysis period was 3 days, the surface temperature differences were in the range of approximately 15.5–24.8 °C, and the air temperature differences were in the range of approximately 17.1–27.7 °C. These temperature differences gradually became similar and reached the average temperature differences of the entire test period, namely 19.6 °C and 21.8 °C, because of the influence of the cumulative average as the analysis periods increased. These temperature differences across the test wall were considerably higher than the temperature condition—that is, 10.0 °C—recommended by the ISO 9869-1 standard [4].

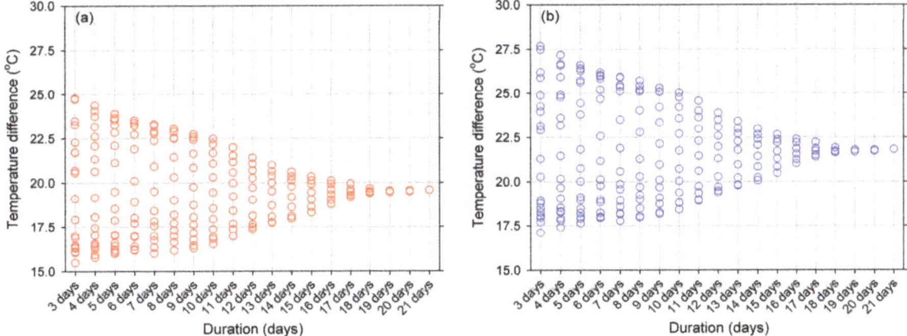

Figure 8. Average (**a**) surface and (**b**) air temperature differences between inside and outside environments for different analysis periods.

The average temperature differences and the corresponding *R*-values and *U*-values for different analysis periods are plotted in Figures 9 and 10, respectively. In the cases where the average surface temperature difference was more than 19.6 °C, the *R*-values were mostly within ±5% of the value derived at the end of the test. On the other hand, in the cases where the average surface temperature difference was less than 19.6 °C, the *R*-values that deviate by more than ±5% from the final value were more frequently found than in the other cases. For example, for an analysis period of 3 days, there were nine cases where the average surface temperature difference was greater than 19.6 °C, and only two of these cases showed deviations greater than ±5% from the final value. In contrast, six out of the 10 cases with an average surface temperature difference below 19.6 °C had deviations outside the ±5% range. As can be observed in Figure 9a,b, this tendency is commonly confirmed in cases where the analysis period is relatively short. Therefore, the results show that if the test duration is the same, the larger the surface temperature difference and the greater the possibility of causing a lower deviation.

The *R*-values for all the cases with an analysis period of 13 days or more converged within ±5% of the final value without a large influence of the surface temperature difference between the inside and outside environments. Therefore, if the surface temperature difference is higher than a certain temperature difference—that is, 10 °C—recommended in previous studies [7,33] and the ISO 9869-1 standard [4], the convergence of the *R*-values is affected more by the length of the analysis period than by the surface temperature difference. These results are also seen in the relationship between the *U*-values and the air temperature differences between the inside and outside environments (Figure 10). However, these analysis results still have limitations in that measurement data for 21 days were used; thus, further research based on long-term measurements is needed.

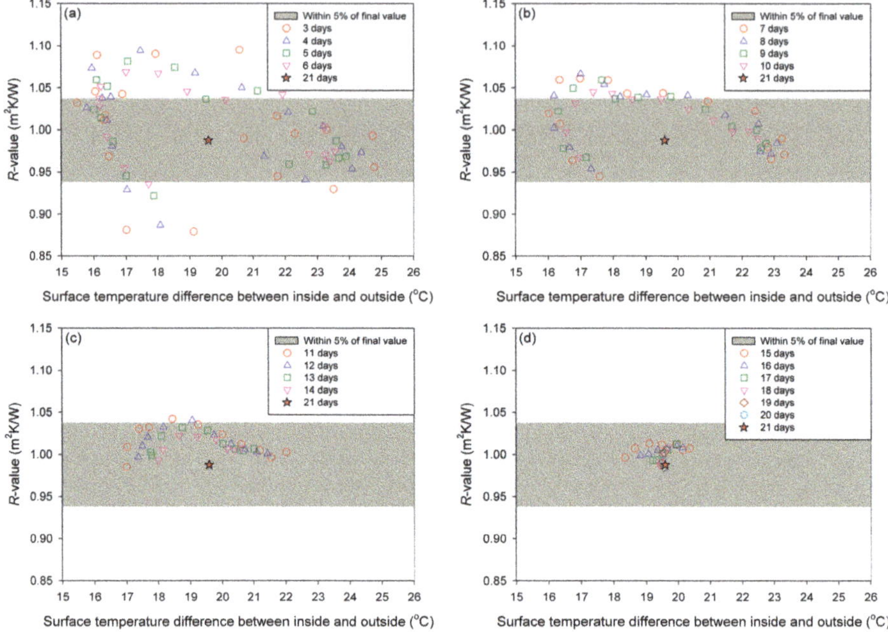

Figure 9. Relationship between average surface temperature differences and final R-values evaluated for different analysis periods. For ease of understanding, the plots are divided into four sub-figures based on the length of the analysis period: (**a**) 3–6 days; (**b**) 7–10 days; (**c**) 11–14 days; (**d**) 15–20 days.

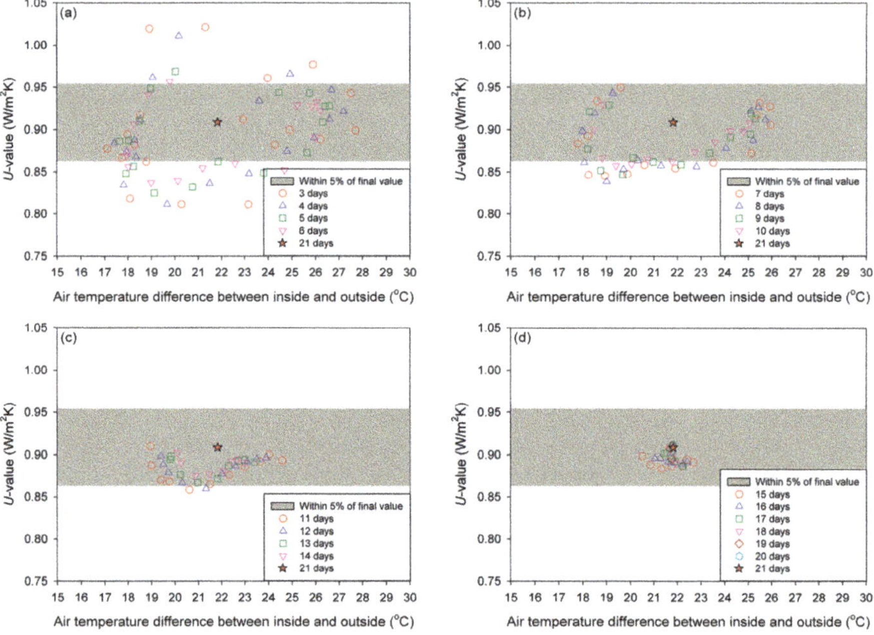

Figure 10. Relationship between average air temperature differences and final U-values evaluated for different analysis periods. For ease of understanding, the plots are divided into four sub-figures based on the length of the analysis period: (**a**) 3–6 days; (**b**) 7–10 days; (**c**) 11–14 days; (**d**) 15–20 days.

4. Conclusions

This study evaluated the convergence characteristics of the in situ R-value and U-value analyzed using the standardized average method. The convergence characteristics were analyzed according to the convergence criteria regulated by ISO 9869-1 [4]. Onsite measurement was conducted on the northwest-facing external wall for over 21 days in winter under fairly stable environmental conditions, as recommended by ISO 9869-1 [4] and the literature [7,12–14,33]. To analyze the effect of the length of the analysis period and the temperature difference on the convergence characteristics of the in situ R-value and U-value, datasets for different analysis periods were created from the onsite measurement data for 21 consecutive days.

Our results show that in situ R-values and U-values that were within ±5% of the values obtained across a full test duration were obtained starting from the 7th day, but the convergence conditions were satisfied only from the 17th day. This is because the length of the overlap period and the periods used for comparing the deviations are different between the second and third convergence conditions. The overlap period for the second condition increases proportionally as the measurement period becomes longer, but, in the third condition, the overlap period does not exceed 50% of the comparison period. This result indicates that the convergence according to the ISO 9869-1 standard largely depends on the third condition. Therefore, to obtain reliable in situ R-values and U-values in a short test duration, it is necessary to keep the environmental variables constant throughout the entire test duration, or an appropriate duration should be selected.

Our results also show that when the test duration is relatively short, the larger the temperature difference and the smaller the deviation for the convergence conditions. However, when the test duration is longer (approximately 2 weeks or more in this study), the effect of the temperature difference on the convergence of the in situ R-value and U-value decreases gradually because of cumulative averaging. Therefore, if the temperature difference is higher than the recommended value—that is, 10 °C—the convergence of the in situ R-value and U-value is affected more by the length of the test duration than by the temperature difference.

In addition, our findings indicate that for the in situ R-value and U-value, although the deviation values for the convergence conditions are symmetrical, other aspects such as the fulfillment of the convergence conditions and the proportion of cases for which the convergence conditions are fulfilled are very similar for the two values. Therefore, it is found that there is no difference between the use of the R-value and U-value in determining the end of the test.

In this study, it is assumed that many measurements were conducted on the same test wall by creating datasets for different analysis periods from a single onsite measurement dataset for 21 consecutive days. Thus, we intend to conduct further research by increasing the number of test walls and using onsite measurement data for longer periods. Furthermore, we intend to investigate the selection of an appropriate test duration and how the duration should be shortened.

Author Contributions: Conceptualization, D.S.C. and M.J.K.; methodology, M.J.K.; validation, D.S.C.; formal analysis, D.S.C. and M.J.K.; investigation, M.J.K.; data curation, M.J.K.; writing—original draft preparation, M.J.K.; writing—review and editing, D.S.C. and M.J.K.; visualization, M.J.K.; supervision, D.S.C.; project administration, D.S.C.; funding acquisition, D.S.C.

Funding: This research was supported by an Academic Research Fund of Chungwoon University in 2017 and Basic Science Research Program through the National Research Foundation of Korea (NRF) funded by the Ministry of Education (NRF-2017R1D1A1B03034281).

Conflicts of Interest: The authors declare no conflict of interest.

Nomenclature

D_T	Test duration, days
f	Functional relationship between measurand Y and input quantities X_i on which measurand Y depends
INT	Integer part
j	Individual measurements
n	Number of measurement data points
N	Number of input quantities X_i on which measurand Y depends
q_j	Density of heat flow rate, W/m^2
R_{AM}	Thermal resistance value evaluated by average method, m^2·K/W
R_{si}	Interior surface resistance, m^2·K/W
R_{se}	Exterior surface resistance, m^2·K/W
$t.$	Material thickness, m
$T_{e,j}$	Exterior air temrature, K
$T_{i,j}$	Interior air temperature, K
$T_{se,j}$	Exterior wall surface temperature, K
$T_{si,j}$	Interior wall surface temperature, K
$u_c(y)$	Combined standard uncertainty
$u(x_i)$	Standard uncertainty of input estimate x_i
U_{AM}	Thermal transmittance value evaluated by average method, W/m^2·K
U_D	Thermal transmittance value evaluated by calculation method, W/m^2·K
X_i	ith input quantity on which measurand Y depends
x_i	Estimate of input X_i
Y	Measurand
y	Estimate of measurand Y
λ	Thermal conductivity, W/m·K

References

1. Ministry of Land, Infrastructure and Transport (MOLIT). *Energy Saving Design Standards for Building*; Ministry of Land, Infrastructure and Transport: Sejong, Korea, 2017. Available online: http://www.molit.go.kr/USR/I0204/m_45/dtl.jsp?idx=15270 (accessed on 19 February 2019).
2. ISO 6946:2007—Building Components and Building Elements—Thermal Resistance and Thermal Transmittance—Calculation Method. Available online: https://www.iso.org/standard/65708.html (accessed on 19 February 2019).
3. Choi, D.S.; Ko, M.J. Comparison of various analysis methods based on heat flowmeters and infrared thermography measurements for the evaluation of the in situ thermal transmittance of opaque exterior walls. *Energies* **2017**, *10*, 1019. [CrossRef]
4. ISO 9869–1:2014. Building Elements—In-Situ Measurement of Thermal Resistance and Thermal Transmittance—Part 1: Heat Flow Meter Method. Available online: https://www.iso.org/standard/59697.html (accessed on 19 February 2019).
5. Adhikari, R.S.; Lucchi, E.; Pracchi, V. Experimental measurements on thermal transmittance of the opaque vertical walls in the historical buildings. In Proceedings of the 28th International PLEA Conference on Sustainable Architecture + Urban Design, Lima, Peru, 7–9 November 2012.
6. Cabeza, L.F.; Castell, A.; Medrano, M.; Martorell, I.; Perez, G.; Fernandez, I. Experimental study on the performance of insulation materials in mediterranean construction. *Energy Build.* **2010**, *42*, 630–636. [CrossRef]
7. Asdrubali, F.; D'Alessandro, F.; Baldinelli, G.; Bianchi, F. Evaluating in situ thermal transmittance of green buildings masonries—A case study. *Case Stud. Constr. Mater.* **2014**, *1*, 53–59. [CrossRef]
8. Evangelisti, L.; Guattari, C.; Gori, P.; Vollaro, R.D.L. In situ thermal transmittance measurements for investigating differences between wall models and actual building performance. *Sustainability* **2015**, *7*, 10388–10398. [CrossRef]

9. Baker, P. *U-Value and Traditional Buildings: In Situ Measurements and Their Comparisons to Calculated Values*; Historic Scotland Technical Paper: Edinburgh, UK, 2011.

10. Rye, C.; Scott, C. The SPAB Research Report 1: U-Value Report. Society for the Protection of Ancient Buildings, London. 2012. Available online: https://www.spab.org.uk/sites/default/files/documents/MainSociety/Advice/SPABU-valueReport.Nov2012.v2.pdf (accessed on 19 February 2019).

11. Bros-Williamson, J.; Carnier, C.; Currie, J.I. A longitudinal building fabric and energy performance analysis of two homes built to different energy principles. *Energy Build.* **2016**, *130*, 578–591. [CrossRef]

12. Gaspar, K.; Casals, M.; Gangolells, M. In situ measurement of façades with a low U-value: Avoiding deviations. *Energy Build.* **2018**, *170*, 61–73. [CrossRef]

13. Ahmad, A.; Maslehuddin, M.; Al-Hadhrami, L.M. In situ measurement of thermal transmittance and thermal resistance of hollow reinforced precast concrete walls. *Energy Build.* **2014**, *84*, 132–141. [CrossRef]

14. Ficco, G.; Iannetta, F.; Ianniello, E.; d'Ambrosio Alfano, F.R.; Dell'Isola, M. U-value in situ measurement for energy diagnosis of existing buildings. *Energy Build.* **2015**, *104*, 108–121. [CrossRef]

15. Deconinck, A.H.; Roels, S. Comparison of characterization methods determining the thermal resistance of building components from onsite measurements. *Energy Build.* **2016**, *130*, 309–320. [CrossRef]

16. Gaspar, K.; Casals, M.; Gangolells, M. Review of criteria for determining HFM minimum test duration. *Energy Build.* **2018**, *176*, 360–370. [CrossRef]

17. ASTM C 1155. Standard Practice for Determining Thermal Resistance of Building Envelope Components from the In-Situ Data. Available online: https://www.astm.org/Standards/C1155.htm (accessed on 19 February 2019).

18. Haralambopoulos, D.A.; Paparsenos, G.F. Assessing the thermal insulation of old buildings—The need for in situ spot measurements of thermal resistance and planar infrared thermography. *Energy Convers. Manag.* **1998**, *39*, 65–79. [CrossRef]

19. Balaras, C.A.; Argiriou, A.A. Infrared thermography for building diagnostics. *Energy Build.* **2002**, *34*, 171–183. [CrossRef]

20. Ocana, S.M.; Guerrero, I.C.; Requena, I.G. Thermographic survey of two rural buildings in Spain. *Energy Build.* **2004**, *36*, 515–523. [CrossRef]

21. Kalamees, T. Air tightness and air leakages of new lightweight single-family detached houses in Estonia. *Build. Environ.* **2007**, *42*, 2369–2377. [CrossRef]

22. Taylor, T.; Counsell, J.; Gill, S. Energy efficiency is more than skin deep: Improving construction quality control in new-build housing using thermography. *Energy Build.* **2013**, *66*, 222–231. [CrossRef]

23. Albatici, R.; Tonelli, A.M. Infrared thermovision technique for the assessment of the thermal transmittance value of opaque building elements on site. *Energy Build.* **2010**, *42*, 2177–2183. [CrossRef]

24. Albatici, R.; Tonelli, A.M.; Chiogna, M. A comprehensive experimental approach for the validation of quantitative infrared thermography in the evaluation of building thermal transmittance. *Appl. Energy* **2015**, *141*, 218–228. [CrossRef]

25. Fokaides, P.A.; Kalogirou, S.A. Application of infrared thermography for the determination of the overall heat transfer coefficient (U-value) in building envelopes. *Appl. Energy* **2011**, *88*, 4358–4365. [CrossRef]

26. Dall'O, G.; Sarto, L.; Panza, A. Infrared screening of residential buildings for energy audit purposes: Results of a field test. *Energies* **2013**, *6*, 3859–3878. [CrossRef]

27. Nardi, I.; Sfarra, S.; Ambrosini, D. Quantitative thermography for the estimation of the U-value: State of the art and a case study. In Proceedings of the 32nd UIT (Italian Union of Thermo-fluid-dynamics) Heat Transfer Conference, Pisa, Italy, 23–25 June 2014. [CrossRef]

28. Nardi, I.; Ambrosini, D.; Rubeis, T.D.; Sfarra, S.; Perilli, S.; Pasqualoni, G. A comparison between thermographic and flow-meter methods for the evaluation of thermal transmittance of different wall constructions. In Proceedings of the 33th UIT (Italian Union of Thermo-fluid-dynamics) Heat Transfer Conference, L'Aquila, Italy, 22–24 June 2015. [CrossRef]

29. Biddulph, P.; Gori, V.; Elwell, C.; Scott, C.; Rye, C.; Lowe, R.; Oreszczyn, T. Inferring the thermal resistance and effective thermal mass of a wall using frequent temperature and heat flux measurements. *Energy Build.* **2014**, *78*, 10–16. [CrossRef]

30. Gori, V.; Marincioni, V.; Biddulph, P.; Elwell, C.A. Inferring the thermal resistance and effective thermal mass distribution of a wall from in situ measurements to characterize heat transfer at both the interior and exterior surfaces. *Energy Build.* **2017**, *135*, 398–409. [CrossRef]

31. Iglesias, M.; Sawlan, Z.; Scavino, M.; Tempone, R.; Wood, C. Bayesian inferences of the thermal properties of a wall using temperature and heat flux measurements. *Int. J. Heat Mass Transf.* **2018**, *116*, 417–431. [CrossRef]
32. ISO/IEC Guide 98–9:2008. Uncertainty of Measurement—Part 3: Guide to the Expression of Uncertainty in Measurement. Available online: https://www.iso.org/standard/50461.html (accessed on 19 February 2019).
33. Desogus, G.; Mura, S.; Ricciu, R. Comparing different approaches to in situ measurement of building components thermal resistance. *Energy Build.* **2011**, *43*, 2613–2620. [CrossRef]

Article

Study on the Thermal Performance of a Hybrid Heat Collecting Facade Used for Passive Solar Buildings in Cold Region

Xiaoliang Wang [1,2,*], Bo Lei [2], Haiquan Bi [2] and Tao Yu [2]

[1]   Architecture and Urban Planning College, Southwest Minzu University, Chengdu 610041, China
[2]   School of Mechanical Engineering, Southwest Jiaotong University, Chengdu 610031, China;
lbswjtu@163.com (B.L.); bhquan@163.com (H.B.); yutao196@gmail.com (T.Y.)
*   Correspondence: xiao_liangwang@126.com; Tel.: +86-028-8552-2056

Received: 25 January 2019; Accepted: 12 March 2019; Published: 18 March 2019

Abstract: Passive solar technologies are traditionally considered as cost-effective ways for the building heating. However, conventional passive solar buildings are insufficient to create a relatively stable and comfortable indoor thermal environment. To further increase the indoor air temperature and reduce the heating energy consumption, a hybrid heat collecting facade (HHCF) is proposed in this paper. To analyze the thermal performance of the HHCF, a heat transfer model based on the heat balance method is established and validated by experimental results. Meanwhile, the energy saving potential of a room with the HHCF is evaluated as well. When the HHCF is applied to places where heating is required in the cold season while refrigeration is unnecessary in hot season, the HHCF can reduce the heating need by 40.2% and 21.5% compared with the conventional direct solar heat gain window and the Trombe wall, respectively. Furthermore, a series of parametric analyses are performed to investigate the thermal performance of the room with HHCF under various design and operating conditions. It is found that the thermal performance of the HHCF mainly depends on the window operational schedule, the width and the absorptivity of heat collecting wall, and the thermal performance of the inner double-glass window. The modeling and the parametric study in this paper are beneficial to the design and the optimization of the HHCF in passive solar buildings.

Keywords: hybrid heat collecting facade (HHCF); passive solar building; heat transfer model; thermal performance

---

1. Introduction

Passive solar building is a type of low-energy buildings exploiting solar energy to create a relatively comfortable environment in buildings [1,2]. Thermal energy collection and storage in building envelops may be enhanced by integrating some passive solar measures [3–7]. It was reported by measurements that passive solar buildings can save more than 25% of total primary energy consumption than the same buildings without passive solar measures [8–10]. As two efficient technologies used in passive solar houses, direct solar heat gain window and thermal storage wall (e.g., Trombe wall) are widely used [11]. For the direct solar heat gain window, sunlight passes through the window and enters into the indoor space, solar energy is absorbed in the floor or the interior wall during the daytime due to the effect of thermal mass. Then the stored thermal energy is gradually released into the indoor space when the room temperature falls down at night. While for the thermal storage wall, solar energy is initially absorbed and stored by the heat collecting wall (such as the exterior wall) in the daytime, rather than directly coming into the room. The thermal energy stored in the heat collecting wall is released at night to improve the indoor temperature as well. Both of these two methods are beneficial to improve the thermal performance of the passive solar building in the cold region [12].

In fact, due to the high U-value of the window, the direct solar heat gain strategy is more easily influenced by the outdoor temperature. Meanwhile, the indoor air temperature fluctuation of the direct solar heat gain room is relatively large [13], which limits the further improvement of the indoor thermal performance. The Trombe wall not only collects and stores solar energy for the indoor heating, but also reduces the heat loss from the indoor to the outdoor environment due to the outside glass cover [14]. Meanwhile, for a specific facade, the larger area of the Trombe wall decreases the maximum window to wall ratio, resulting in less solar energy coming into the indoor space [15,16]. Based on the above descriptions, if the direct solar heat gain window and the Trombe wall are combined well with each other in the practical application, the thermal performance of the passive solar building can be further improved. Therefore, we developed a hybrid passive solar energy utilization form, named hybrid heat collecting façade (HHCF).

Although there are very few directly related studies about the HHCF, plenty of investigations about the thermal performance of passive solar houses with the direct solar gain window [17–22] or the thermal storage wall [23–26] could provide effective research techniques for analyzing the thermal performance of the HHCF. Through literature research, it can be found that experiments and numerical simulations are two widely used methods. Based on experiments and simulations, researchers have shown that the direct solar heat gain strategy is quite effective to improve the indoor thermal performance [27–30]. For instance, numerical simulations have been carried out by Gong et al. [31] and it was found that the direct solar heat gain house could significantly reduce the annual thermal load of the building. A study conducted by M.C. Ruiz [32] also showed that the overall thermal consumption of a building was reduced by almost 13% through optimizing the building direction, the window–wall ratio, and the insulation of the direct solar heat gain building. These analytical methods are helpful for carrying out the thermal performance of the HHCF.

In this paper a novel hybrid heat collecting façade (HHCF) is proposed, integrating the advantages of both the direct solar heat gain window and the traditional Trombe wall. A heat transfer simulation model is developed for analyzing the thermal performance of the building with the proposed HHCF. Experiments are conducted to validate the accuracy of the heat transfer model. Meanwhile, in order to explore the energy saving potential of the HHCF, the thermal performance of a building with the HHCF is compared with those of a conventional building with the direct solar heat gain and with the Trombe wall, respectively. During the comparison, the numerical simulation method is adopted and the cell buildings are with the same geometry except for the solar system. In addition, a parametric study is employed to analyze the influences of various factors on the thermal performance of the HHCF. With the model established in this paper, the thermal performance of the building with the HHCF can be analyzed and optimized, which will provide a simulation tool for designing the HHCF in regions where heating is required in the cold season while refrigeration is unnecessary in hot season.

2. Materials and Methods

2.1. Principle of HHCF

Previous studies have shown that the direct solar heat gain window and the thermal storage wall are effective to improve the indoor thermal performance. However, the application of the direct solar heat gain window is liable to decrease the indoor temperature at night, since the high U value of the window largely increases the heat loss from the indoor space to the outdoor environment. The Trombe wall is opaque and prevents the sunlight into the indoor space. Thus, in order to ensure the necessary daylight for the building with the Trombe wall, a transparent window is dispensable but it leads to the extra heat loss from the room to the outside environment. To solve this problem, a novel hybrid heat collecting facade is proposed in this paper.

The schematic diagram of the HHCF is shown in Figure 1. The HHCF is normally mounted on the south wall, and mainly consists of a single-glass window, a double-glass window, and a heat collecting wall. In the HHCF, there are two functional spaces including the heat collecting space and the heat transfer space. The heat collecting space is surrounded by the single-glass window and the heat collecting wall, which behaves as the "Trombe Wall" and can be used to convert solar energy into thermal energy for heating the air. The heat transfer space is surrounded by the single-glass window and the double-glass window. Since the double-glass window can be opened and closed acting as a door, the heat transfer space allows the solar energy to directly transmit into the indoor space, as well as increasing the thermal resistance between the indoor space and the outdoor environment. For indoor heat exhaust ventilation and to prevent indoor overheating that may occur in the hot season, a transparent window is inlaid in the upper part of the single glass window and it can be opened and closed. In this paper, we mainly focus on places where heating is required in the cold season while refrigeration is unnecessary in the hot season and for this situation the inlaid window is closed.

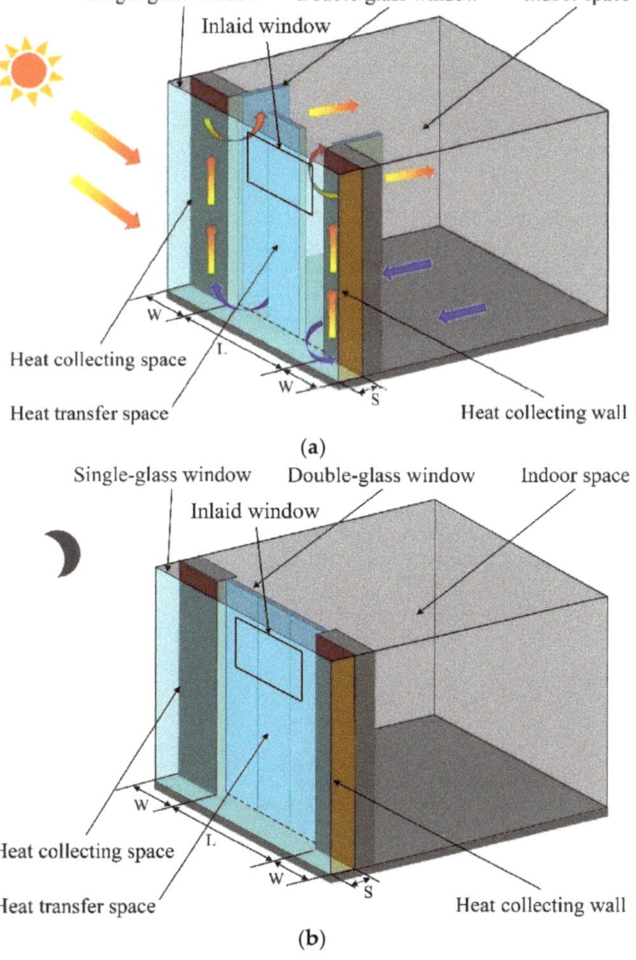

Figure 1. Schematic diagram of the hybrid heat collecting facade (HHCF). (**a**) with double-glass window open; (**b**) with double-glass window closed.

The operation principle of the HHCF can be seen in Figure 1:

1. In a sunny day, solar energy is absorbed by the exterior surface of the heat collecting wall and converted into thermal energy for heating the air convectively within the heat collecting space.

With the inner double-glass window open, the heated air in the heat collecting space rises up and reaches the top of the heat collecting space under the effect of the buoyancy force. Then, the heated air gathers at the top of the heat collecting space and increases the air pressure at the top of the heat collecting space. Finally, under the effect of the pressure difference, the heated air passes through the heat transfer space and enters into the indoor space along the horizontal direction. Meanwhile, at the bottom of the heat collecting space, the low-pressure cavity is supplied by the indoor air along the horizontal direction. In this way, thermal energy is transferred into the room space and the indoor temperature rises. Moreover, sunlight can also penetrate the windows into the indoor space directly and the solar energy is stored in the interior building construction.

2. At night or during a cloudy day, the inner double-glass window should be kept closed, and it increases the thermal resistance between the indoor space and the outdoor environment. Consequently, the heat loss from the indoor space to the outdoor environment can be effectively reduced, especially at night.

2.2. Modeling of the Building with HHCF

To evaluate the thermal performance of the building with HHCF, the simulation model is established in this section and necessary assumptions are made for simplifying the simulation.

2.2.1. Assumptions

For simplifying the analysis, the following assumptions are made:

(1) Thermal properties of the building materials are kept constant.
(2) Heat transfer processes through walls, floor, roof, and windows are considered as one-dimensional.
(3) The heat storage of glass is ignored.
(4) Air in each zone is well-mixed.
(5) Mean air flow rate between the heat collecting space and the heat transfer space is identical with that between the heat transfer space and the indoor space.

2.2.2. Energy Balance Equations

For solving the heat transfer processes of a building with HHCF, energy balance equations for HHCF, indoor air and building constructions have been established as displayed in Figure 2.

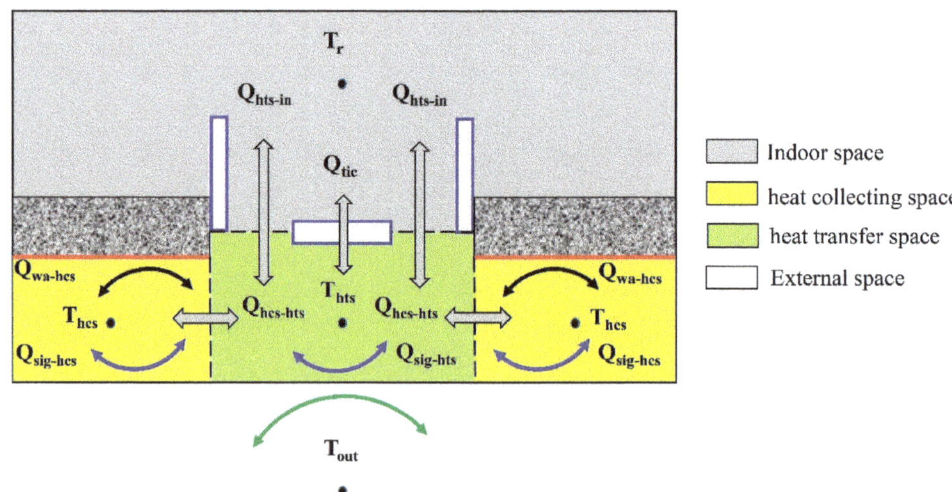

Figure 2. Heat transfer processes of HHCF.

- **Heat transfer of HHCF**

Heat transfer of HHCF can be considered as two heat processes as shown in Figure 2: (1) heat transfer in the heat collecting space and (2) heat transfer in the heat transfer space. Heat balance equations are established for these two spaces of the HHCF.

(1) For the heat collecting space,

$$c_a \rho_a V_{hcs} \frac{dT_{hcs}(\tau)}{d\tau} = Q_{wa\text{-}hcs}(\tau) + Q_{sig\text{-}hcs}(\tau) - Q_{hcs\text{-}hts}(\tau) \tag{1}$$

In Equation (1), $Q_{wa\text{-}hcs}$ is the convective heat transfer rate from the heat collecting wall to the air in the gap. $Q_{sig\text{-}hcs}$ is the heat transmission rate from the outdoor to the heat collecting space through the single-glass window, which considers the effect of the heat conduction of the single-glass window and the heat convection of both the interior surface and the exterior surface of the single-glass window. $Q_{hcs\text{-}hts}$ is the convective heat transfer rate from the heat collecting space to the heat transfer space. $Q_{wa\text{-}hcs}$, $Q_{sig\text{-}hcs}$, $Q_{hcs\text{-}hts}$ can be obtained by the following equations:

$$Q_{wa\text{-}hcs}(\tau) = h_{wa\text{-}hcs} A_{con,out}(T_{con,out}(\tau) - T_{hcs}(\tau)) \tag{2}$$

$$Q_{sig\text{-}hcs} = U_{sig} WH(T_{out}(\tau) - T_{hcs}(\tau)) \tag{3}$$

$$Q_{hcs\text{-}hts}(\tau) = c_a m(\tau)(T_{hcs}(\tau) - T_{hts}(\tau)) \tag{4}$$

The mean air flow rate between the heat collecting space and the heat transfer space (or between the heat transfer space and the indoor space) $m(\tau)$ is obtained by Bernoulli's equation [33], as given in Equation (5).

$$m(\tau) = \rho_a A_{cs} \sqrt{\frac{2gH}{\left(\xi_1\left(\frac{A_{cs}}{A_{va}}\right)^2 + \xi_2\right)} \cdot \frac{(T_{hcs}(\tau) - T_{hts}(\tau))}{|T_{hcs}(\tau)|}} \tag{5}$$

where the term $\left(\xi_1(A_{cs}/A_{va})^2 + \xi_2\right)$ represents the pressure loss of the heat collecting space. The ratio $(A_{cs}/A_{va})^2$ indicates the difference between the air velocity in the vents and the air velocity in the air gap.

(2) For the heat transfer space,

$$c_a \rho_a V_{hts} \frac{dT_{hts}(\tau)}{d\tau} = 2Q_{hcs\text{-}hts}(\tau) + Q_{sig\text{-}hts}(\tau) - 2Q_{hts\text{-}in}(\tau) - Q_{tic}(\tau) \tag{6}$$

where $Q_{sig\text{-}hts}$ is the heat transfer rate from the outdoor to the heat transfer space passing through the single-glass window.

$$Q_{sig\text{-}hts}(\tau) = U_{sig} LH(T_{out}(\tau) - T_{hts}(\tau)) \tag{7}$$

$Q_{hts\text{-}in}$ is the convective heat transfer rate from the heat transfer space to the indoor space passing through each open double-glass window, Q_{tic} is the conductive heat transfer rate from the heat transfer space to the indoor space through the closed double-glass window.

(a) When the double-glass window is closed, $Q_{hts\text{-}in}$, Q_{tic} can be expressed as

$$Q_{hts\text{-}in}(\tau) = 0 \tag{8}$$

$$Q_{tic}(\tau) = U_{doub} LH(T_{hts}(\tau) - T_r(\tau)) \tag{9}$$

(b) When the double-glass window is open, $Q_{hts\text{-}in}$, Q_{tic} can be expressed as

$$Q_{hts\text{-}in}(\tau) = c_a m(\tau)(T_{hts}(\tau) - T_r(\tau)) \tag{10}$$

$$Q_{tic}(\tau) = U_{doub}L_{clo}H(T_{hts}(\tau) - T_r(\tau)) \tag{11}$$

where $L = L_{clo} + L_{ope}$.

- **Heat transfer of building construction**

For the building construction, the heat transfer schematic is illustrated in Figure 3 and the implicit finite difference technique is employed to model the heat transfer process of building envelopes by solving the one dimensional transient heat conduction equation [34]:

$$\rho_{con}c_{con}\frac{\partial T_{con}(x,\tau)}{\partial \tau} = \lambda_{con}\frac{\partial^2 T_{con}(x,\tau)}{\partial x^2} \tag{12}$$

Subjected to the boundary conditions:

$$\text{at } x = 0, \; q_{out}(\tau) + h_{out}(T_{out}(\tau) - T_{con,out}(\tau)) = -\lambda_{con}\frac{\partial T_{con}(\tau)}{\partial x} \tag{13}$$

$$\text{at } x = D, \; q_{lw}(\tau) + q_{ts}(\tau) + h_{in}(T_r(\tau) - T_{con,in}(\tau)) = -\lambda_{con}\frac{\partial T_{con}(\tau)}{\partial x} \tag{14}$$

where q_{out} is the total thermal radiation heat flux from the outdoor. For example, the exterior surface of the heat collecting wall is heated by the solar energy, and the solar radiation heat flux can be calculated by

$$q_{out}(\tau) = \alpha \times SHGC_{sig}I(\tau) \tag{15}$$

For general building constructions, q_{out} could be expressed as

$$q_{out}(\tau) = \alpha \times I(\tau) \tag{16}$$

q_{lw} represents the long-wave radiation exchange from surrounding surfaces. The gray interchange model is used to calculate the thermal radiation heat exchange between interior surfaces. Meanwhile, the radiosity concept developed by Hottel and Sarofim is adopted [35]. The net radiative heat transfer at a surface can be determined by Equation (17).

$$q_{lw,i}(\tau) = \frac{A_i\varepsilon_i}{1 - \varepsilon_i}(\sigma T_i(\tau)^4 - J_i(\tau)) \tag{17}$$

where the radiosity, J, represents the sum of the gray body radiation of temperature and the incident radiation, and it can be expressed as Equation (18).

$$J_i(\tau) = \varepsilon_i\sigma T_i(\tau)^4 + (1 - \varepsilon_i)H_i(\tau) \tag{18}$$

The incident radiation, Ir, is normally unknown. If a certain surface i is hit by radiation from another surface j, the radiation heat energy incident on surface i [36] can be described as Equation (19).

$$Ir_i(\tau) = \frac{\sum_{j=1}^{N}F_{ji}A_jJ_i(\tau)}{A_i} \tag{19}$$

where F_{ji} is the view factor from surface j to i.

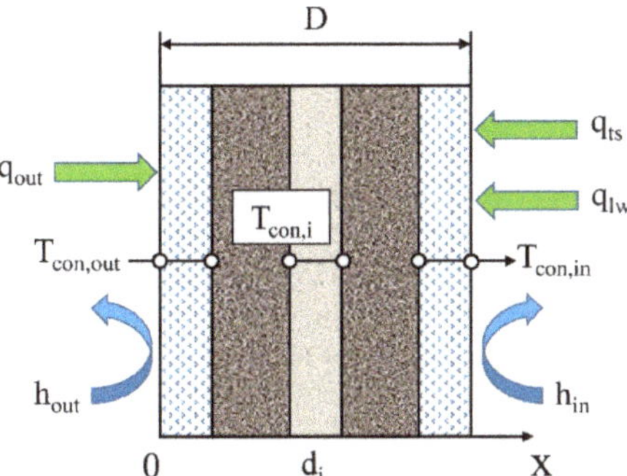

Figure 3. Heat transfer process of the building construction.

q_{ts} is the transmitted solar radiation through the fenestration absorbed by the interior surface. In this model, a simplified interior solar distribution model developed by Benjamin and Moncef Krarti [37] is employed to calculate the transmitted solar radiation absorbed by the floor and other interior surfaces, respectively.

The transmitted solar radiation absorbed by the floor can be expressed as

$$q_{ts,floor}(\tau) = \frac{\sum Q_{tsf,diffuse}(\tau) + (1 - \alpha_{floor})\sum Q_{tsf,direct}(\tau)}{\sum_{j=1}^{N} A_j} + \frac{\alpha_{floor}\sum Q_{tsf,direct}(\tau)}{A_{floor}} \tag{20}$$

The transmitted solar radiation absorbed by other interior surfaces excluding the floor can be expressed as

$$q_{ts,j}(\tau) = \frac{\sum Q_{tsf,diffuse}(\tau) + (1 - \alpha_{floor})\sum Q_{tsf,direct}(\tau)}{\sum_{j=1}^{N} A_j} \tag{21}$$

where Q_{tsf} is the transmitted solar radiation through the window.

Convective heat transfer coefficients of the exterior surface and the interior surface of the building construction can be calculated by the following equations [38]:

$$h_{out} = \frac{0.023}{\lambda_a}\left(\frac{V_m}{\nu_a}\right)^{0.891} l^{-0.109} \tag{22}$$

$$h_{in} = \begin{cases} 1.22(\rho_a^2 \Delta t/l)^{1/4}, & 10^{-1} < GrPr < 10^9 \\ 0.28(\rho_a^2 \Delta t/l)^{1/6} + 1.13(\rho_a^2 \Delta t)^{1/3}, & 10^9 < GrPr < 10^{12} \end{cases} \tag{23}$$

where Δt is the temperature difference between the air and the construction surface. V_m is the outside average wind velocity.

- **Heat balance of indoor air**

The indoor air heat balance model establishes the interaction among walls, the indoor air and the hybrid heat collecting facade, which can be written as

$$c_a \rho_a V_r \frac{dT_r(\tau)}{d\tau} = \sum_{i=1}^{N} Q_{con,in}^i(\tau) + Q_{win}(\tau) + Q_{leak}(\tau) + Q_{interheat}(\tau) + 2Q_{hts-in}(\tau) + Q_{tic}(\tau) \tag{24}$$

where $Q_{con,in}$ is the convective heat transfer rate from the interior surface of the building construction to the indoor air, Q_{win} is the heat transfer rate through other windows excluding the HHCF, Q_{leak} is the heat transfer rate by air leakage, $Q_{interheat}$ is the heat transfer rate from indoor heat sources including lighting, occupants, and equipment. $Q_{con,in}$, Q_{win}, and Q_{leak} are calculated by the following equations:

$$Q_{con,in}(\tau) = h_{in} \cdot (T_{con,in}(\tau) - T_r(\tau)) \cdot A_{con,in} \tag{25}$$

$$Q_{leak}(\tau) = c_a \rho_a V_r \cdot ACH \cdot (T_{out}(\tau) - T_r(\tau))/3600 \tag{26}$$

$$Q_{win}(\tau) = h_{win} \cdot (T_{out}(\tau) - T_r(\tau)) \cdot A_{win} \tag{27}$$

where $T_{con,in}$ is the interior surface temperature of the wall, the ceiling or the floor, $A_{con,in}$ is the interior surface area, ACH is the air change rate of the room, h_{win} and A_{win} are the heat transfer coefficient and the area of the window, respectively.

Figure 4 gives the heat transfer mechanisms and solving methods for various surfaces and spaces. In the developed model, the implicit finite difference method (FDM) is adopted to calculate the hourly thermal performance of the room with the HHCF [39]. The heat balance equations established for the HHCF, building constructions and the indoor air are converted into algebraic equations using the central difference scheme. To reduce the consumption of computational memory and the computational time, the Gauss-Seidel iteration method is adopted to calculate the indoor air temperature and surface temperatures as described in the reference [38]. It should be noted that for calculating the dynamic heating load of the building, the indoor air temperature will be fixed to the set-point temperature in the procedure.

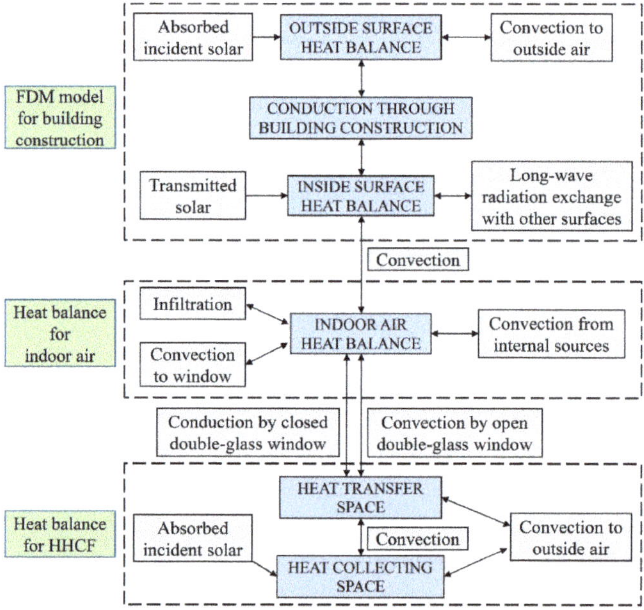

Figure 4. Heat balance solution procedure for the building with the HHCF.

2.2.3. Validation of Heat Transfer Model

Experiments are employed to validate the simulation model established in this paper. Figure 5 shows a 3-storey dormitory building with a HHCF in Ruoergai, China. In order to reduce the influence of outdoor conditions and the ground on the indoor air temperature, the middle room on the second floor of the building marked in Figure 5 was selected as the experimental room. Figure 6 shows the geometry of the test room, with the dimensions of 3.3 m (width) × 6.4 m (depth) × 3.0 m (height). The HHCF faces south, where two heat collecting walls with a width of 0.5 m and one double-glass window with a width of 2.3 m are located. The percentage of double-glass window area compared to the whole wall area is 69.7%. One door faces north with the size of 2.0 m (height) × 1.0 m (width). Detailed constructions of the room and thermal properties of the material are listed in Table 1.

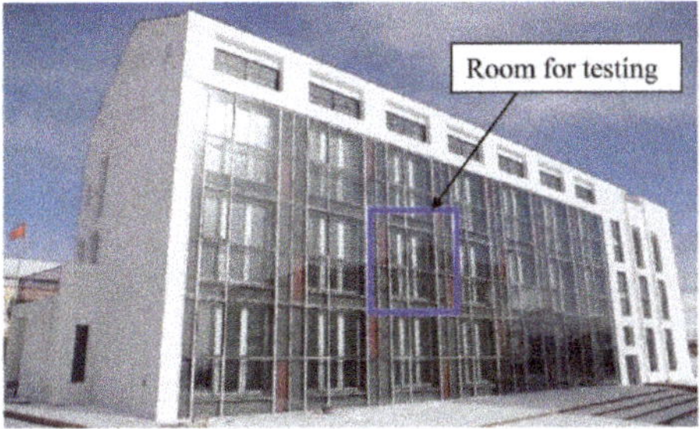

Figure 5. Test room of the dormitory building.

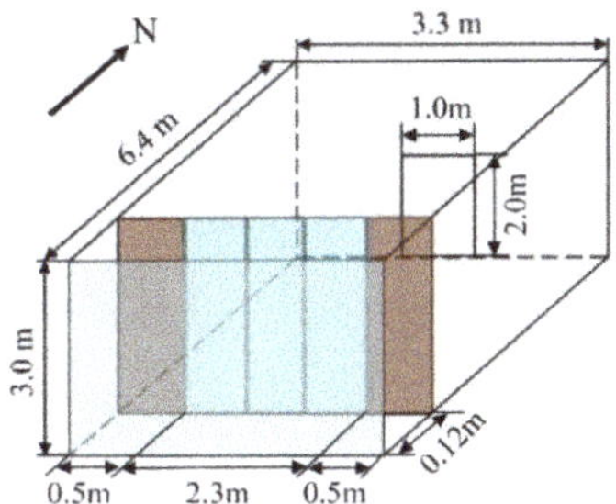

Figure 6. Geometry of the room studied.

Table 1. Constructions of the building.

Construction	Material	Thermal Conductivity (W/(m·K))	Heat Capacity (J/(kg·K))	Density (kg/m^3)
Exterior wall	80 mm polyurethane	0.033	1380	40
	240 mm brick wall	0.89	1000	1800
	20 mm Cement mortar	0.93	1050	1800
Interior wall	20 mm Cement mortar	0.93	1050	1800
	240 mm brick wall	0.89	1000	1800
	20 mm Cement mortar	0.93	1050	1800
Ceiling	20 mm Cement mortar	0.93	1050	1800
	180 mm concrete	1.74	920	2500
	20 mm Cement mortar	0.93	1050	1800
Door	60 mm wood	0.15	1630	608
HHCF	3 mm single-glass window, U = 5.56 W/(m^2·K); SHGC = 0.9; 120 mm-thickness air gap; 12 mm double-glass window, U = 2.83 W/(m^2·K); SHGC = 0.76; 0.5 m-width for each heat collecting wall (W = 0.5 m); Materials of the heat collecting wall is the same with the exterior wall Absorptivity of the heat collecting wall is 0.8; 2.3 m-width for double-glass window (L = 2.3 m)			

For the validation of the heat transfer model established above, indoor air temperature of the test room, air temperature of the heat transfer space, and interior surface temperature of the west interior wall were measured with the double-glass window open and closed. Meanwhile, environmental parameters including outdoor temperature, solar radiation intensity, and outdoor wind speed were also recorded for the input data in the simulation model. In addition, exterior surface temperatures of five interior constructions (three interior walls, the floor, and the ceiling) contacting with adjacent rooms were monitored, which would be used as boundary conditions in the simulation model.

Experiments were performed from 27 March to 2 April 2016. During the experimental period, the door was closed all the time, while the double-glass window was closed during the first five days and in the last two days it was open at 9:00–17:00 for each day. In the test room, no other space heating system is considered, and there is no internal heat gain from the lighting, occupants, and equipment.

To conduct the experimental test, thermocouples were used to monitor the indoor air temperature and various surface temperatures as seen in Figure 7. While outdoor temperature, solar radiation intensity and outdoor wind speed were measured by a small weather station as shown in Figure 8. The K-T method proposed by Klien and Theilacker [40] was adopted to convert the measured horizontal solar radiation into the incident solar radiation on the south wall.

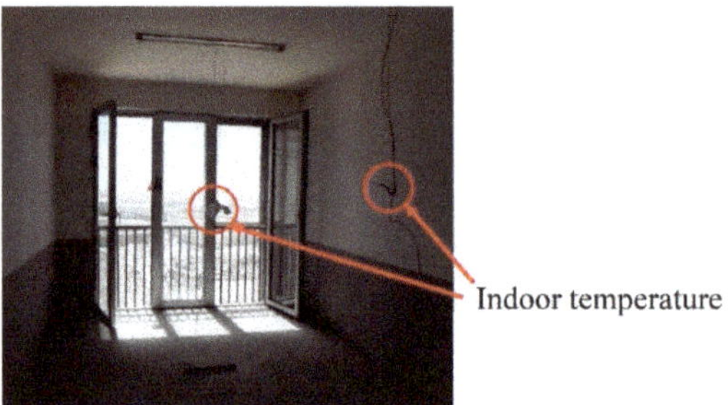

Figure 7. Indoor temperature test.

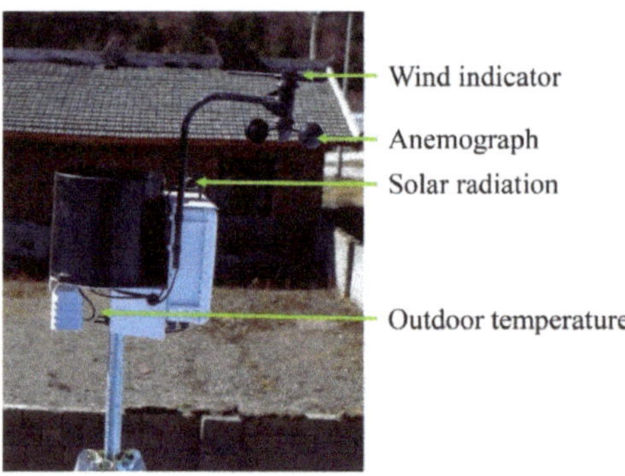

Figure 8. Meteorological data measurement.

In the transient simulation, the initial thermal inertia and initial conditions of the building have significant influences on the simulated results. In order to eliminate these influences, the monitored experimental data of the first three days are used for making the simulation stable and test results for the last four days are validated against results from the simulation. Figure 9 gives the comparative analysis of results of the test room obtained from the experimental test and the simulation. Comparing the mean indoor air temperature (Figure 9a), the air temperature of the heat transfer space (Figure 9b), and the interior surface temperature of the west wall (Figure 9c) under two different conditions (with double-glass window closed all the time and with it open at 9:00–17:00), it can be seen that the simulated results agree quite well with the experimental results under both conditions, which means the heat balance model established in this paper is enough accurate for evaluating the thermal performance of building with HHCF.

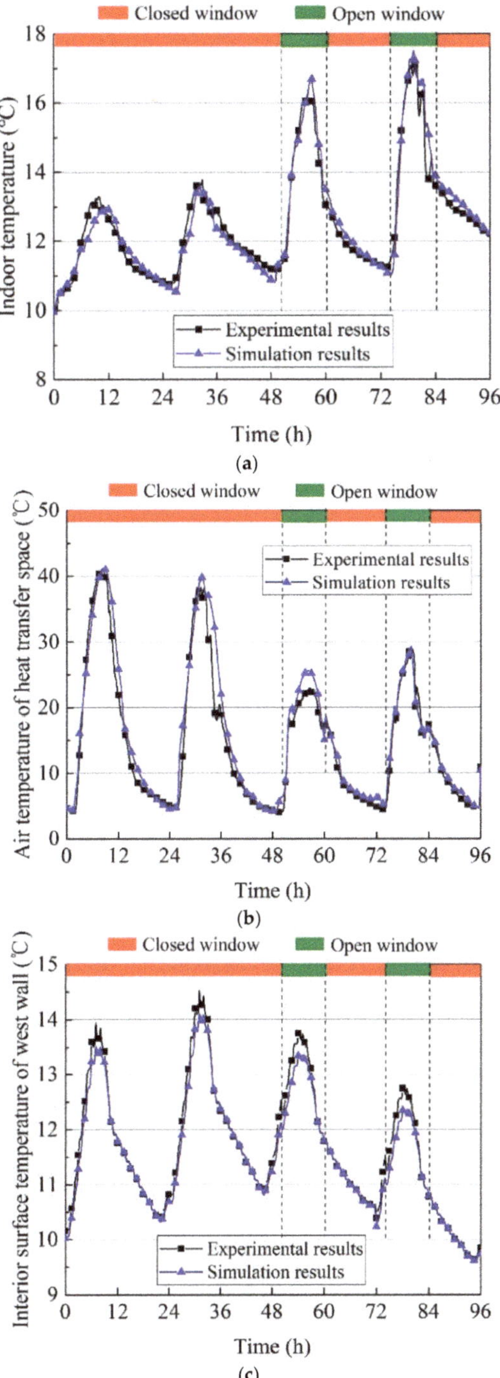

Figure 9. Comparison of Results from experiments and simulations. (**a**) Indoor air temperature; (**b**) Air temperature of heat transfer space; (**c**) Interior surface temperature of west wall.

2.2.4. Energy Saving Comparison of HHCF

In order to illustrate the energy saving potential of the proposed HHCF, the thermal performances of the above studied dormitory room using the HHCF and using other typical passive solar measures are compared. All the physical dimensions and thermal properties of building materials are the same as the experimental room except the passive solar design on the south wall. This typical room is assumed to be located in Ruoergai of China and the weather data are obtained from the typical year weather data of China based on the past 30 years' climate data [41].

Three cases used in the comparison are illustrated in Figure 10. Case 1 is the room with the direct solar heat gain window. In this case the sunlight directly passes through the window and enters into the indoor space to improve the thermal performance of the room. Case 2 is the room with the traditional Trombe wall. The heated air in the Trombe wall rises up with the effect of buoyancy effect and enters into the indoor space through the hole at the top, while the indoor air is sucked into the Trombe wall through the hole at the bottom. Case 3 is the room with the proposed HHCF in this paper.

For Case 3, the design of the HHCF is the same with the test room. In Cases 1 and 2, the direct solar heat gain window is the same with the double-glass window used in the HHCF, and the percentage of the direct solar heat gain window area compared to the exterior wall area is 69.7%. Unlike Case 1, the Trombe wall in Case 2 is installed on both sides of the south wall, each side with a width of 0.5 m. For the Trombe wall, the inlet and the outlet have the same dimensions of 0.25 m (width) × 0.15 m (height).

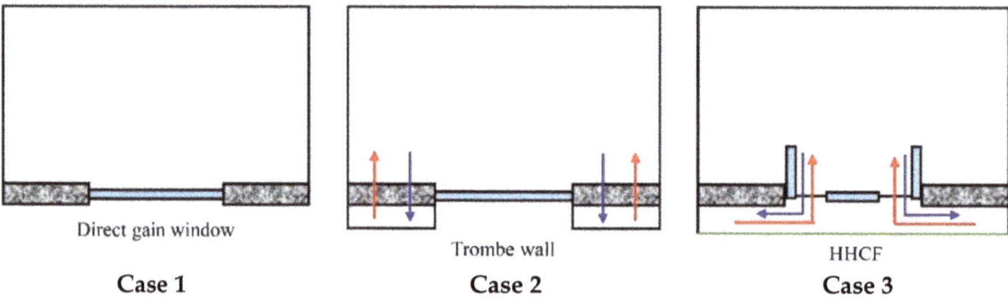

Figure 10. Three cases studied.

In order to compare the thermal performance of these three cases, both the indoor air temperature and the heating need of those cases are compared. It should be noted that Case 3 is investigated with the proposed simulation model and the other cases are simulated by TRNSYS [42]. The TRNSYS model for Cases 1 and 2 are shown in Figure 11a,b, respectively. In Figure 11a, Type56 building model is used to simulate the direct solar heat gain window. In Figure 11b, Type36 and Type56 are used to separately simulate the Trombe wall and the building. Meanwhile, only the heating season is considered and the heating season is from November 1st to March 31st of the next year. The heating schedule is from 18:00 to 8:00 of the next day during weekdays (from Monday to Friday) and the heating set-point temperature is 18.0 °C. While during the daytime of weekdays (from 8:00 to 18:00) and the weekend, no heating system runs and the indoor thermal performance largely depends on the passive solar measures such as the direct solar heat gain window, the Trombe wall and the HHCF.

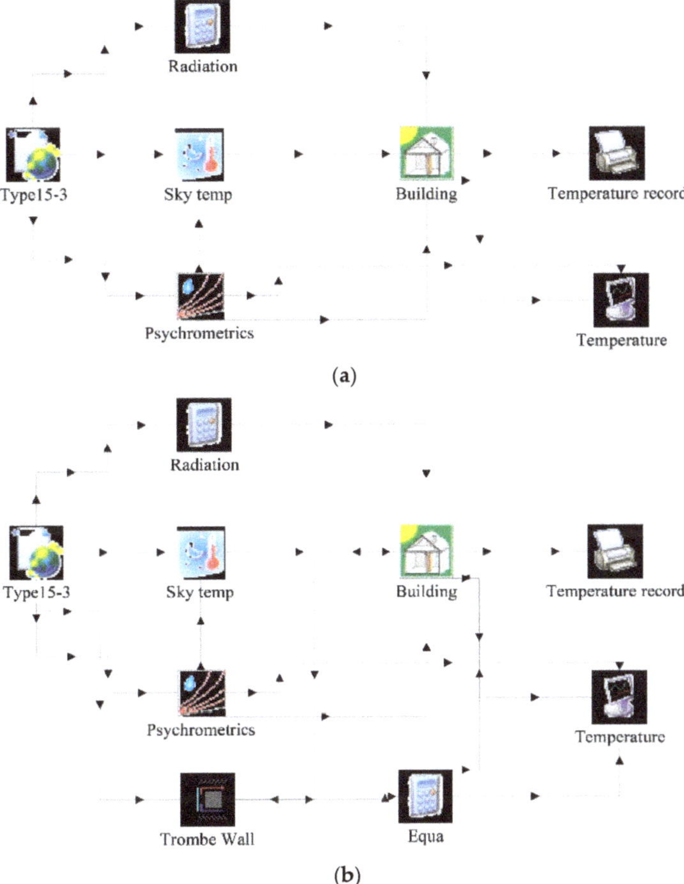

Figure 11. TRNSYS model for Case 1 and Case 2. (**a**) TRNSYS model for Case 1; (**b**) TRNSYS model for Case 2.

3. Results and Discussion

3.1. Energy Saving Potential of HHCF

Figure 12 shows the indoor air temperature and the heating need on the heating design day (21 January) predicted by the simulations. As seen in Figure 12, the studied room with the proposed HHCF (Case 3) has a higher temperature and a lower heating need than the other two passive measures (Cases 1 and 2). Table 2 summarizes the indoor air temperature and the total heating need of the studied room during the heating season. As shown in Table 2, during the heating season, the mean indoor air temperature for the studied room with the proposed HHCF is 18.6 °C, which is 1.1 °C and 0.4 °C higher than those of the direct solar heat gain window and the Trombe wall, respectively. Meanwhile, the total heating need of the studied room with the proposed HHCF in the heating season is 28.7 kWh/m^2. Compared with the conventional direct solar gain window, the HHCF reduces the total heating need of the room by 19.2 kWh/m^2 and the energy-saving efficiency reaches 40.2%. Even in contrast to the conventional Trombe wall, the HHCF also decreases the total heating need by 21.5%. The comparison results show that the HHCF proposed in this paper has very high energy saving potential.

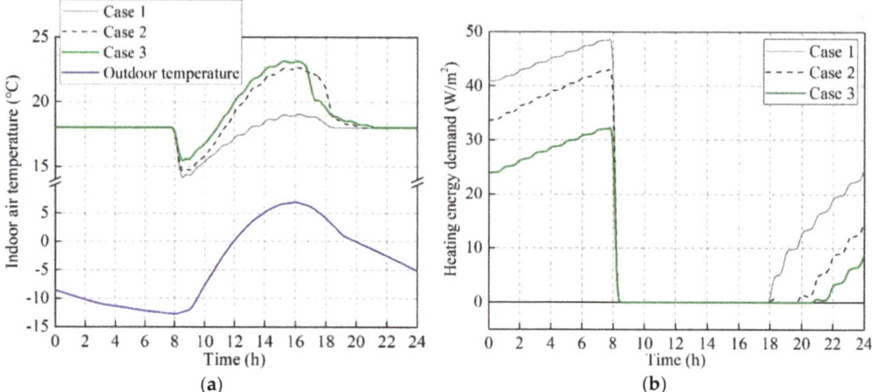

Figure 12. Comparisons of indoor air temperature and heating energy demand on heating design day. (a) Indoor air temperature for studied cases; (b) Heating energy demand for studied cases.

Table 2. Summary of indoor air temperature and total heating energy demand in heating season.

	Indoor Air Temperature (°C)			Total Heating Energy Demand (kWh/m^2)
	Mean	**Minimum**	**Maximum**	
Case 1	17.5	10.4	21.9	47.9
Case 2	18.2	10.5	26.7	36.6
Case 3	18.6	11.0	27.1	28.7

3.2. Parametric Study on the Thermal Performance of a Room with a HHCF

In order to analyze the thermal performance of the HHCF in a passive solar building, the dormitory located in Ruoergai as mentioned in Section 3 is still used as the simulated case. The weather data used in simulations are obtained from the typical year weather data of China [41]. Similarly, no other space heating system is considered in this building, and the internal heat gain from the lighting, occupants, and equipment are neglected.

The thermal performance of a room with a HHCF facing south is influenced by various factors, such as window operational schedule, absorptivity of heat collecting wall, thickness of air gap, window to wall ratio, solar heat gain coefficient and U-value of both single-glass windows and double-glass windows, etc. To analyze the effect of each factor on the thermal performance, a parametric study is carried out. Each factor is changed while the others kept constant. Simulated results are presented in this section.

3.2.1. Effects of Window Operational Schedule

For given climatic conditions, the operational schedule of inner double-glass windows of the HHCF is crucial to improve the building thermal performance. Opening inner double-glass windows too early or too late in the daytime will increase the heat loss of the single-glass window. Certainly, if inner double-glass windows are always closed in the daytime, the heat absorbed by the HHCF cannot be transferred to the indoor space effectively. Therefore, for a specific region, an optimal window operational schedule for the HHCF exists in order to maximize the indoor air temperature.

Through transient simulations, the optimal window schedule is determined by choosing the highest indoor air temperature among cases with different window operational schedules as showed in Table 3. The optimal window schedule for the plateau region located in western Sichuan Province is opening double-glass windows at 9:00 and closing them at 17:00.

Table 3. Indoor air temperature ranges under different window schedules.

	Schedule A	Schedule B	Schedule C	Schedule D
Time to open windows	With windows closed all day	8:00	9:00	11:00
Time to close windows		18:00	17:00	15:00
Indoor air temperature (°C)	6.7~13.2	8.2~19.0	8.6~19.0	7.8~17.5

Figure 13 shows the indoor air temperature under different window schedules. It can be found that the indoor air temperature with the double-glass window open is higher than that with the double-glass window closed, since the closed windows prevent the hot air in the air gap from flowing into the indoor space. Similarly, if opening double-glass windows too late or closing them too early (such as Schedule D), the heat transferred into the indoor space will be reduced as well. Furthermore, comparing Schedule C with Schedule B, it can be found that reducing the window opening hours decreases the minimum indoor temperature. The reason is that in the early morning or the late afternoon the solar radiation is very weak, the solar heat gain is less than the heat loss from the single-glass window to the outdoor.

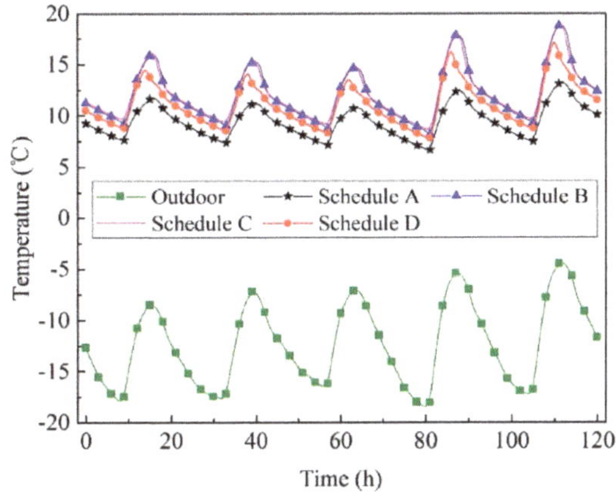

Figure 13. Air temperatures under diverse window schedules.

3.2.2. Effects of Width of Heat Collecting Wall

When the height of the collecting wall is constant, the area of heat collecting wall is determined by its width. In the south wall of the passive dormitory, the area of heat collecting wall increases with the decreasing area of the double-glass window. For the south wall, the solar energy can be used in two ways: one way is going through windows into the indoor space directly, the other way is being absorbed by the heating collecting wall and then heating the air in the gap. Different widths of the heating collecting wall are simulated to calculate the indoor air temperature in winter as showed in Figure 14. It can be seen that when the width of heat collecting wall is about 0.4 m, the mean indoor air temperature is higher than others. Furthermore, when the width of heat collecting wall is less than 0.4 m, the heat collecting wall is enough effective to capture more heat than windows. When the width of the heat collecting wall exceeds 0.4 m, the decrease of window area will largely reduce the solar energy transmitted directly into the indoor space, resulting in the drop of indoor air temperature.

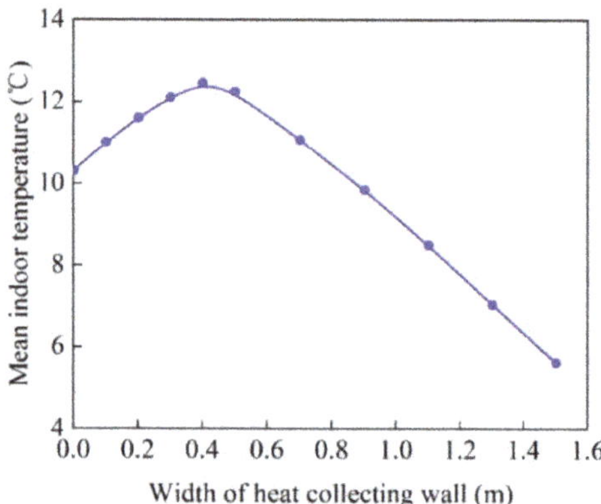

Figure 14. Mean indoor air temperature under different widths of heat collecting wall.

3.2.3. Effects of Absorptivity of Heat Collecting Wall

Absorptivity of the heat collecting wall is another important factor affecting the thermal performance of the HHCF. To illustrate the influence of absorptivity on the heating collecting wall, four optional surface coatings including the black paint, the blue paint, the brown paint and the green paint are selected for the comparison. The corresponding values of absorptivity for these four paints are 0.92, 0.88, 0.84, and 0.74 according to the reference [43], respectively. Figure 15 displays the indoor air temperature under different surface coatings of the heat collecting wall. It can be seen that when the absorptivity of the heat collecting wall increases by 0.04, the mean indoor air temperature has a rise of 0.18 °C. This is because the increase of absorptivity of heat collecting wall leads to more solar energy converted into thermal energy and subsequently the indoor air is heated.

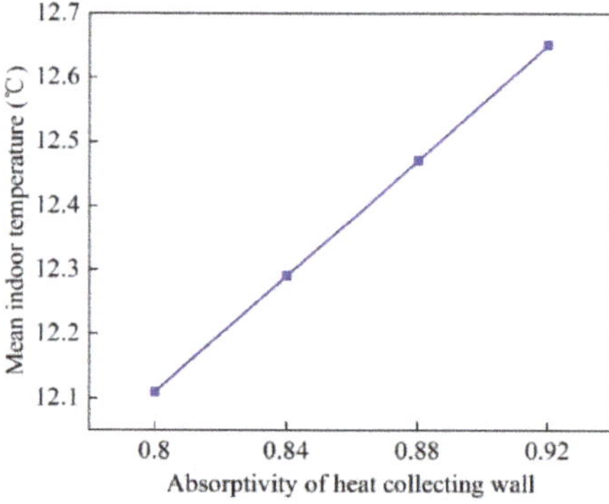

Figure 15. Indoor air temperature for different values of absorptivity of heat collecting wall.

3.2.4. Effects of Thermal Performance of Inner Double-Glass Window

For the HHCF, the thermal performance of the inner double-glass window plays an important role in influencing the indoor thermal environment of the building. For double-glass windows, both U-value and solar heat gain coefficient are the dominant parameters affecting the thermal performance of the building.

- **U-value**

For analyzing the influence of the U-value of the inner double-glass window, some optional real double-glass windows in the window database of TRNSYS [42] are used for simulations. Figure 16 displays the change of the mean indoor air temperature with the U-value of the inner double-glass window. The inner double-glass window with different solar heat gain coefficients of 0.298, 0.333, 0.440, 0.586, and 0.623 are studied as well. As depicted in Figure 16, when the solar heat gain coefficient is constant, the indoor temperature decreases with the increasing U-value. If the U-value increases by 0.1 W/(m^2·K), the mean indoor temperature will have a drop of 0.3 °C.

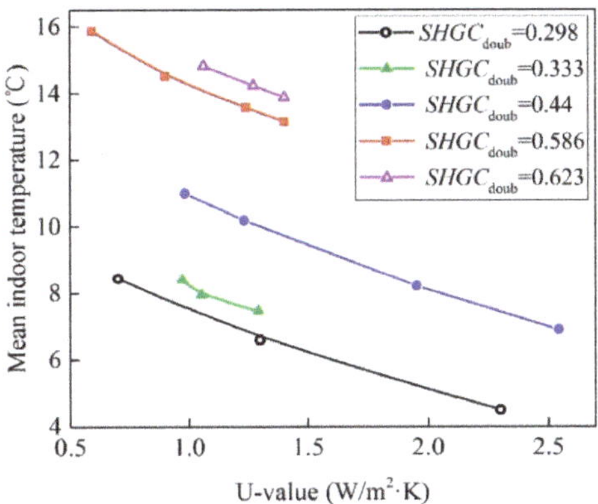

Figure 16. Mean indoor air temperature for various U-values of double-glass window.

● **Solar heat gain coefficient**

Figure 17 displays the change of the mean indoor temperature with the solar heat gain coefficient of the inner double-glass window, under the different U-values of 0.59, 1.05, 1.24, and 1.27 W/(m^2·K). As depicted in Figure 17, when the U-value is constant, the indoor temperature increases with the solar heat gain coefficient. If the solar heat gain coefficient increases by 0.1, the mean indoor air temperature has a rise of 2.1 °C. In practical applications, to reduce the heating energy use for the building, both a higher solar heat gain coefficient and a lower U-value are necessary for the double-glass window.

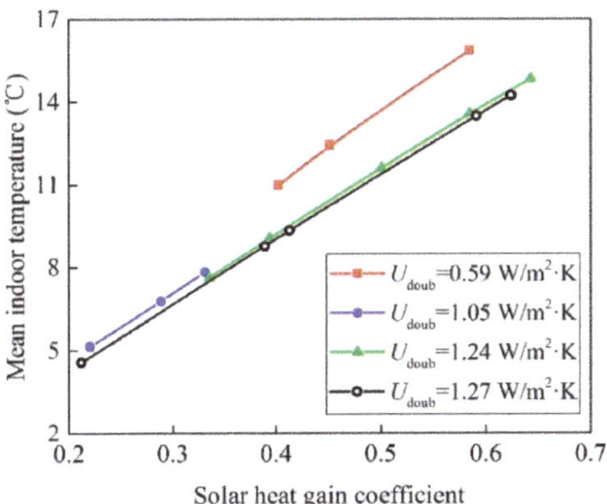

Figure 17. Mean indoor temperature under various solar heat gain coefficients.

3.2.5. Effects of Outer Single-Glass Window

For single-glass windows, both the U-value and the solar heat gain coefficient are still two important parameters affecting the thermal performance of the building. Generally, the U-value and the solar heat gain coefficient of single-glass windows are higher than those of double-glass windows. Compared with the double-glass window, the single-glass window allows more solar energy transferring into the building and meanwhile increases the heat transmission from the indoor to the outdoor. Figure 18 shows the change of the mean indoor air temperature with the U-value of the outer single-glass window. As shown in Figure 18, when the solar heat gain coefficient is constant, the indoor air temperature decreases with the increasing U-value. If the U-value increases by 1.0 W/(m²·K), the mean indoor temperature has a drop of about 0.17 °C. Similarly, if the solar heat gain coefficient increases by 0.1, the mean indoor temperature rises about 0.4 °C.

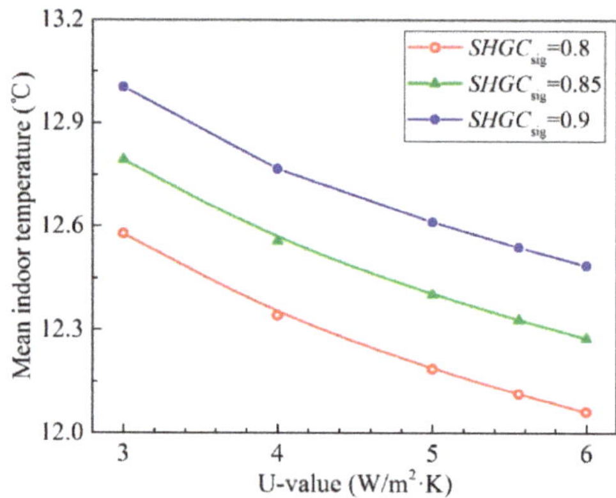

Figure 18. Mean indoor air temperature under various U-values of single-glass window.

3.2.6. Effects of Air Gap Thickness

Thickness of the air gap between the outer single-glass window and the inner double-glass window also has influences on the thermal performance of the HHCF. Figure 19 displays the relationship between the mean indoor air temperature and the thickness of the air gap. It can be seen that when the thickness of the air gap exceeds 30 mm, increasing the air gap thickness by 270 mm results in a drop of 1.24 °C for the indoor air temperature, which means the increase of air gap thickness decreases the thermal performance of the HHCF. Thus, in practical applications of the HHCF, we should try to reduce the air gap thickness as long as the building structure allows. However, according to the requirement of the building structure, the minimum thickness of the air gap should be more than 30 mm. Therefore, in order to improve the thermal performance of the HHCF, the thickness of the air gap should be designed as close as possible to 30 mm.

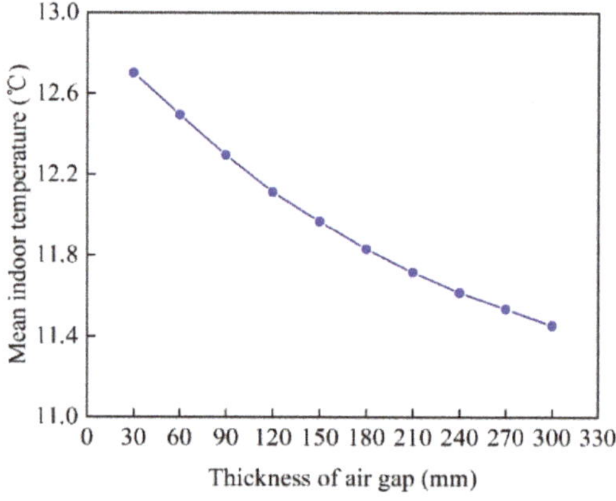

Figure 19. Mean indoor air temperature under various thicknesses of air gap.

3.3. Summary of the Presented Results and Discussion

Energy saving potential analysis mentioned above has shown that the HHCF is more energy efficient than the conventional direct solar gain window and the Trombe wall, which provides a novel efficient passive solar energy utilization form. It should be noticed that the HHCF is originally developed for making the most use of solar energy to reduce the heating demand for buildings located in the regions where it is cold in winter and mild in summer, such as the Qinghai–Tibet plateau of China. In these regions, building heating is necessary during the heating season, while in other seasons reasonable natural ventilation can meet the indoor comfort requirement due to the mild outdoor climate and there is no need for building cooling throughout the year.

In order to illustrate the performance of the HHCF, two similar studies performed by Evangelos Bellos et al. [44] and Zou Huifen [45] have been selected for comparsion. Evangelos Bellos et al. proposed an innovative Trombe wall as a passive heating system for a building in Athens. The predicted results shown that the mean indoor temperature for the building with the innovative Trombe wall is about 0.5 °C higher than that with the traditional Trombe wall. In the present study, the indoor temperature increase is about 0.4 °C, a value which is in the same level. The structure of the double-skin façade studied by Zou Huifen et al. is similar to the HHCF in this paper, and it can also decrease the heating energy consumption. Even compared with the plating Low-e film insulating glass curtain wall, the double-skin façade will decrease the energy consumption by 16.0%, while the HHCF in this paper can decrease the total heating need by 21.5%.

4. Conclusions

To improve the thermal performance of passive solar buildings, a hybrid heat collecting façade (HHCF) is proposed in this paper. The heat transfer model for analyzing the thermal performance of the building with the HHCF is established and validated by the experimental results. In order to illustrate the energy saving potential, the heating need of the building with the HHCF is compared with those of the conventional direct solar gain window and the Trombe wall, and results show that the HHCF can reduce the heating energy demand by 40.2% and 21.5%, respectively. A parametric study is performed to determine the thermal performance of the building with a HHCF under various design and operating conditions. It is found that the thermal performance of the HHCF mainly depends on the window operational schedule, the width and absorptivity of heat collecting wall, and the thermal performance of the inner double-glass window. However, other parameters (such as the thermal performance of outer single-glass window and the air gap thickness) of the HHCF could only increase the indoor air temperature by no more than 1.0 °C even within a wide range of these parameters. Apparently, to further improve the indoor air temperature, the U-value of the single-glass window and the thickness of the air gap should be designed as small as possible.

The HHCF mentioned in this paper is suitable for regions where it is cold in winter and mild in summer. If the HHCF is used in hot region, energy consumption performance should be analyzed on an annual basis, since reducing the heating consumption in winter might increase the refrigeration consumption in summer. At the same time, the HHCF in hot regions may cause indoor overheating due to higher outdoor temperature and more solar heat gain in summer. This needs further research to discuss the practicality of the HHCF in other climatic zones.

Author Contributions: Conceptualization, X.W.; Data curation, X.W.; Formal analysis, B.L. and T.Y.; Investigation, X.W. and H.B.; Methodology, X.W.; Project administration, B.L.; Supervision, B.L. and H.B.; Validation, X.W.; Writing—original draft, X.W.; Writing—review & editing, T.Y.

Funding: This research was supported by the Fundamental Research Funds for the Central Universities, Southwest Minzu University, grant number 2018NQN57.

Nomenclature

A	building envelope area (m^2)	Subscripts and superscripts	
ACH	air change per hour (1/h)	a	air
C	specific heat capacity (J/(kg·K))	con	building construction
D	total thickness of building construction (m)	con, in	interior surface of building construction
g	gravity acceleration (m/s^2)	con, out	exterior surface of building construction
Gr	Grashof number	clo	closed double-glass window
h	convective heat transfer coefficient (W/(m^2·K))	cs	horizontal cross section of the heat collecting space
H	height of HHCF (m)	diffuse	diffuse radiation
I	intensity of solar radiation (W/m^2)	direct	direct radiation
Ir	incident radiation (W/m^2)	doub	double-glass window
L	width of double-glass window (m)	hcs	heat collecting space
l	characteristic length (m)	hcs-hts	convection from heat transfer space to heat collecting space
m	mass flow rate of air in the heat collecting space (kg/s)	hts	heat transfer space
Pr	Prandtl number	hts-in	convection from heat transfer space to indoor space
q	heat flux (W/m^2)	in	indoor
q$_{ts}$	solar radiation absorbed by the interior surface (W/m^2)	interheat	indoor heat source of building
q$_{tsf}$	transmitted solar radiation through the fenestration (W/m^2)	leak	air leak from room
Q	heat transfer rate (W)	lw	long-wave radiation exchange
S	thickness of air gap (m)	ope	open double-glass window
SHGC	solar heat gain coefficient	out	outdoor
T	temperature (°C)	r	room
U	U-value of window (W/(m^2·K))	sig	single-glass window
V	volume (m^3)	sig-hcs	convection from single-glass single-glass window to air in the heat collecting space
V$_m$	average wind velocity (m/s)	sig-hts	convection from single-glass single-glass window to air in the heat transfer space
W	width of heat collecting wall (m)	tic	conduction through closed double-glass window from heat transfer space to indoor space
Greek symbols		va	outlet area of the vent
α	absorptivity of building surface	wa-hcs	convection from heat collecting wall to air in the gap
ρ	density (kg/m^3)	win	window
λ	thermal conductivity (W/(m·K))		
τ	time (s)		
ν	dynamic viscosity (Pa·s)		
ε	emissivity of building surface		
ξ$_1$	vent pressure loss coefficient		
ξ$_2$	air gap pressure loss coefficient		

References

1. Gil-Baez, M.; Padura, A.B.; Huelva, M.M. Passive actions in the building envelope to enhance sustainability of schools in a Mediterranean climate. *Energy* **2019**, *167*, 144–158. [CrossRef]
2. Wang, R.Z.; Zhai, X.Q. *Handbook of Energy Systems in Green Buildings*; Springer Nature: Berlin/Heidelberg, Germany, 2018.

3.	Athienitis, A.K.; Barone, G.; Buonomano, A.; Palombo, A. Assessing active and passive effects of façade building integrated photovoltaics/thermal systems: Dynamic modelling and simulation. *Appl. Energy* **2018**, *209*, 355–382. [CrossRef]

4.	Duan, S.P.; Jing, C.J.; Zhao, Z.Q. Energy and exergy analysis of different Trombe walls. *Energy Build.* **2016**, *126*, 517–523. [CrossRef]

5.	Barea, G.; Ganem, C.; Esteves, A. The multi-azimuthal window as a passive solar system: A study of Heat gain for the rational use of energy. *Energy Build.* **2017**, *144*, 251–261. [CrossRef]

6.	Schnieders, J.; Feist, W.; Rongen, L. Passive Houses for different climate zones. *Energy Build.* **2015**, *105*, 71–87. [CrossRef]

7.	Nguyen, A.T.; Tran, Q.B.; Tran, D.Q.; Reiter, S. An investigation on climate responsive design strategies of vernacular housing in Vietnam. *Build. Environ.* **2011**, *46*, 2088–2106. [CrossRef]

8.	Zhang, T.T.; Tan, Y.F.; Yang, H.X.; Zhang, X.D. The application of air layers in building envelopes: A review. *Appl. Energy* **2016**, *165*, 707–734. [CrossRef]

9.	Lebied, M.; Sick, F.; Choulli, Z.; Bouardi, A.E. Improving the passive building energy efficiency through numerical simulation—A case study for Tetouan climate in northern of Morocco. *Case Stud. Therm. Eng.* **2018**, *11*, 125–134. [CrossRef]

10.	Shen, C.; Li, X.T. Solar heat gain reduction of double glazing window with cooling pipes embedded in venetian blinds by utilizing natural cooling. *Energy Build.* **2016**, *112*, 173–183. [CrossRef]

11.	Egolf, P.W.; Amacker, N.; Gottschalk, G.; Courret, G.; Noume, A.; Hutter, K. A translucent honeycomb solar collector and thermal storage module for building façades. *Int. J. Heat Mass Transf.* **2018**, *127*, 781–795. [CrossRef]

12.	Stevanović, S. Optimization of passive solar design strategies: A review. *Renew. Sustain. Energy* **2013**, *25*, 177–196. [CrossRef]

13.	Alfredo, F.G. Analysis of the thermal performance and comfort conditions produced by five different passive solar heating strategies in the United States Midwest. *Sol. Energy* **2007**, *81*, 581–593.

14.	Sebald, A.V.; Clinton, J.R.; Langenbacher, F. Performance effects of Trombe wall control strategies. *Sol. Energy* **1979**, *23*, 479–487. [CrossRef]

15.	Kara, Y.A. Diurnal performance analysis of phase change material walls. *Appl. Therm. Eng.* **2016**, *102*, 1–8. [CrossRef]

16.	Jaber, S.; Ajib, S. Optimum design of Trombe wall system in Mediterranean region. *Sol. Energy* **2011**, *85*, 1891–1898. [CrossRef]

17.	Ihm, P.; Krartib, M. Design optimization of energy efficient residential buildings in Tunisia. *Build. Environ.* **2012**, *58*, 81–90. [CrossRef]

18.	Naylor, D.; Foroushani, S.S.M.; Zalcman, D. Free convection heat transfer from a window glazing with an insect screen. *Energy Build.* **2017**, *138*, 206–214. [CrossRef]

19.	Lou, S.W.; Li, D.H.W.; Lam, J.C.; Lee, E.W.M. Estimation of obstructed vertical solar irradiation under the 15 CIE Standard Skies. *Build. Environ.* **2016**, *106*, 123–133. [CrossRef]

20.	Parvin, S.; Ahmed, S.; Chowdhury, R. Effect of Solar Irradiation on Forced Convective Heat Transfer through a Nanofluid Based Direct Absorption Solar Collector. In Proceedings of the 7th BSME International Conference on Thermal Engineering, Dhaka, Bangladesh, 22–24 December 2016.

21.	Aksoy, U.T.; Inalli, M. Impacts of some building passive design parameters on heating demand for a cold region. *Build. Environ.* **2006**, *41*, 1742–1754. [CrossRef]

22.	Marinoski, D.L.; Guths, S.; Pereiara, F.O.R.; Lamberts, R. Improvement of a measurement system for solar heat gain through fenestrations. *Energy Build.* **2007**, *39*, 478–487. [CrossRef]

23.	Zhou, G.B.; Pang, M.M. Experimental investigations on the performance of a collector–storage wall system using phase change materials. *Energy Convers. Manag.* **2015**, *105*, 178–188. [CrossRef]

24.	Tian, Z.C.; Zhang, X.K.; Jin, X.; Zhou, X.; Si, B.H.; Shi, X. Towards adoption of building energy simulation and optimization for passive building design: A survey and a review. *Energy Build.* **2018**, *158*, 1306–1316. [CrossRef]

25.	Wei, H.; Hu, Z.T.; Luo, B.Q.; Hong, X.Q.; Sun, W.; Ji, J. The thermal behavior of Trombe wall system with venetian blind: An experimental and numerical study. *Energy Build.* **2015**, *104*, 395–404.

26. Gou, S.Q.; Nik, V.M.; Scartezzini, J.L.; Zhao, Q.; Li, Z.R. Passive design optimization of newly-built residential buildings in Shanghai for improving indoor thermal comfort while reducing building energy demand. *Energy Build.* **2018**, *169*, 484–506. [CrossRef]

27. Harkouss, F.; Fardoun, F.; Biwole, P.H. Passive design optimization of low energy buildings in different climates. *Energy* **2018**, *165*, 591–613. [CrossRef]

28. Larsen, S.F.; Rengifo, L.; Filippin, C. Double skin glazed façades in sunny Mediterranean climates. *Energy Build.* **2015**, *102*, 18–31. [CrossRef]

29. Samuelson, H.; Claussnitzer, S.; Goyal, A.; Chen, Y.J.; Alejandra, R.C. Parametric energy simulation in early design: High-rise residential buildings in urban contexts. *Build. Environ.* **2016**, *101*, 19–31. [CrossRef]

30. Owrak, M.; Aminy, M.; Jamal-Abad, M.T.; Dehghan, M. Experiments and simulations on the thermal performance of a sunspace attached to a room including heat-storing porous bed and water tanks. *Build. Environ.* **2015**, *92*, 142–151. [CrossRef]

31. Gong, X.Z.; Akashi, Y.; Sumiyoshi, D. Optimization of passive design measures for residential buildings in different Chinese areas. *Build. Environ.* **2012**, *58*, 46–57. [CrossRef]

32. Ruiz, M.C.; Romero, E. Energy saving in the conventional design of a Spanish house using thermal simulation. *Energy Build.* **2011**, *43*, 3226–3235. [CrossRef]

33. Klein, S.A.; Beckman, W.A.; Mitchell, J.W.; Duffie, J.A.; Duffie, N.A.; Freeman, T.L. *TRNSYS Manual*; University of Wisconsin-Madison: Madison, WI, USA, 2007.

34. Pantankar, S.V. *Numerical Heat Transfer and Fluid Flow*; Hemisphere: New York, NY, USA, 1980.

35. Hottel, H.C.; Sarofim, A.F. *Radiative Transfer*; McGraw-Hill: New York, NY, USA, 1967.

36. Cengel, Y.A.; Ghajar, A.J. *Fundamentals of Thermal-Fluid Sciences*; McGraw Hill: New York, NY, USA, 2001.

37. Park, B.; Krarti, M. Development of a simulation analysis environment for ventilated slab systems. *Appl. Therm. Eng.* **2015**, *87*, 66–78. [CrossRef]

38. Xiao, W. *Study of the Direct-Gain Solar Heating in Remote Southwest Tibet*; Tsinghua University: Beijing, China, 2010.

39. He, L.Q.; Ding, L.X. *The Calculation Basis of Thermal Physics for SOLAR Building*; Press of University of Science and Technology of China: Hefei, China, 2010.

40. Klien, S.A.; Theilacker, J.C. An algorithm for calculating monthly-average radiation on inclined surfaces. *J. Sol. Energy Eng.* **1981**, *103*, 29–33. [CrossRef]

41. Meteorological information center of China, Department of building technology science of Tsinghua university. *Special Meteorological Data Sets for the Analysis of the Thermal Environment of Chinese Buildings*; China building industry press: Beijing, China, 2005.

42. TRNSYS 18 Manuals, Component Mathematical Reference. Available online: http://sel.me.wisc.edu/trnsys/user18-resources/index.html (accessed on 1 November 2018).

43. Zhu, Y.X.; Yan, Q.S. *Built Environment*, 2nd ed.; China building industry press: Beijing, China, 2007.

44. Bellos, E.; Tzivanidis, C.; Zisopoulou, E.; Mitsopoulos, G. An innovative Trombe wall as a passive heating system for a buildingin Athens—A comparison with the conventional Trombe wall and theinsulated wall. *Energy Build.* **2016**, *133*, 754–769. [CrossRef]

45. Zou, H.F.; Fei, Y.C.; Yang, F.H.; Tang, H.; Zhang, Y.; Ye, S. Mathematical Modeling of Double-Skin Facade in Northern Area of China. *Math. Probl. Eng.* **2013**, *2013*. [CrossRef]

Article

Thermally Anisotropic Composites for Improving the Energy Efficiency of Building Envelopes [†]

Kaushik Biswas *, Som Shrestha, Diana Hun and Jerald Atchley

Oak Ridge National Laboratory, One Bethel Valley Road, Oak Ridge, TN 37831, USA; shresthass@ornl.gov (S.S.); hunde@ornl.gov (D.H.); atchleyja@ornl.gov (J.A.)
* Correspondence: biswask@ornl.gov; Tel.: +1-86-5574-0917

† This manuscript has been authored by UT-Battelle, LLC, under Contract No. DE-AC05-00OR22725 with the U.S. Department of Energy. The United States Government retains and the publisher, by accepting the article for publication, acknowledges that the United States Government retains a non-exclusive, paid-up, irrevocable, world-wide license to publish or reproduce the published form of this manuscript, or allow others to do so, for United States Government purposes. DOE will provide public access to these results of federally sponsored research in accordance with the DOE Public Access Plan (http://energy.gov/downloads/doe-public-access-plan).

Received: 9 September 2019; Accepted: 29 September 2019; Published: 5 October 2019

Abstract: This article describes a novel application of thermal anisotropy for improving the energy efficiency of building envelopes. The current work was inspired by existing research on improved heat dissipation in electronics using thermal anisotropy. Past work has shown that thermally anisotropic composites (TACs) can be created by the alternate layering of two dissimilar, isotropic materials. Here, a TAC consisting of alternate layers of rigid foam insulation and thin, high-conductivity aluminum foil was investigated. The TAC was coupled with copper tubes with circulating water that acted as a heat sink and source. The TAC system was applied to a conventional wood-framed wall assembly, and the energy benefits were investigated experimentally and numerically. For experimental testing, large scale test wall specimens were built with and without the TAC system and tested in an environmental chamber under simulated diurnal hot and cold weather conditions. Component-level and whole building numerical simulations were performed to investigate the energy benefits of applying the TAC system to the external walls of a typical, single-family residential building.

Keywords: thermal anisotropy; building envelope; thermal management; energy efficiency; peak load reduction

---

1. Introduction

Globally, buildings consume about 40% of the total energy, and are responsible for 30% of carbon dioxide emissions [1]. In building envelope systems, thermal management is important from both energy conservation and thermal comfort perspectives [1–3]. Thermal management to reduce unwanted heat flows through the opaque building envelope sections (walls, roof, and foundation) has traditionally been done via insulation materials. Alternative methods that have been proposed include thermal mass, solar control and shading, ventilation, etc. [2,3]. Phase change materials (PCMs), used as latent thermal storage technologies, have shown the potential for reductions in envelope-generated heating and cooling loads [4,5], but systematic studies evaluating the benefits of PCMs in large-scale, real building applications are missing. Vacuum insulation panels (VIPs) and aerogels are among the new generation of high-performance insulation materials being investigated for building envelope applications [6–8], but suffer from high cost and/or durability-related questions.

This study investigates the feasibility of applying thermal anisotropy [9] for improved thermal management in building envelopes. Unlike isotropic materials, the thermal conductivity of anisotropic materials is different in different directions, enabling preferential heat transfer in one direction compared to another. Thermal anisotropy allows heat to dissipate in a preferential direction, and has been investigated for improved heat dissipation and hot spot remediation in electronics [10–13]. Suszko and El-Genk [11] numerically investigated thermally anisotropic composite heat spreaders comprised of two 0.5 mm thick copper (Cu) laments separated by a thin (0.25–1.0 mm) layer of graphite to achieve in-plane thermal conductivities of 200–325 W/(mK) and cross-plane conductivities of 5–20 W/(mK). The composite spreaders were predicted to remove up to 429% higher heat than an all-Cu spreader [11]. Conversely, Ren and Lee [13] utilized the high cross-plane conductivity and low in-plane conductivity of holey silicon nanostructures to achieve improved thermoelectric cooling effectiveness compared to high-thermal conductivity bulk silicon.

The goal of the current research was to investigate if preferential heat transfer using thermal anisotropy can be used for a more effective thermal management of building envelopes compared to passive insulation systems. Narayana and Sato [14] and Vemuri and Bandaru [15], among others, proposed that a practical approach to creating anisotropy is to build a stacked composite from macroscopic layers of isotropic materials. The reasoning is that in a composite made of alternating sheets of two materials, the overall conductivity parallel to the layers is the arithmetic mean (AM) of the individual conductivities of the layers, while the overall conductivity perpendicular to the layers is the harmonic mean (HM) of the individual conductivities [15]. Since HM ≤ AM, the resultant composite will exhibit thermal anisotropy, and the degree of anisotropy can be tuned based on the selection of the isotropic materials and their geometric orientation [15].

Buildings, as major energy end users, have great potential to relieve the stress of the power surplus/shortage of the electric grid by shifting their loads from on-peak to off-peak periods [16]. Furthermore, with an increasing focus on net-zero energy buildings and stringent future carbon emission targets, envelopes need to switch from passive to dynamic and "responsive" systems [17,18]. For this study, the authors investigated the efficacy of a thermally anisotropic composite (TAC) comprising of alternating layers of rigid extruded polystyrene (XPS) foam sheets and thin aluminum (Al) foil in reducing the heat transfer through a wall system compared to insulation materials alone. The specific TAC materials were chosen because XPS is a commonly used building insulation material, and Al foil is another low-cost, readily available material. In fact, Al foil-faced insulation materials are commercially available. The system design was guided by some preliminary finite element analysis using COMSOL Multiphysics® (https://www.comsol.com/heat-transfer-module) (version 5.4, COMSOL, Inc., Burlington, MA, USA). The Al foil layers were connected to copper (Cu) tubes circulating water that acted as heat sinks and sources, thus providing a dynamic and controllable system. The presence of the Al foils provides very high in-plane conductance compared to the cross-plane conductance across the thickness of the TAC. Figure 1 graphically describes the concept of a TAC diverting heat to a heat sink.

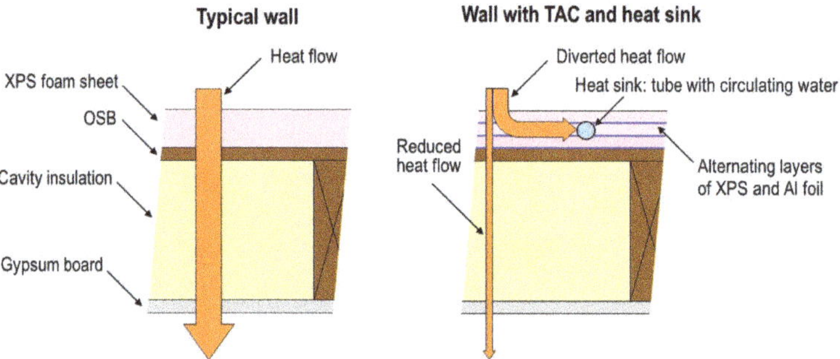

Figure 1. Conceptual operation of a thermally anisotropic composite (TAC)–heat sink combination applied to a wall to reduce heat gains during hot outdoor conditions.

The coupled TAC–heat sink/source concept is promising in terms of reducing energy consumption as well as reducing and/or shifting peak loads. This proof-of-concept research involved experimental evaluations of a full-scale 2.44 × 2.44 m^2 wall in an environmental chamber as well as component-level and whole-building simulations to estimate the energy benefits of the TAC combined with heat sinks/sources. The component-level simulations were performed using COMSOL, and the whole-building simulations were performed with EnergyPlus (E+) (https://energyplus.net/).

2. Materials and Methods

2.1. Test Walls, TAC, and Heat Sink/Source

Three test walls were created for testing and evaluation. The baseline ("Base") test wall consisted of typical wood-framed assemblies that are used in residential buildings in North America. The walls were built with wood studs of 3.8 cm width and 8.9 cm depth that were spaced 0.4 m apart. The wall cavities, i.e., the spaces between the wood studs, were filled with fiberglass insulation and were covered by 1.3 cm thick oriented strand board (OSB) and gypsum board as exterior and interior sheathings, respectively. Next, the baseline wall was upgraded by adding three layers of 1.3 cm thick XPS as exterior insulation to create the "Base + XPS" test wall. Finally, a third test wall ("Base + TAC") was created by adding the TAC and Cu tubes to the baseline wall. Figure 2 shows a COMSOL-generated schematic of the cross-section of the "Base + TAC" wall. For clarity, only the half-width of the wall is shown.

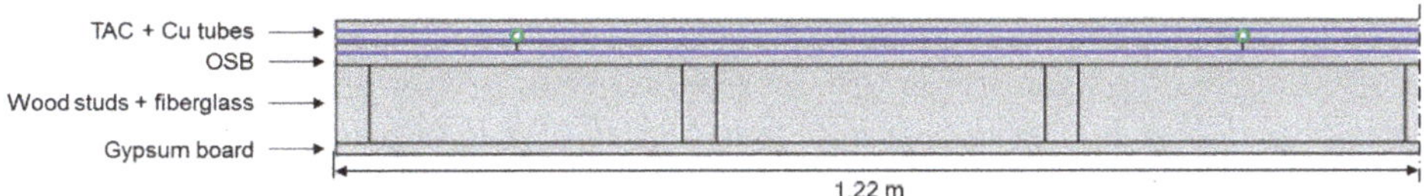

Figure 2. COMSOL rendering of the schematic cross-section of the "Base + TAC" wall.

The TAC consisted of three alternating layers of 1.3 cm thick XPS sheets and 0.13 mm thick Al foil, for a total of six layers. The solid blue lines in Figure 2 indicate the Al foil layers that create the thermal anisotropy and are connected to the Cu tubes, highlighted by the green circles. The external and internal diameter of the Cu tubes were nominally 12.7 mm and 10.9 mm, respectively. Three Cu tubes were installed along alternate wall cavities. Figure 3 shows the physical assembly and construction of the TAC and Cu tubes system. Each Al foil layer was installed and attached to the Cu tubes using Al foil adhesive tape before the corresponding XPS layer was added above. The process was repeated twice to install the three layers each of XPS and Al foil.

2.2. Environmental Chamber and Test Conditions

The experimental evaluations were performed in an environmental chamber called the large scale climate simulator (LSCS). The LSCS consists of three chambers—climate, meter, and guard—as shown in Figure 4. The climate chamber is above ground and simulates outdoor weather conditions of temperature, humidity, irradiance (using infrared or IR lamps), and wind speed. It can maintain steady conditions or simulate diurnal weather conditions. The lower portion of the LSCS contains a guard chamber and a meter chamber. Test specimens are installed at the interface of the climate and meter/guard chambers. The meter chamber is surrounded by the guard chamber on five sides, except for the side facing up, which is exposed to the test specimen. The meter and guard chambers simulate indoor temperature and humidity. The LSCS serves as a guarded hot box apparatus, and tests are performed in accordance with ASTM C1363 [19].

Figure 3. Creating the TAC and Cu tubes system for the test wall. (**a**) Cu tubes installed above the oriented strand board (OSB) that was covered with a water and vapor resistive membrane, (**b**) addition of the first Al foil and rigid extruded polystyrene (XPS) layers, (**c**) second and third XPS layers sandwiching the third Al foil layer, and (**d**) finished TAC and Cu tubes system.

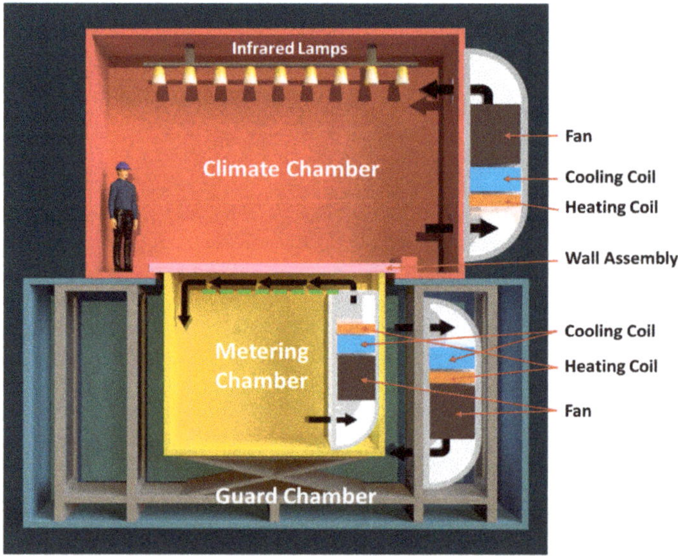

Figure 4. Sketch of the large scale climate simulator (LSCS).

The test wall assemblies were installed horizontally such that their exterior surfaces (OSB or XPS) were exposed to the climate chamber, and their interior surfaces were exposed to the meter chamber. The wall assemblies were supported at the open end of the guard and meter chambers. The meter chamber can be raised and lowered to accommodate the test specimens. During testing, the edge of the meter chamber is sealed against the indoor side of the wall assembly, and provides a measurement of the total heat flow through the 2.44×2.44 m^2 central measurement area of the test walls. The surrounding guard temperature is maintained close to the meter chamber temperature to minimize heat flows across the meter chamber walls. Thus, the heat flow through the test wall can be calculated from an energy balance of the meter chamber, i.e., the heat input to or removal from the meter chamber to maintain the "indoor" temperature. Reduction in the net heat transfer between the climate and meter chambers was evaluated to prove the efficacy of the proposed TAC with heat sink/source concept.

The climate chamber of the LSCS was programmed to impose a diurnally varying cyclic temperature and IR irradiance on the exterior side of the test wall assemblies. The meter chamber was nominally set to an indoor or "room" condition of 23.9 °C. The heaters, chillers, and fans associated with the meter chamber are controlled to maintain the indoor temperature as near as 23.9 °C as possible, while compensating for the heat gain or loss through the test walls. The walls were instrumented to measure the surface temperatures on the exterior and interior surfaces. For the tests with the TAC walls, a water pump and chiller were used to circulate water through the Cu tubes. The tests were performed under assumed summer and winter environmental conditions. Under summer conditions, the water circulation in the Cu tubes acted as heat sinks with an average inlet water temperature of about 27.8 °C. Under winter conditions, the circulating water acted as heat sources with an average inlet temperature of about 20.5 °C. The sink and source temperatures were based on some preliminary numerical simulations using COMSOL that showed that relatively mild sink/source temperatures can enable significant reductions in heat transfer through the wall.

2.3. Numerical Simulation Methods and Tools

To further evaluate the energy benefits of TACs combined with heat sinks/sources, both component-level and whole building numerical simulations were performed. First, geometries matching the LSCS test walls were created using COMSOL, and model validation simulations were performed. Two-dimensional (2D) geometries were created to match the cross-sections of the different test walls (similar to Figure 2) and utilizing appropriate materials properties, as listed in Table 1. In the "Base" wall model, the OSB was the outmost layer, and the "Base + XPS" wall model only contained the XPS sheets as the exterior insulation. The material properties were obtained from the ASHRAE Handbook of Fundamentals [20], except for the properties of Cu and Al, which were taken from the COMSOL material library.

Table 1. Material properties used in the simulations.

Material	Thermal Conductivity [k, W/(m·K)]	Density [ρ, kg/m^3]	Specific Heat [cp, J/(kg·K)]
Gypsum	0.159	640.7	879.2
Wood stud	0.144	576.7	1632.9
Fiberglass	0.039	7.8	837.4
OSB	0.130	656.8	1884.1
XPS	0.029	32.0	1465.4
Aluminum	238	2700	900
Copper	400	8960	385

The model validations were performed using the transient test data from the LSCS. The measured surface temperatures on the exterior wall surface and the meter chamber air temperatures were used as boundary conditions for these simulations. The heat transfer between the interior wall surface and meter chamber air was calculated using a surface heat transfer coefficient ('h_{meter}') for non-reflective horizontal surfaces according to the ASHRAE Handbook of Fundamentals [20]. 'h_{meter}' was 9.26 W/(m^2·K) for heat flow in the upward direction, and 6.13 W/(m^2·K) for heat flow in the downward direction [20]. The measured and calculated heat flows between the climate and meter chambers were compared to validate the models. For the "Base + TAC" model, the heat sink/source represented by the Cu tubes were modeled using an internal convection heat transfer coefficient (h_{int}) and a mean water temperature. The mean water temperature was based on the measured inlet and outlet water temperatures in the Cu tubes. 'h_{int}' was calculated using the following correlation for a laminar, fully developed internal flow under constant heat flux conditions [21].

$$\frac{h_{int}D}{k} = 4.36 \tag{1}$$

In Equation (1), 'D' is the internal tube diameter, and 'k' is the conductivity of the fluid (water). Equation (1) is valid for Reynolds numbers (Re) < 10,000 [21]. For convection heat transfer calculations in internal flows, the assumption of constant heat flux conditions applies to cases where the outer tube surface is uniformly heated or irradiated [21]. This assumption of constant heat flux is deemed reasonable for the current situation, because the exterior surfaces of the test walls were irradiated by an array of uniformly distributed IR lamps.

Following the 2D model validation, annual simulations were performed using COMSOL and E+ using typical climate conditions for two US cities, Phoenix, AZ and Baltimore, MD, which lie in climate zones 2 and 4, respectively, according to ASHRAE 90.1 [22]. Figure 5 illustrates the overall modeling and analysis process used for the annual simulations.

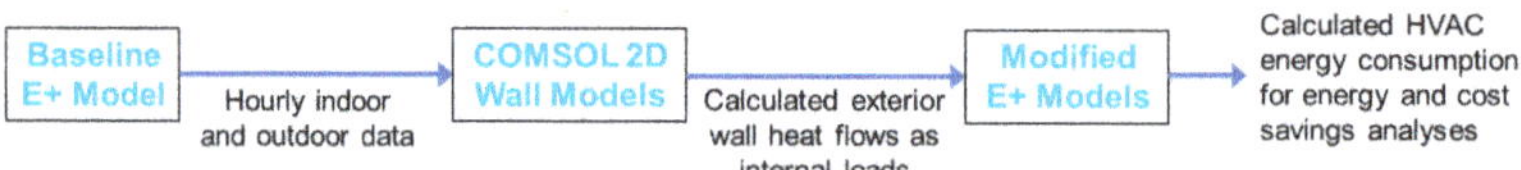

Figure 5. Flowchart illustrating the modeling and analysis process for the annual simulations.

The first step of the annual simulation methodology was to build an E+ baseline whole building model, based on a residential prototype building model [23]. The building used for this analysis is a wood-framed, two-story single-family detached house with a total conditioned floor area of about 223 m². The baseline whole-building E+ model was simulated to generate the indoor and outdoor boundary conditions for the 2D COMSOL models, based on the climate conditions of Phoenix and Baltimore. The exterior surface boundary condition included the impacts of solar irradiance as well as convection and radiation heat transfer with the outdoor environment. The internal boundary conditions included convection and radiation heat transfer with the interior space. The room temperatures for both Phoenix and Baltimore were assumed to be within the heating and cooling set points of 22.2 °C and 23.9 °C, respectively, from the residential prototypes model. The parameters of outdoor and indoor temperatures, solar irradiance, and exterior and interior convection coefficients were generated from the baseline E+ model. An assumption was made that exterior and interior heat transfer coefficients are independent of the wall construction. This is a reasonable approximation based on the research team's experience with whole-building simulations, which has shown that wall construction has a trivial impact on the calculated surface heat transfer coefficients.

Next, the simplified 2D COMSOL models were used to calculate the wall-generated heating and cooling loads from the "Base", "Base + XPS", and "Base + TAC" wall configurations. Figure 6 shows the simplified 2D model of the cross-section of the "Base + TAC" wall for the annual simulations. The simplified models include only two wall cavities and symmetric boundary conditions, i.e., the model assumes that the geometry is repeated symmetrically on either side. The wall components and properties were the same as the 2D validation models.

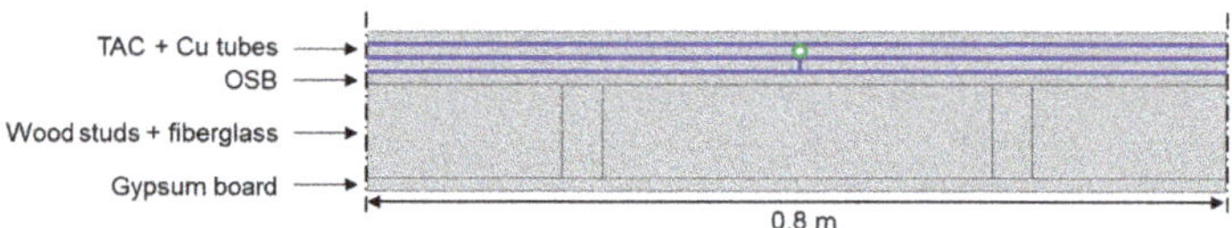

Figure 6. Simplified 2D COMSOL model used for annual simulation with symmetric boundary conditions at the two edges.

In the "Base + TAC" model, a Cu tube was assumed to be installed along the center of one of the two modeled wall cavities, same as the LSCS test wall. The internal convection coefficient within the tube was calculated using Equation (1). The Cu tubes were assumed to switch between heat sinks and heat sources during cooling and heating periods, respectively. The mean water temperature was assumed to be 27.8 °C in heat sink mode and 20 °C in heat source mode. For the current work, the switch from heat sink to heat source was assumed when the outdoor ambient temperature fell below 12.8 °C and vice versa. For this study, 12.8 °C was assumed to be the balance point temperature (BPT) for the modeled building. The BPT varies based on building and climate conditions [24], but 12.8 °C is deemed a suitable approximation for the current proof-of-concept simulations.

Finally, the hourly heat flows between the walls and the indoor space calculated from COMSOL were inserted into modified residential prototype E+ models as internal loads. The residential prototype building model used for the E+ simulations was a single-family detached house that complied with International Energy Conservation Code (IECC) 2006. The coefficient of performance (COP) of the cooling system was assumed to be 3.97 at standard design conditions. A gas furnace was modeled to provide heating, and the gas burner efficiency is assumed to be 0.78.

The E+ models were modified to make the opaque wall sections adiabatic, while the other aspects of the prototype models remained the same. The COMSOL-calculated heat gains and losses through the 2D model of the opaque wall sections were treated as cooling and heating loads to the corresponding conditioned space, respectively, in the modified E+ models. This modification was necessary, as E+ utilizes a one-dimensional conduction heat transfer for wall assemblies, and it is not possible to capture the 2D nature of the directional heat flows in the TAC in E+. Thus, this modified approach was used to calculate the heat flows through the opaque wall sections using 2D COMSOL models, and then incorporate them within modified E+ models as wall-generated cooling or heating loads to evaluate the energy benefits of using TACs.

3. Results

3.1. LSCS Experimental Evaluations

The climate chamber conditions were programmed to create 24-h diurnal cycles of air temperature and IR irradiance (IRR) for assumed summer and winter conditions. The goal of these proof-of-concept tests was to demonstrate the potential of TAC to reduce the heat transfer through the test wall. Each test was performed over multiple diurnal cycles, and data from 72 h (or three 24-h cycles) were used for evaluation and comparison. Figures 7 and 8 show the diurnally varying climate chamber air temperature and the IRR on the exterior surfaces, respectively, of the test walls for the assumed summer and winter conditions. The LSCS was programmed to create the same diurnal conditions for the three test walls as much as possible, but some differences were observed between the different test walls. Figure 9 shows the average inlet and outlet temperatures of the water circulating through the three Cu tubes in "Base + TAC" wall. The water circulation rate varied between 0.053 and 0.057 kg/s through each Cu tube.

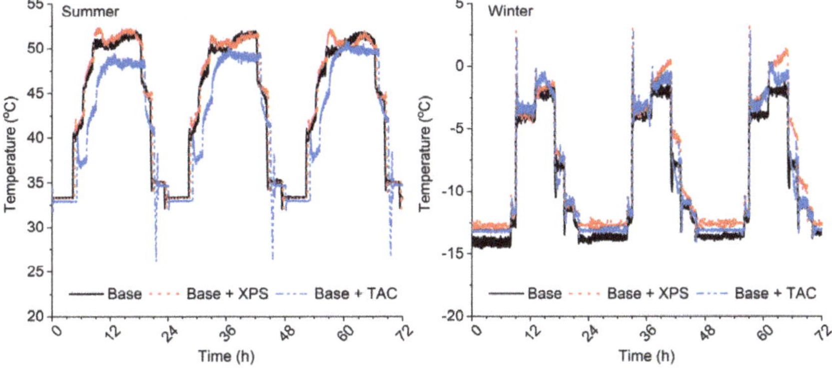

Figure 7. Air temperature in the climate chamber.

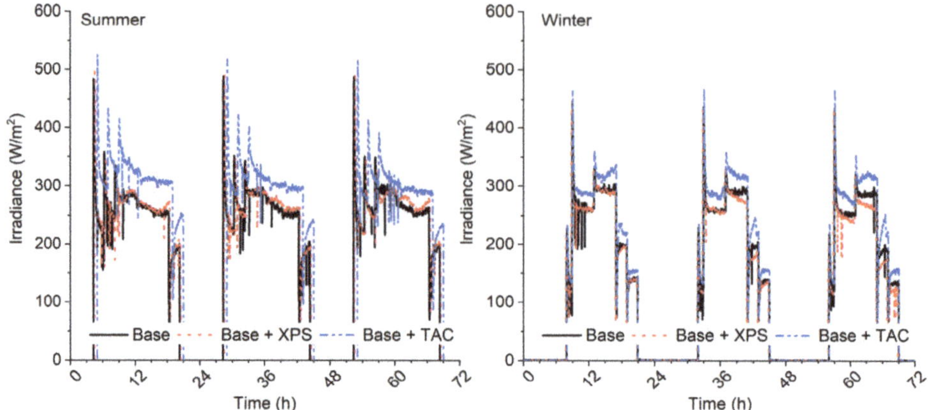

Figure 8. IR irradiance on the exterior wall surfaces.

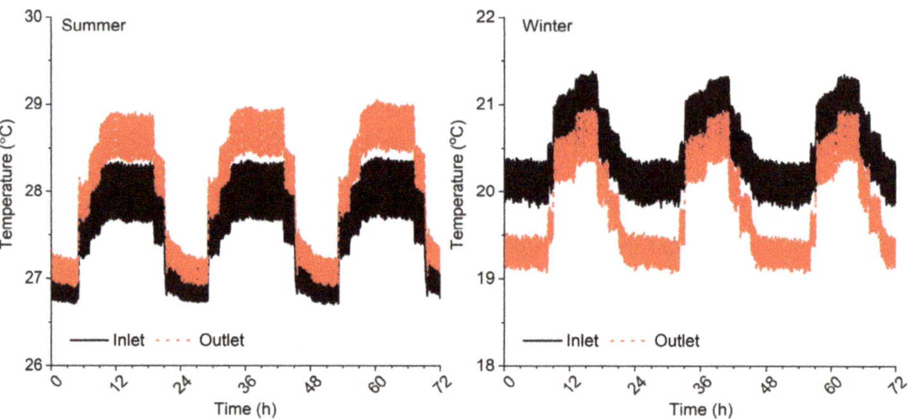

Figure 9. Average inlet and outlet water temperatures in the Cu tubes for the "Base + TAC" tests.

Finally, Figure 10 shows the net heat transfer between the climate and meter chamber through the different test walls, which is measured by the net heating or cooling power (Q_{meter}) needed to maintain the meter chamber at or near the "room" temperature of 23.9 °C. A positive Q_{meter} indicates net heat flow from the climate to meter chamber (heat gain) or a cooling load, while negative Q_{meter} indicates heat loss from the meter chamber or a heating load. Figure 10 clearly shows the effectiveness of the TAC and heat sink/source system in reducing both peak cooling and peak heating loads compared to both the baseline wall and "Base + XPS" wall. Under summer conditions, the peak cooling loads were reduced by 43.4% with the "Base + XPS" wall and by 79.5% with the "Base + TAC" wall compared to the "Base" wall. Under winter conditions, the reductions in peak heating loads were 42.1% and 63.7% with the "Base + XPS" and "Base + TAC" walls, respectively.

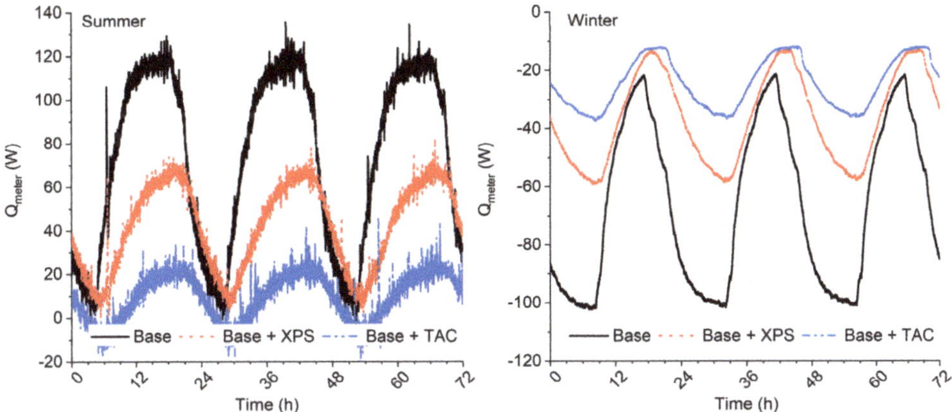

Figure 10. Measured heat gains to and heat losses from the meter chamber.

Table 2 compares the integrated cooling and heating loads over the 72-h summer and winter periods as well as the percent reductions in the loads with the addition of XPS only and the TAC + heat sink/source system to the baseline wall. The LSCS test results showed that using thermal anisotropy combined with a heat sink and sources can significantly outperform insulation materials of similar thickness. Under summer conditions, "Base + TAC" doubled the decrease in cooling loads compared to "Base + XPS"; under winter conditions, "Base + TAC" increased the reduction in heating loads by 40% compared to "Base + XPS". A future task is to include the energy penalty due to water circulation in the performance evaluation of the "Base + TAC" wall. However, given the low flow rates and mild water temperatures utilized, the energy penalty is expected to be low.

Table 2. Comparison of integrated cooling and heating loads.

	Test Parameter	Base	Base + XPS	Base + TAC
Summer	Cooling load (Wh)	5310	3021	770
	% difference	-	−43.1%	−85.5%
Winter	Heating load (Wh)	4781	2629	1760
	% difference	-	−45.0%	−63.2%

3.2. COMSOL Validation Results

The 2D models matching the test wall cross-sections were created, and the simulated heat flows through the test walls were compared to the LSCS measurements. Figure 11 compares the calculated heat flows through the different test walls with the LSCS measurements. For all cases, the measured exterior surface temperatures of the test walls and meter chamber air temperatures were used as the boundary conditions. The meter-facing or interior surface heat transfer was calculated using 'h_{meter}' of 6.13 W/(m^2·K) for summer conditions (heat flow downwards from climate to meter chamber) and 9.26 W/(m^2·K) for winter conditions (heat flow upwards from meter to climate chamber). For the "Base + TAC" summer and winter cases, transient 'h_{int}' values were calculated using Equation (1). With the flow rates of 0.053–0.057 kg/s and the interior Cu tube diameter of 10.9 mm, the resultant Reynolds numbers were in the range 6200–8200. The 'k' of water was obtained from the average of the inlet and outlet temperatures of the water circulating in the Cu tubes. As shown in Figure 9, the difference between the inlet and outlet temperatures were about 1 °C or less, so using an average temperature for 'h_{int}' calculations is reasonable. The resultant values of 'h_{int}' were 240–246 W/(m^2·K).

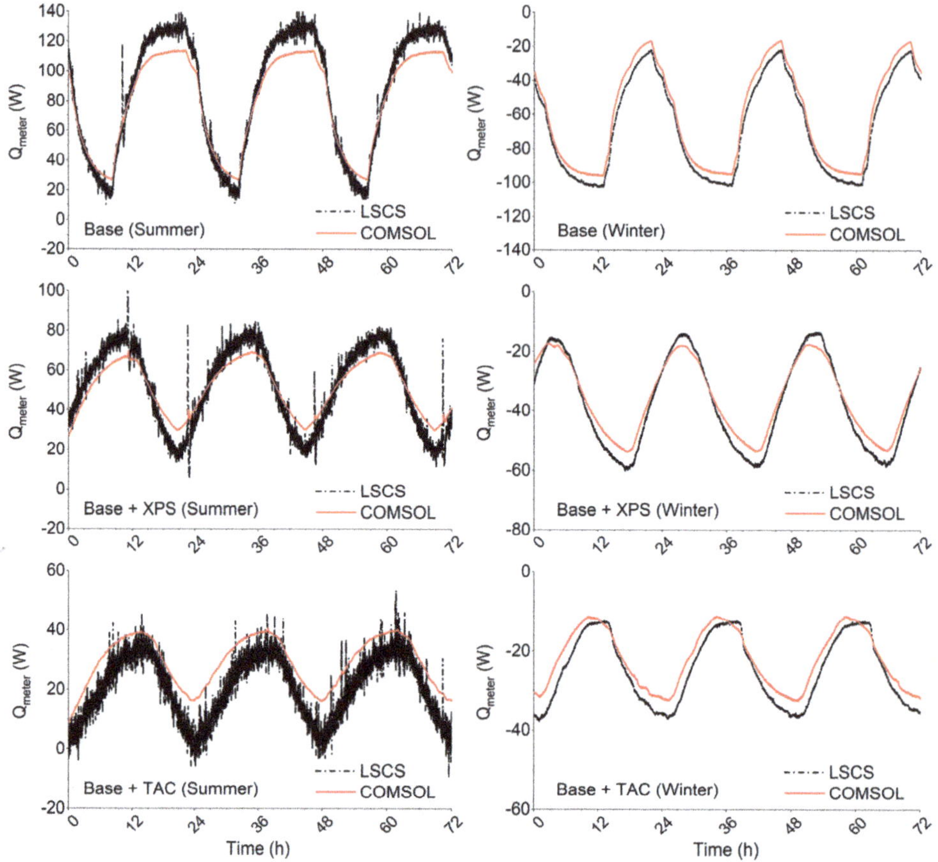

Figure 11. Comparison of measured and calculated heat gains to and heat losses from the meter chamber.

Overall, the 2D models were able to capture the thermal behavior of all the walls under both summer and winter conditions. Under summer conditions, the calculated peak cooling loads were within 11% of the measurements for the "Base" and "Base + XPS" cases, and within 15% for the "Base + TAC" case. Under winter conditions, the calculated peak heating loads were within 6%, 8%, and 11% of the measurements for the "Base", "Base + XPS", and "Base + TAC" cases, respectively.

3.3. Annual Simulation Results

The annual simulations were based on typical weather conditions in Phoenix and Baltimore; the former is in a cooling-dominated climate zone, and the latter has a heating-dominated climate. For the simplified 2D COMSOL models, the exterior and interior boundary conditions were generated using E+ simulations of the residential prototype building model [23]. For the "Base + TAC" model, a flow rate of 0.056 kg/s was assumed in the Cu tube, and the water temperature was assumed to be 27.8 °C in the heat sink mode and 20.0 °C in the heat source mode. The resultant 'h_{int}' for the heat sink and heat source modes were 245.7 and 241.2 W/(m^2·K), respectively. The switch between heat sink and heat sources happened when the outdoor temperature went above or below the assumed BPT of 12.8 °C. Hourly heat flows through the opaque wall sections were calculated from the 2D COMSOL models and used as cooling and heating loads in the modified E+ models with the adiabatic walls. COMSOL simulations were performed for walls oriented in all four cardinal directions—north, east, south, and west. Figure 12 shows a comparison of the integrated monthly heat gains through a south-facing wall in Phoenix and monthly heat losses through a north-facing wall in Baltimore. Table 3 presents the calculated annual heat gains and losses for the different cases.

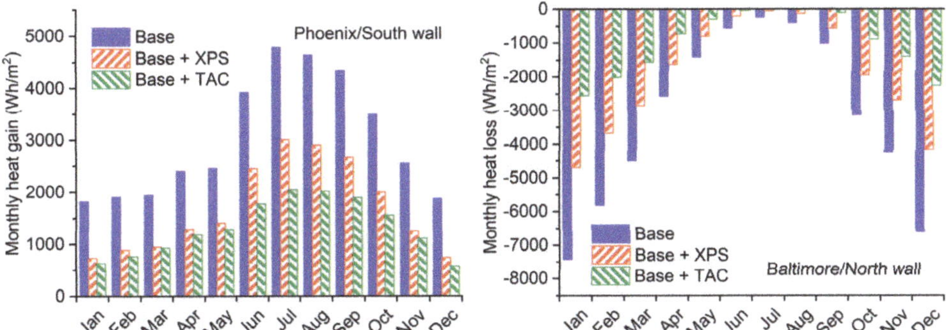

Figure 12. Calculated monthly heat gains through a south-facing wall in Phoenix and heat losses through a north-facing wall in Baltimore.

Table 3. Comparison of calculated annual wall heat gains and losses and percent difference with respect to the "Base" wall.

Wall Type	Performance Metric	North	East	South	West
	Phoenix				
Base	Heat gain (kWh/m²)	19.0	31.6	36.1	32.4
Base + XPS	Heat gain (kWh/m²)	11.4	18.1	20.3	18.9
	% difference	−40.1%	−42.6%	−43.9%	−41.6%
Base + TAC	Heat gain (kWh/m²)	10.6	14.3	15.8	14.9
	% difference	−44.2%	−54.7%	−56.4%	−53.8%
Base	Heat loss (kWh/m²)	−14.9	−10.9	−10.1	−12.7
Base + XPS	Heat loss (kWh/m²)	−8.9	−5.7	−4.2	−6.3
	% difference	−40.0%	−48.1%	−58.8%	−50.3%
Base + TAC	Heat loss (kWh/m²)	−3.6	−2.5	−2.2	−3.1
	% difference	−75.8%	−77.0%	−78.1%	−75.6%
	Baltimore				
Base	Heat gain (kWh/m²)	5.1	10.4	12.1	10.6
Base + XPS	Heat gain (kWh/m²)	2.6	5.1	5.2	5.4
	% difference	−49.3%	−50.7%	−57.1%	−48.8%
Base + TAC	Heat gain (kWh/m²)	4.2	5.7	5.9	6.0
	% difference	−17.6%	−45.2%	−51.1%	−43.3%
Base	Heat loss (kWh/m²)	−37.9	−32.9	−30.1	−34.4
Base + XPS	Heat loss (kWh/m²)	−23.4	−19.7	−16.7	−20.4
	% difference	−38.3%	−40.3%	−44.6%	−40.7%
Base + TAC	Heat loss (kWh/m²)	−11.9	−10.4	−9.1	−10.9
	% difference	−68.6%	−68.5%	−69.6%	−68.3%

In addition to the reductions in overall heat gains and losses, the potential of the 'Base + TAC' case to reduce and shift peak loads was also explored. Table 4 lists the COMSOL-calculated average reduction in peak heat gains and average time shift in the peak heat gains in Phoenix during the month of July with the 'Base + XPS' and 'Base + TAC' cases compared to the 'Base' case. While both the

'Base + XPS' and 'Base + TAC' cases showed similar potential for peak load shifting, the reduction in peak loads was much higher with the 'Base + TAC' case.

Table 4. Calculated average peak heat gain reduction and average peak shift during July in Phoenix.

Wall Type	Performance Metric	North	East	South	West
Base + XPS	Peak reduction (%)	40.6	53.0	46.5	48.7
	Peak shift (h)	0.7	4.0	3.8	1.2
Base + TAC	Peak reduction (%)	58.6	69.3	63.6	66.6
	Peak shift (h)	0.6	5.0	3.6	1.1

It should be noted that all heat gains for real buildings don't directly translate to cooling energy consumption and vice versa for heat losses. For example, heat gains during winter can reduce heating energy consumption without adding to cooling energy use. Therefore, whole building simulations are needed. Using the COMSOL-calculated hourly wall heat flows for the different wall configurations, E+ simulations were performed to estimate their impacts on whole-building cooling and heating energy consumption. Table 5 lists the E+ calculated cooling, heating, and fan energy use for the different wall configurations. The percent reductions in energy use are smaller than the reductions in heat flows, which is expected because overall energy use is also impacted by heat flows through other envelope sections (for example, roofs and windows), indoor loads, etc. In general, the "Base + TAC" case provided greater reductions in energy use than the "Base + XPS" case, except for the cooling energy use in Baltimore. It should be noted that the operating parameters chosen for the "Base + TAC" case for the current simulations were not optimized. Further investigations are ongoing for tuning the TAC and heat sink/source configuration, heat sink/source temperatures, etc., to maximize the peak load reductions and/or energy savings with TACs.

Table 5. Comparison of calculated annual whole building cooling and heating-related energy use and percent difference with respect to the "Base" case.

Wall Type	Cooling and Fan Energy Use (kWh)	% Difference	Heating Energy Use (kWh)	% Difference
		Phoenix		
Base	11,278	-	4855	-
Base + XPS	10,268	−9.0%	3804	−21.6%
Base + TAC	9998	−11.3%	3342	−31.2%
		Baltimore		
Base	4158	-	21,945	-
Base + XPS	3801	−8.6%	19,244	−12.3%
Base + TAC	3838	−7.7%	17,413	−20.6%

The annual simulation presented here was done to support the proof-of-concept work and show the potential of TACs coupled with heat sink/source to reduce wall-generated heating and cooling loads. The simulation controls and algorithm have not been optimized yet. In Tables 3 and 4, it is observed that under Baltimore weather conditions, the 'Base + XPS' case is more effective at reducing the heat gains and cooling energy use compared to 'Base + TAC'. For the 'Base + TAC' case, it was assumed that the water circulation is always on, and the switch between heat sink and heat source happens when the outdoor temperature crosses 12.8 °C. This can lead to scenarios when the outdoor temperature is lower than the room temperature (22.2–23.9 °C) but greater than 12.8 °C, so the circulating water is assumed to be at 27.8 °C. Thus, during these times, there is a greater temperature gradient for inward heat flow with the 'Base + TAC' case compared to the 'Base + XPS' case. Figure 13 shows a five-day period in April, when the outdoor temperature ('*Tout*') was predominantly lower than the room temperature but higher than 12.8 °C. As seen from the calculations, there were no heat gains (i.e., heat flows greater than zero) through the 'Base + XPS' wall, but there were significant heat gains through the 'Base + TAC' wall due to the water circulation at 27.8 °C.

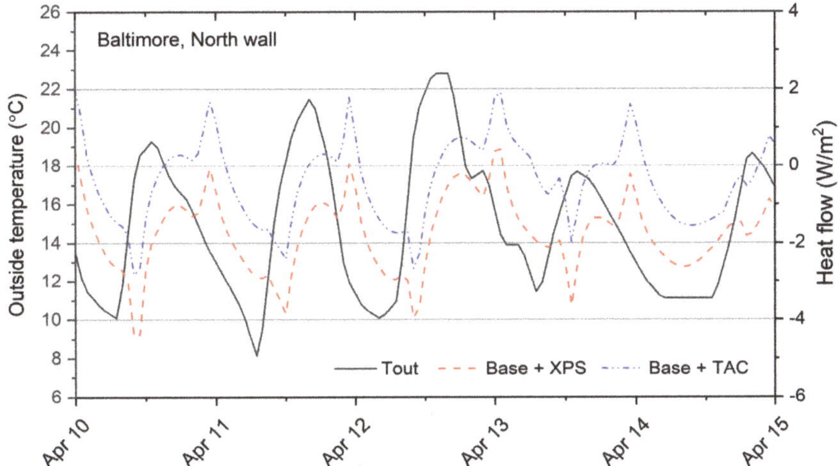

Figure 13. Outdoor temperature (left axis) and calculated heat flows (right axis) through a north-facing wall in Baltimore.

The preliminary annual simulations utilized a simple binary approach to switch between heat sink and source based on a single, constant outdoor temperature value. This was deemed reasonable for this proof-of-concept work. However, a more logical approach would be to use the exterior wall surface and interior temperatures to control the switching between heat sink and source modes, and even turn off the water circulation at times. Such algorithms that would maximize the benefits of the TAC system are being developed, and those results will be published in the near future.

4. Conclusions and Future Work

Here, the development and implementation of a novel TAC-based active envelope system is described. The potential of a TAC to reduce heat flows through building envelopes was experimentally and numerically investigated. A TAC coupled with a heat sink/source was shown to be more effective in reducing both cooling and heating loads and peak cooling loads compared to foam insulation of the same thickness. The TACs consisted of alternating layers of foam insulation and aluminum foil, both of which are commonly available materials. The TAC was connected to copper tubes circulating water for the experimental evaluations, and were able to reduce cooling and heating loads by 86% and 63% compared to a baseline wall with only cavity insulation. Preliminary numerical simulations were performed under two different climate conditions, cooling-dominated and heating-dominated. The TAC system was predicted to reduce cooling energy use by 11% under cooling-dominated climate and heating energy use by 21% in the heating-dominated climate.

The current study was a proof-of-concept investigation. Further optimization of the TAC system is needed to maximize its benefits with respected to peak load reduction and shifting, as well as overall reduction in heating and cooling energy use. By actively controlling the heat sink/source, the building envelope can be tuned to interact with the electric grid and provide benefits to the energy suppliers via peak load reduction and load shifting. The algorithm to switch between heat sink and source as well as turning off the water circulation will be optimized for different building and climate types to maximize the benefits of the TAC system. Finally, the energy penalty of creating a heat sink and source will also be considered.

Author Contributions: K.B. was primarily responsible for creating this manuscript. D.H., K.B., and S.S. developed the preliminary concept of thermal management using anisotropy. K.B. performed the COMSOL simulations. S.S. developed the coupled COMSOL and E+ simulation methodology and performed the E+ simulations. K.B., S.S. and J.A. were jointly responsible for the overall experimental design, testing and data analysis. S.S., D.H. and K.B. were responsible for acquiring funding support for this research.

Funding: This work was funded by the Building Technologies Office (BTO) of the US Department of Energy (DOE), under Contract No. DE-AC05-00OR22725. We gratefully acknowledge the support from Sven Mumme, the responsible Technology Manager of the DOE's BTO.

Acknowledgments: The authors would like to acknowledge the contribution of Daniel Howard, a post-Bachelor researcher at ORNL, who created the updated graphic for Figure 4.

Conflicts of Interest: The authors declare no conflict of interest.

References

1. Yang, L.; Yan, H.; Lam, J.C. Thermal comfort and building energy consumption implications—A review. *Appl. Energy* **2014**, *115*, 164–173. [CrossRef]
2. Sadineni, S.B.; Madala, S.; Boehm, R.F. Passive building energy savings: A review of building envelope components. *Renew. Sustain. Energy Rev.* **2011**, *15*, 3617–3631. [CrossRef]
3. Bhamare, D.K.; Rathod, M.K.; Banerjee, J. Passive cooling techniques for building and their applicability in different climatic zones-The State of Art. *Energy Build.* **2019**, *198*, 467–490. [CrossRef]
4. Kosny, J.; Fallahi, A.; Shukla, N.; Kossecka, E.; Ahbari, R. Thermal load mitigation and passive cooling in residential attics containing PCM-enhanced insulations. *Sol. Energy* **2014**, *108*, 164–177. [CrossRef]
5. Biswas, K.; Shukla, Y.; Desjarlais, A.; Rawal, R. Thermal characterization of full-scale PCM products and numerical simulations, including hysteresis, to evaluate energy impacts in an envelope application. *Appl. Therm. Eng.* **2018**, *138*, 501–512. [CrossRef]
6. Baetens, R.; Jelle, B.P.; Thue, J.V.; Tenpierik, M.J.; Grynning, S.; Uvslokk, S.; Gustavsen, A. Vacuum insulation panels for building applications: A review and beyond. *Energy Build.* **2010**, *42*, 147–172. [CrossRef]
7. Baetens, R.; Jelle, B.P.; Gustavsen, A. Aerogel insulation for building applications: A state-of-the-art review. *Energy Build.* **2011**, *43*, 761–769. [CrossRef]
8. Biswas, K. Development and Validation of Numerical Models for Evaluation of Foam-Vacuum Insulation Panel Composite Boards, Including Edge Effects. *Energies* **2018**, *11*, 2228. [CrossRef]
9. Termentzidis, K. Thermal conductivity anisotropy in nanostructures and nanostructured materials. *J. Phys. D-Appl. Phys.* **2018**, *51*, 094003. [CrossRef]
10. Huang, S.R.; Bao, J.; Ye, H.; Wang, N.; Yuan, G.J.; Ke, W.; Zhang, D.S.; Yue, W.; Fu, Y.F.; Ye, L.L.; et al. The Effects of Graphene-Based Films as Heat Spreaders for Thermal Management in Electronic Packaging. In Proceedings of the 17th International Conference on Electronic Packaging Technology, Wuhan, China, 16–19 August 2016; pp. 889–892.
11. Suszko, A.; El-Genk, M.S. Thermally anisotropic composite heat spreaders for enhanced thermal management of high-performance microprocessors. *Int. J. Therm. Sci.* **2016**, *100*, 213–228. [CrossRef]
12. Cometto, O.; Samani, M.K.; Liu, B.; Sun, S.X.; Tsang, S.H.; Liu, J.; Zhou, K.; Teo, E.H.T. Control of Nanoplane Orientation in voBN for High Thermal Anisotropy in a Dielectric Thin Film: A New Solution for Thermal Hotspot Mitigation in Electronics. *ACS Appl. Mater. Interfaces* **2017**, *9*, 7456–7464. [CrossRef] [PubMed]
13. Ren, Z.Q.; Lee, J. Thermal conductivity anisotropy in holey silicon nanostructures and its impact on thermoelectric cooling. *Nanotechnology* **2018**, *29*, 045404. [CrossRef] [PubMed]
14. Narayana, S.; Sato, Y. Heat Flux Manipulation with Engineered Thermal Materials. *Phys. Rev. Lett.* **2012**, *108*, 214303. [CrossRef] [PubMed]
15. Vemuri, K.P.; Bandaru, P.R. Geometrical considerations in the control and manipulation of conductive heat flux in multilayered thermal metamaterials. *Appl. Phys. Lett.* **2013**, *103*, 133111. [CrossRef]
16. Xue, X.; Wang, S.W. Interactive Building Load Management for Smart Grid. In Proceedings of the 2012 Power Engineering and Automation Conference, Wuhan, China, 14–16 September 2012; pp. 371–375.
17. Perino, M.; Serra, V. Switching from static to adaptable and dynamic building envelopes: A paradigm shift for the energy efficiency in buildings. *J. Facade Des. Eng.* **2015**, *3*, 143–163. [CrossRef]
18. Lufkin, S. Towards dynamic active facades. *Nat. Energy* **2019**, *4*, 635–636. [CrossRef]

19. ASTM. *ASTM C1363-11, Standard Test Method for Thermal Performance of Building Materials and Envelope Assemblies by Means of a Hot Box Apparatus*; ASTM International: West Conshohocken, PA, USA, 2011.

20. ASHRAE. Handbook—Fundamentals. 2013. Available online: https://www.ashrae.org/technical-resources/ashrae-handbook (accessed on 4 October 2019).

21. Incropera, F.P.; DeWitt, D.P. *Fundamentals of Heat and Mass Transfer*, 4th ed.; John Wiley & Sons, Inc.: Hoboken, NJ, USA, 1996.

22. *ANSI/ASHRAE/IES Standard 90.1-2016, in Energy Standard for Buildings Except Low-Rise Residential Buildings*; ASHRAE: Atlanta, GA, USA, 2016.

23. Residential Prototype Building Models. Available online: https://www.energycodes.gov/development/residential/iecc_models (accessed on 4 October 2019).

24. Krese, G.; Lampret, Z.; Butala, V.; Prek, M. Determination of a building's balance point temperature as an energy characteristic. *Energy* **2018**, *165*, 1034–1049. [CrossRef]

Article

Envelope Thermal Performance Analysis Based on Building Information Model (BIM) Cloud Platform—Proposed Green Mark Collaboration Environment

Ziwen Liu [1,2,3,*], Qian Wang [4], Vincent J.L. Gan [5] and Luke Peh [2]

[1] Digitalization Department, Built Environment Research and Innovation Institute (BERII), Building and Construction Authority, Zero Energy Building (ZEB), 200 Braddell Road, Singapore 579700, Singapore

[2] School of Science and Technology, Singapore University of Social Science, 463 Clementi Road, Singapore 599494, Singapore; lukepehlc@suss.edu.sg

[3] BCA Academy, 200 Braddell Road, Singapore 579700, Singapore

[4] Department of Building, School of Design and Environment, National University of Singapore, 4 Architecture Drive, Singapore 117566, Singapore; bdgwang@nus.edu.sg

[5] Department of Civil and Environmental Engineering, The Hong Kong University of Science and Technology, Clear Water Bay, Kowloon, Hong Kong 999077, China; vincentgan@ust.hk

* Correspondence: zwliu001@suss.edu.sg or liu_ziwen@bcaa.edu.sg

Received: 19 November 2019; Accepted: 21 January 2020; Published: 27 January 2020

Abstract: Building Information Modeling (BIM) and sustainable buildings are two future cornerstones of the Architectural, Engineering and Construction (AEC) industry. In Singapore's context, the Green Mark (GM) scoring system is prevalently used to assess the sustainability index of green buildings. BIM provides the semantic and geometry information of buildings, which is proliferated as the technological and process backbone for the green building assessment. This research, through vast literature reviews, identified that the current procedure of achieving a Green Mark score is tedious and cumbersome, which hampers productivity, especially in the calculation of building envelope thermal performance. Furthermore, the project stakeholders work in silos, in a non-collaborative, manual and 2D-based environment for generating relevant documentation to achieve the requisite green mark score. To this end, a cloud-based BIM platform was developed, with the aim of encouraging project stakeholders to collaboratively generate the project's green mark score digitally in accordance with the regulatory requirements. Through this research, the authors have validated the Envelope Thermal Transfer Value (ETTV) calculation, which is one of the prerequisite criteria to achieve a Green Mark score, through a case study using the developed cloud-based BIM platform. The results indicated that using the proposed platform enhances the productivity and accuracy as far as ETTV calculation is concerned. This study provides a basis for future research in implementing the proposed platform for other criteria under the Green Mark Scheme.

Keywords: Building Information Modeling; cloud-based; envelope thermal performance; green buildings; Green Mark; Integrated Digital Delivery (IDD)

---

1. Introduction

In the recent decade, Building Information Modeling (BIM) has widely been adopted in the Architectural, Engineering, and Construction (AEC) industry and completely upended the way we build [1]. Building Information Modeling (BIM) has been identified by the Building and Construction Authority (BCA) of Singapore as one of the key drivers to improve productivity in the Architectural, Engineering, and Construction (AEC) industry. In 2010, the first BIM Roadmap was implemented and had since achieved progressive and positive results in various facets of building design [2]. The one-stop BIM submission to a number of agencies is one of the transformative policies introduced to the construction industry that made Singapore the first country in the region to accept and approve building plan submissions with BIM models [3]. With the increasing benefits of BIM, firms have been challenged to extend their BIM capabilities from modeling to design and coordination. Design integration of multiple disciplines with BIM has been improving steadily with better coordination and interfacing between different discipline models.

While BIM continues to gain momentum, especially in building plan submissions and coordination, it is also important to look at the other aspects of BIM, such as BIM for sustainability. There is an increasing demand for developing sustainable buildings because of rising energy costs and environmental impacts. Using BIM during the early stages of a project can facilitate complex building performance analyses, especially if it is used during the early design and pre-construction phases [4]. The traditional CAD-based practice often leads to retroactive modification of the design to achieve building performance requirements [5]. The information required for sustainable design, analyses, and certification can be readily and routinely made available by using BIM during the early design stages.

There are challenges in both developing a project in compliance with the Green Mark requirements and assessing the level of sustainability with partial and fragmented data. Project development, along with the emergence of integrated digital delivery (IDD) within the collaborative environment, could potentially ease these processes.

In this paper, a cloud-based platform for automated analysis and digital analysis technology, to be used for Green Mark envelope thermal performance analysis based on a BIM and 3D graphic environment is proposed. The calculations and data have been validated for accuracy.

2. Literature Review

2.1. Building Information Modeling (BIM)

The concept of BIM originated in the late 1970s with Professor Charles Eastman from Georgia Tech [6]. Since its development, different definitions have been given by scholars. To date, the most authoritative and highly recognized by the AEC industry is the National Building Information Modeling Standard (NBIM) published by the National Institute of Building Sciences (NIBS): *"A BIM is a digital representation of physical and functional characteristics of a facility. As such it serves as a shared knowledge resource for information about a facility forming a reliable basis for decisions during its lifecycle from inception onward* [7]." From the definition of BIM by NIBS, it can be seen that BIM includes a Building Information Model, Building Information Modeling and Building Information Management, which are interrelated but independent, the details are as follows:

- From a static perspective, BIM is a digital model that integrates all information. This model can be used for virtual simulation and information sharing, that is, a building information model.
- From a dynamic perspective, BIM is a behavioral process that continuously updates, inputs and extracts information. In this process, the model was gradually perfected to meet the needs of all stakeholders, namely Building Information Modeling.
- From a management perspective, BIM provides a collaborative working environment, which integrates work and management processes, facilitates information sharing and connects stakeholders on the same project throughout the building life cycle. It allows timely identifying problems, making correct decisions and improving management efficiency, that is, Building Information Management.

In the above three perspectives of BIM, Building Information Model is the foundation, Building Information Modeling is the core and Building Information Management is the basic guarantee to achieve a Building Information Model, which is complementary and indispensable. The reasonable coupling of the three functions enables BIM technology to have the characteristics of operation visualization and integration of information completeness, coordination and interoperability, etc. BIM technology can provide basic support for improving the productivity of the construction industry, increasing professional communication and reducing resource waste.

2.2. Green Building Assessment (GBA) Systems

In order to better regulate the development of green buildings and improve the living environment of human beings, countries around the world have started to develop relevant assessment standards. The most common green building assessment (GBA) systems worldwide include but are not limited to Leadership in Energy and Environmental Design (LEED), Building Research Establishment Environmental Assessment Method (BREEAM), Building Environmental Assessment Method (BEAM) Plus, Green Star, Green Building Index (GBI) and Green Mark. All these GBA systems provide a quantitative and comprehensive assessment of the sustainability level of building that is influenced by energy efficiency, water efficiency, environmental protection, indoor environment quality (IEQ) and life cycle assessment (LCA) [8]. Table 1 compares the various green building certifications.

Table 1. Comparison of various green building certifications.

Scheme & Developed	LEED V4 By USGBC	BREEAM By BRE	BEAM Plus By HKGBC	Green Star By GBCA	GBI By PAM/ACEM	Green Mark By BCA
Reference	[9]	[10]	[11]	[12]	[13]	[14]
Origin & Launch Year	USA, 1998	UK, 1990	Hong Kong, 1996	Australia, 2002	Malaysia, 2009	Singapore, 2005
No. of Countries/ Regions adopted	160	77	1	2	1	15
Application	Various Climate Condition	Various Climate Condition	Various Climate Condition	Tropical/Subtropical Climate	Tropical Climate	Tropical/ Subtropical Climate
Certified by	GBCI (*Green Business Certification Inc.*)	BREEAM Assessor (*BRE Global monitor the assessment quality process of the assessor*)	BEAM Plus TRC (*Technical Review Committee*)	Green Star Certified Assessor	GBIAP (*GBI Accreditation Panel*)	BCA Green Mark Department
Envelope Thermal Performance Requirement	Building envelope opaque: roofs, walls, floors, slabs, doors, etc. Building envelope glazing: vertical fenestration	The design analysis must cover peripheral microclimate conditions, building form, layout, building envelope, and thermal mass, as well as daylighting and ventilation strategies.	Building Envelope: Overall Thermal Transfer Value (OTTV) calculation	Credit considered in energy modeling	OTTV $\leq$ 50, RTTV $\leq$ 25. Submit calculations using the BEIT software or other GBI approved software(s)	Energy Efficiency (1-1: Thermal Performance of Building Envelope)

It can be observed from Table 1 that most green building assessments have clear requirements on the thermal building envelope.

BCA Green Mark Scheme

The BCA's Green Mark scheme, launched in 2005, is an internationally recognized green building rating system for tropical climates. Green Mark provides a meaningful differentiation and stratification of buildings, in terms of how green and sustainable they are, in the real estate market with a benchmarking scheme, which incorporates internationally recognized best practices in environmental design and performance [15]. It encompasses a range of assessment standards for different types of buildings, which covers new buildings, existing buildings, user-centric spaces (office interior, retail, supermarket) and beyond buildings (districts, parks and infrastructure).

Green Mark for Non-Residential Building (NRB 2015) was launched in 2016. The NRB 2015 includes significant amendments to the previous rating system. For example, Envelope Thermal Transfer Value (ETTV) had been tightened for all Non-Residential projects, with 45 W/m^2 being set as the new baseline instead of the previous 50 W/m^2.

2.3. Building Envelope Thermal Performance

The building envelope is composed of walls, roofs, windows, shading devices and thermal insulation materials, which greatly affects the indoor environment quality in terms of daylight, the energy consumption of HVAC systems and thermal comfort [16]. Existing studies [17–20] have proved that the building envelope has an extremely important influence on thermal performance and is one of the critical factors in obtaining green building certifications. Bressch and Janssens [17] applied a standardized regression coefficient (SRC) to identify the factors that had the greatest impact on thermal comfort on the south side of a typical Belgian office building. They observed that air tightness and heat gains were the two most important variables. Natephra et al. [19] proposed a method (integrating time-stamped 3D thermal data in the BIM) to collect environmental and thermal data and to integrate them with BIM to be used for various applications. Furthermore, Natephra et al. [20] presented the integration of BIM and visual scripting to automatically extract thermal properties from BIM to computing thermal transfer value.

Envelope Thermal Transfer Value (ETTV)

Since 1979, the Building Control Regulation of Singapore had stipulated that the overall thermal transfer value (OTTV) of air-conditioned buildings shall not be more than 45 W/m^2 of the design value. In 2004, OTTV was replaced by the Envelope Thermal Transfer Value (ETTV) because the OTTV formula tended to underestimate the solar radiation gain component through the fenestration system and did not fully account for the full extent of heat gain through the building envelope [21].

The ETTV is a prerequisite or mandatory requirement for BCA Green Mark Non-Residential Building, which means that project stakeholders have to meet minimum requirements in order to qualify for Green Mark certification.

The ETTV is similar to OTTV in that it considers the three basic components of heat gain through the external walls and windows of a building [21]. These three components are:

- Heat conduction through opaque walls;
- Heat conduction through glass windows;
- Solar radiation through glass windows.

The benchmarking of ETTV enforces the optimizing of the design of the building envelop to reduce external heat gain and hence reduce the cooling load for the air-conditioned building [22]. The ETTV concept was extended to cover residential buildings in 2008. As air conditioners in residential buildings are usually turned on in the night, the envelope thermal performance standard for residential buildings is given the name Residential Envelope Transmittance Value (RETV) to differentiate it from ETTV, which is meant for buildings that operate air conditioning systems during or throughout the day [21]. The ETTV formula is as follows:

$$ETTV = 12\,(1 - WWR)\,Uw + 3.4(WWR)U_f + 211(WWR)(CF)(SC) \tag{1}$$

where ETTV: envelope thermal transfer value (W/m^2); WWR: window to wall ratio (fenestration area/gross area of exterior wall); U_w: thermal transmittance of opaque wall (W/m^{2}K); U_f: thermal transmittance of fenestration (W/m^{2}K); CF: correction factor for solar heat gain through fenestration; SC: shading coefficients of fenestration.

Chua et al. [23] studied the diverse building parameters that affect the energy performance of commercial buildings in Singapore and found that there are two key indexes to measure building energy performance, namely, envelope thermal transfer value (ETTV) and the annual cooling energy requirements. The authors then performed a relative ranking of the ETTV functional parameters to evaluate their effectiveness in lowering the ETTV of a building. These parameters are the shading coefficient, window-to-wall ratio, the U-values of the wall and windows, and the absorptance of the opaque wall.

A case study to identify the relationship between Singapore's Green Mark Scheme (GMS) and the Buildable Design and Appraisal System (BDAS)'s requirements for building envelopes was conducted by Singhaputtangkul et al. [24]. Singhaputtangkul et al. [24] discovered that the lengths of windows and walls and the associated materials influence the Green Mark (GM) score of the building envelope and the buildability score of the wall system. In addition, they further concluded that varying the window-to-wall (WWR) ratio has a stronger effect on the GM score, in terms of the building envelope in comparison with the buildability score of the wall system. To illustrate more detail, WWR shows a negative relationship with the GM score of the building envelope as when WWR increases from 0.151 to 0.510, the GM score of the building envelope decreases from 15 points to 0 points [24].

2.4. BIM for Green Building Assessment

Existing studies [8,25–27] have demonstrated that BIM supported the green building assessment. Lu et al. [8] summarized that BIM could support the green building assessment in three aspects. (1) BIM assists stakeholders in choosing effective strategies to achieve a green building, (2) BIM interprets the credits of GBAs, and (3) BIM facilitates documentation management. Ansah et al. [26] conducted an in-depth review of the bibliographic related to BIM for various Green Building Assessments, including LEED, BREEAM, BEAM Plus, Green Mark and GBI and highlighted that cloud-based BIM and GBAS will be the future direction. BIM could assist in automating the Green Mark process, and it was observed that 31 Green Mark items could be attained through BIM software and building performance analysis (BPA) tools [25]. Moakher et al. [27] provided an overview of how purpose-built BIM solutions and integrated analysis tools can help to assess building performance, prioritize investments and evaluate proposals to reduce operational costs, conserve energy, reduce water consumption and improve building air quality, helping to meet sustainability and energy-efficiency goals.

BIM for Envelope Thermal Performance

BIM is able to capture project information and generate documents and the advent of BIM technology, particularly anecdotal evidence of its widespread use in green building projects, has led professionals and researchers to envision the integration of the BIM and green building certification processes. In particular, the application of BIM in envelope thermal performance measurement has attracted attention, such as project Helios [28], Integrated Environmental Solutions Virtual Environment (IES VE) Green Mark Navigator [29], ETTV Assessment through BIM and VPL [22] and a BIM-based OTTV Calculation [20].

Project Helios aimed to develop an add-on application that utilizes the parameters captured within the digital building model created by Autodesk Revit [28]. Although the concept looked promising for ETTV calculation from Autodesk Revit, there seems to be little to no information available on this project after its inception in the year 2009. In addition, the project stakeholders must have a Revit license, which is an additional cost.

A BIM-based building performance analysis (BPA) software was developed by IES VE to calculate ETTV/RTTV by using a solar tracking technique to help the calculation of shading effect from structures with more complex geometry. IES VE aims to provide a step-by-step guide to assess performance elements of the Green Mark scoring system [29]. Envelope thermal performance simulation can be evaluated by importing the extracted information from BIM models with the defined data exchange schema, such as IFC and the gbXML schema [22]. However, Moon et al. [30] pointed out that there are varying levels of interoperability between existing BIM-based BPA software and BIM authoring software, and additional effort and modification to BIM model are ineluctable when transferring information from BIM to BIM-based BPA software. Furthermore, Chen et al. [31] summarized the interoperability issue from BIM to mainstream BPA software into the following six categories: building geometry, space composition, building construction, internal loads, operation schedules and HVAC systems.

Recently, researchers from Universiti Teknologi Malaysia developed a BIM-VPL based tool for building envelope design and assessment support. The research demonstrated the importance and potential of BIM and VPL integration for ETTV assessment, and how it is an important step to automate ETTV calculation [22]. However, the study required additional data extraction and management and had limitations. For instance, curtain walls or tilt walls could not be handled, and the effective shading coefficient of external shading devices (SC2) were not taken into consideration. Furthermore, the author suggested developing a plug-in or application programming interface (API) for Revit to measure thermal envelope performance.

Recently, Natephra et al. [20] presented a BIM approach integrating with scripting to automatically extract thermal parameters from a database and provide an instant OTTV calculation.

3. Research Gaps

According to Inhabit Group [32], *"ETTV compliance presents design challenges for architects, builders and suppliers in Singapore's building and construction market. Recently introduced benchmarks mean that ETTV has to be considered more closely when undertaking any design decisions that influence the façade outcome"*.

It is well known that generating an envelope thermal performance calculation report is a cumbersome and time-consuming process for project stakeholders. The current manual calculation method is time-consuming, and human error can occur throughout the completion of the calculation process [20]. In addition, the current envelope thermal calculation method does not consider some of the contributions by non-conventional shading devices, shading of opaque construction, and shading from surrounding buildings [29]. There is an information delay if the architect initiates a change, such as changing the glazing size and U values. Figure 1 illustrates the current process to calculate the ETTV.

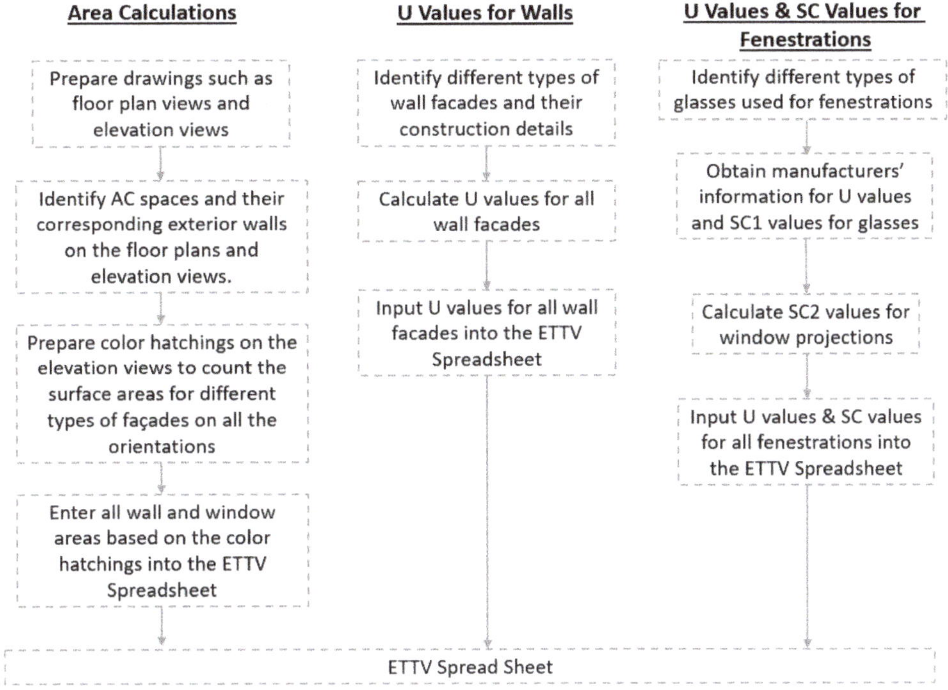

Figure 1. Traditional Envelope Thermal Transfer Value (ETTV) calculation process.

4. Proposed Green Mark Collaboration Environment

The proposed methodology comprises several main steps, as shown in Figure 2. The proposed method starts by creating a design model in BIM authoring tools. In order to streamline the process in computing the ETTV/RETV, the design BIM model in native format follows certain detailed modeling rules, which are provided in Section 4.1. The geometric and semantic information in the design BIM model is then extracted and converted into energy models (in IFC 2×3 format) using a customized program embedded in BIM software. As most of the designers are not able to assign the thermal properties to a BIM model during the early design stage, a common construction material library with all the necessary thermal properties of building materials is provided to enable designers to specify the material property for ETTV/RETV calculation at the early stage. Next, a conventional method is leveraged and integrated with the semantic model to facilitate ETTV/RETV calculation. In addition, a performance-based ETTV/RETV calculation method is also proposed in Section 4.3. The fourth step is to compute ETTV/RETV. The fifth step is to evaluate the ETTV/RETV result. If the results fail to comply, then the Green Mark Collaboration Environment provides functions to change the new material with better thermal properties. It is worth mentioning that the solar data are taken from Table C8–C11 from code on Envelope Thermal Performance for Buildings [33]. The solar data from Table C8–C11 has been built into the proposed platform to facilitate the building envelope thermal performance computation.

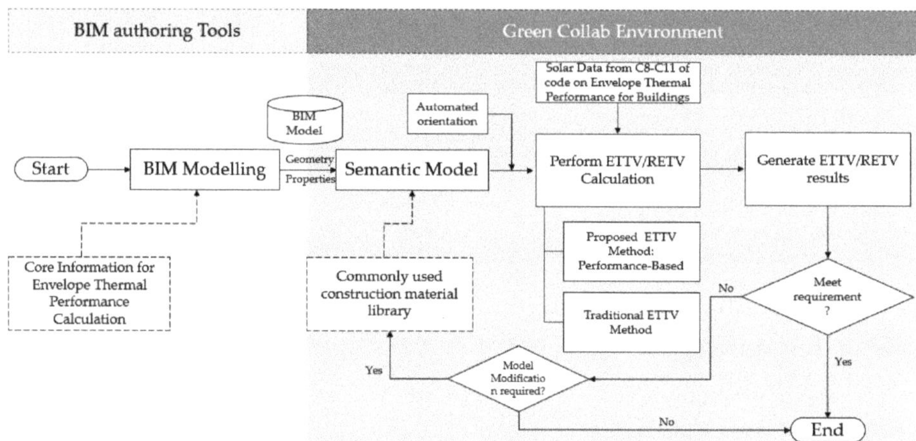

Figure 2. Proposed workflow for Building Information Model (BIM)-based ETTV computation.

4.1. Modeling Requirements and Core Information for ETTV/RETV Computation

Most parameters needed for the envelope thermal performance analysis can be obtained in the BIM model. The modeling requirements for thermal envelope performance calculation in the BIM model are listed in Table 2.

Table 2. Core information for thermal envelope performance calculation.

Element	Modeling Requirements	Core Information
Wall	Ensure that overall partition thickness, including finishes, is correct as per design intent. Model the walls individually for each story/level. Terminate height of wall at soffit of beam and slab, or to exact intended design height.	Build Type Material Thickness Area Volume
Window	Define the window "TYPE" clearly. Ensure that locations and counts are correct per type. Do not over-model. Ensure that overall window size is correct. Window details may refer to 2D typical details. Window to be coordinated with structural openings if it is in the structural wall.	Build Type Material Thickness Area
Door	Define the door type clearly. Ensure that locations and counts are correct per type. Location shall be from room or to room.	Dimensions Location
Column	If the column is a structural column with an architectural cladding, model the cladding finish as the wall. Model the structural column individually for each storey.	Dimension Area Volume
Beam	Model the beams from and to the center of the column For CIP structures, join beams to slabs (clear the connections).	Dimension Area Volume
Slab	Model the structural floors at SFL to be coordinated with Architectural FFL. Model slabs as individual elements from beams.	Thickness Grade

Table 2. *Cont.*

Element	Modeling Requirements	Core Information
Roof	For flat roofs, only the roof finish, i.e., screed must be modeled in the architectural model. The roof slab must be modeled in the structural model. Ensure both finish and slab are aligned and not overlapping.	Thickness Area

If the core information in the information table above is not completed in the original BIM model, the missing information can be patched to the semantic model.

4.2. Features of Green Mark Collaboration Environment

4.2.1. Auto Building Orientation

Project Orientation from the BIM Authoring Tool will automatically be captured when the model is imported to the Green Mark Collaboration Environment. However, the orientation amendment can be easily done in the Green Mark Collaboration Environment, and everything will be followed accordingly, including the calculated results.

4.2.2. Multi-Element Selection

The Green Mark Collaboration Environment is capable of selecting model elements with the same properties and applying the green data to it. Figure 3 shows the selecting and assigning parameters for the elements with the same properties. This would largely reduce the time and procedures of preparing the model for calculation.

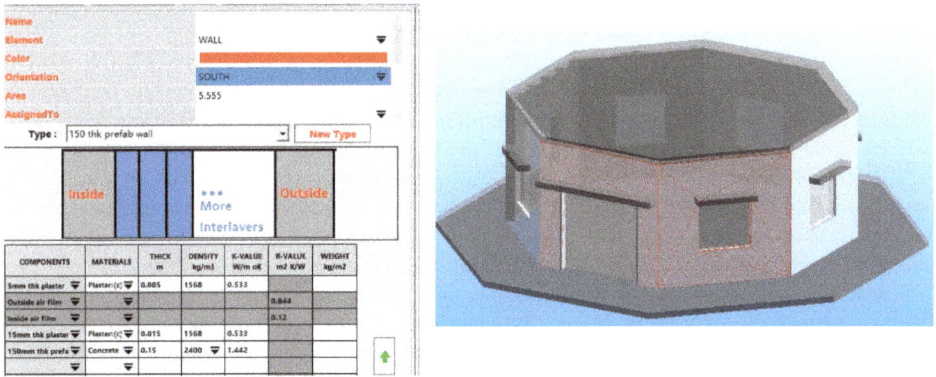

COMPONENTS	MATERIALS	THICK m	DENSITY kg/m3	K-VALUE W/m oK	R-VALUE m2 K/W	WEIGHT kg/m2
5mm thk plaster	Plaster:(c)	0.005	1568	0.533		
Outside air film					0.044	
Inside air film					0.12	
15mm thk plaster	Plaster:(c)	0.015	1568	0.533		
150mm thk prefa	Concrete	0.15	2400	1.442		

Figure 3. Selecting and assigning parameters for the elements with the same properties.

4.2.3. Model Element-Equation Mapping

Model elements, such as windows, walls, beams, columns and others necessary for thermal performance calculation, can be mapped to the Green Mark Collaboration Environment. Manufacturer's data, such as glassing information in Microsoft format or PDF, can also append to the element type. Stakeholders can retrieve the data of the elements whenever verification is needed. As shown in Figure 4, elements with the same thermal properties will have their own assigned color code as per the Green Mark Submission Requirement once it has been mapped with the green building data.

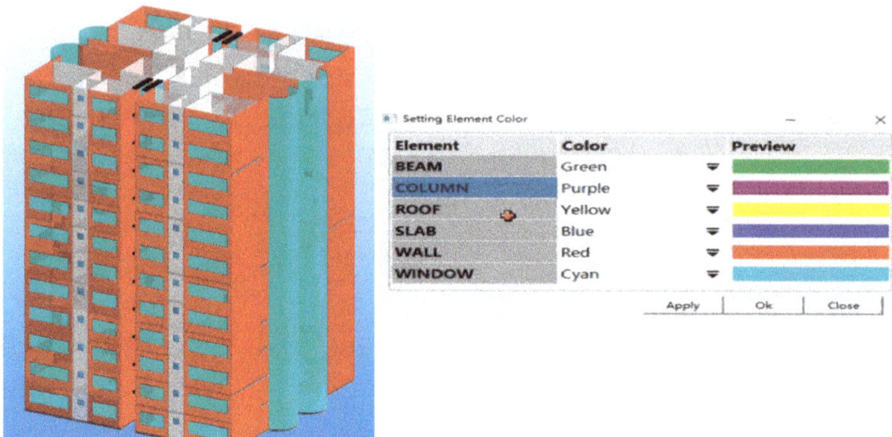

Figure 4. Model elements are mapped with the equation.

4.2.4. Report Generation

The Proposed Green Mark Collaboration Environment has a built-in report template. The template will capture all calculated data while restoring its linkage to the building model elements. This will help in assessing the Green Mark information of the project. Any changes made in the model will be automatically reflected in the reports.

4.2.5. Green Mark Collaborative Working Environment

The Green Mark Collaborative environment is to bridge the gaps caused by isolated and fragmented Green Mark data generated by different stakeholders. It comprises several modules and each module focused on a particular requirement specified in Green Mark for Non-Residential Buildings (NRB) 2015. The ETTV/RETV is the first module embedded into the proposed Green Mark Collaboration Environment. Future research will look into other modules in the Green Mark assessment, such as Concrete Usage Index (CUI), Lighting efficiency and Computational Fluid Dynamic (CFD).

The proposed Green Mark Collaboration Environment was capable of efficiently managing the semantic model(s) and hosting the Green Mark project data and documentation generated, updated and consolidated along with the GM project delivery. Semantic model-based communication and collaboration allows architects, Green Mark Consultants, contractors and other parties, or even Green Mark Assessors, to view reports of real-time information. Its emphasis is on streamlining the automated GM documentation generation, managing GM submission, monitoring the GM score and tracking the GM project status. Figure 5 illustrates the framework of the collaborative working environment.

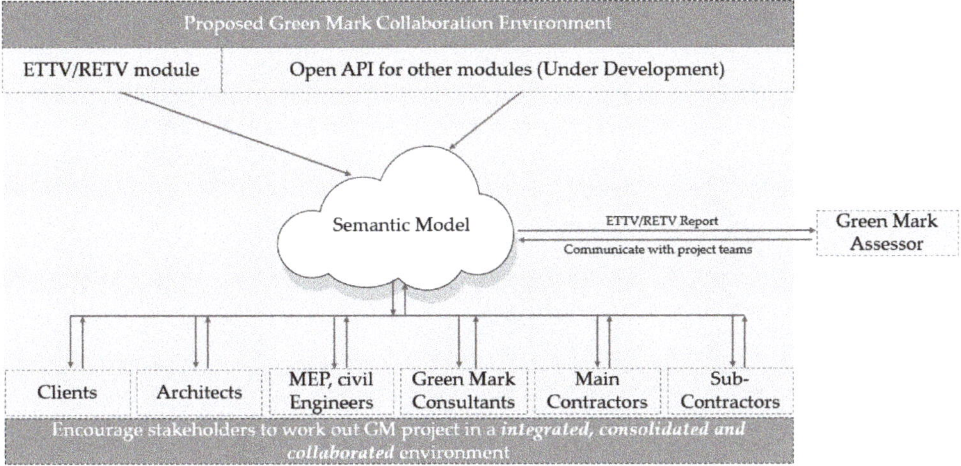

Figure 5. Framework of a collaborative working environment.

To sum up, the collaborative working environment provides various benefits for stakeholders: (1) The collaborative working environment allows remote real-time access and concurrent ETTV analyzing; and (2) integrated GM data acquisition, processing, and management along the project lifecycle with enhanced collaboration among all the stakeholders.

4.3. Performance-Based Method—Proposed Digital Analysis Method for Arbitrarily Shaped Shading Device

The window shading analysis can be very difficult for arbitrarily shaped shading devices. To simplify the analysis process, the tabular reference and analysis sample for the most often used shading device shapes have been provided in the Code on Envelope Thermal Performance for Buildings [33]. However, nowadays, more and more modern buildings have been designed with special shapes and render the simplified analysis process inadequate for envelope thermal transmittance analysis. Therefore, an effective digital analysis technology for envelope thermal transmittance analysis of windows with arbitrarily shaped shading devices is necessary.

This paper hereby proposes a numerical analysis method to analyze envelope thermal transmittance with arbitrarily shaped shading device windows.

Assume the position of the sun can be specified by the angles illustrated in Figure 6, which is adapted from [21,33].

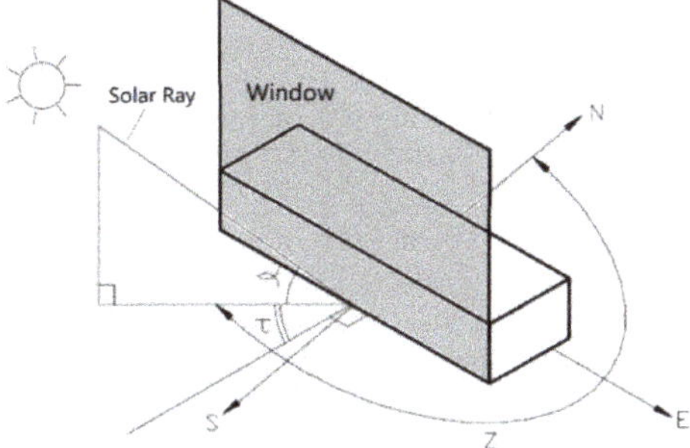

Figure 6. Solar geometry.

By using the finite element approach, the window can be divided into smaller grids. Each small grid will be represented by its central point, as shown in Figure 7.

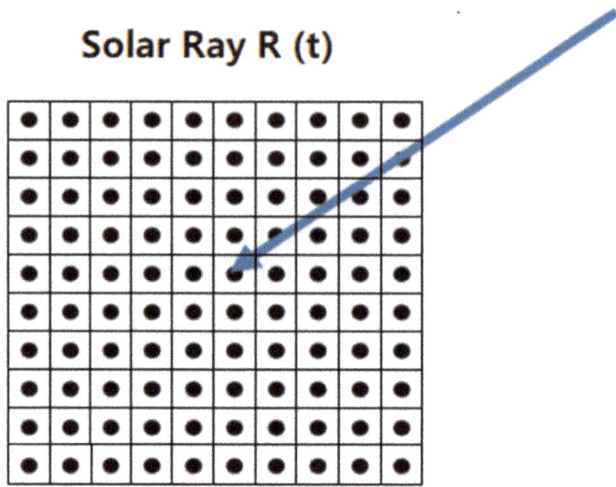

Figure 7. Finite element approach.

A solar ray R(t) with origin O and normalized direction D can be defined as:

$$R(t) = O + tD. \tag{2}$$

Assume the shading device of a window is divided into triangle elements, which is always needed for computer graphical display, as shown in Figure 8.

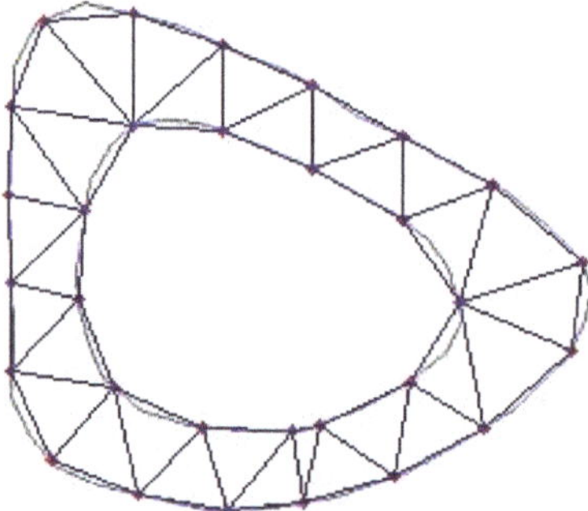

Figure 8. Finite element approach.

Each individual triangle element can be expressed, as illustrated in Figure 9.

Using this approach, the three joints of a triangle element (A, B, C) can also be expressed as vectors V(a), V(b), V(c). By introducing the local parameters (u, v) as shown in Figure 9, any point P within the triangle element can be expressed as:

$$P(u, v) = (1 - u - v)\, V(a) + uV(c) + vV(b) \ (u \geq 0, v \geq 0, u + v \leq 1). \tag{3}$$

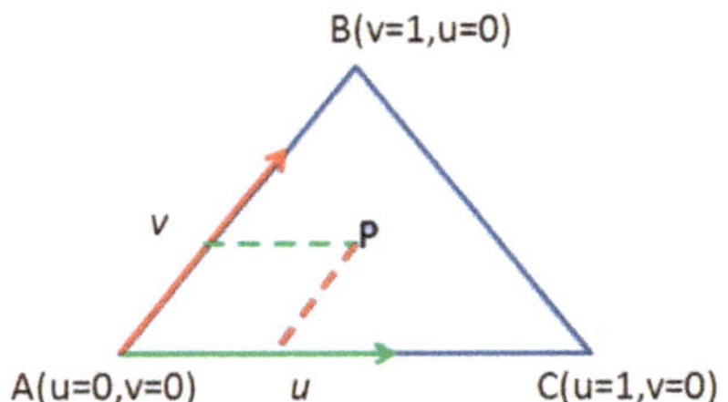

Figure 9. Individual triangle element.

A solar ray in a window point to each individual triangle element can be calculated to determine its intersection, as shown in Figure 10.

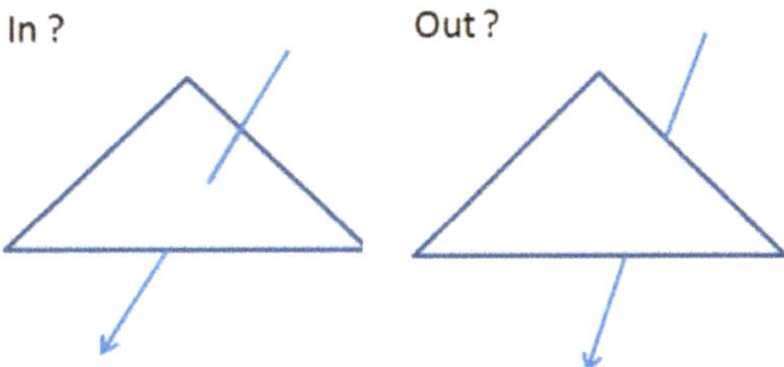

Figure 10. A solar ray in a window.

Computing the intersection between the individual triangle element and solar ray is done by using following equation:

$$O + tD = (1 - u - v)\, V(a) + uV(c) + vV(b) \tag{4}$$

As shown in Figure 7, the window has been divided into several small triangle elements, and the center of each element can be defined as the solar point. Therefore, by referring to each solar point and the shading plan, the above equation can be solved. Inferably, it means that if there is a solution, the solar ray has been shaded; if there is no solution, there is no shading of the solar ray.

To check the results, a cloud-based platform has been developed to provide a 3D view of solar ray shading effects. The user graphic interface is shown in Figure 11a,b. By using the platform, any shape of shading devices and any direction of the solar ray can be calculated and viewed.

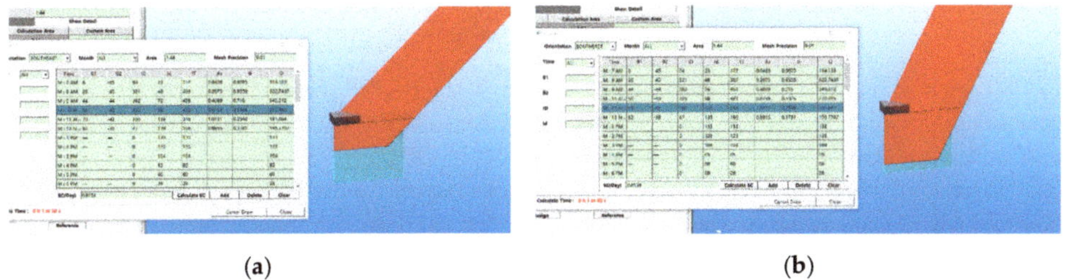

(a)   (b)

Figure 11. Three-dimensional view of solar ray shading effects.

4.4. Analysis of Shading Coefficient

In the ETTV formula as described above, the solar factor has been derived from the annual average of solar radiation transmitted through a 3 mm clear glass window. For other systems of fenestration, the rate of solar heat gain is modified by the shading coefficient of the fenestration system, which is defined as the ratio of solar heat gain through the fenestration system having a combination of glazing and shading device to the solar heat gain through an unshaded 3 mm clear glass. This ratio is a unique characteristic of each type of fenestration system and is represented by the equation:

$$SC = \frac{Solar\ heat\ gain\ of\ any\ glass\ and\ any\ shading\ combination}{Solar\ heat\ gain\ through\ a\ 3\ mm\ unshaded\ clear\ glass} \tag{5}$$

In general, the shading coefficient of any fenestration system can be obtained by multiplying the shading coefficient of the glass (or effective shading coefficient of glass with solar control film where a solar control film is used on the glass) and the shading coefficient of the sun-shading devices as follows:

$$SC = SC1 \times SC2 \tag{6}$$

where SC: shading coefficient of the fenestration system; SC1: shading coefficient of glass or effective shading coefficient of glass with solar control film where a solar control film is used on the glass; SC2: effective shading coefficient of external shading devices.

Therefore, the key part of the analysis is to determine the SC2 parameter. A sample model with different types of windows and shading devices is selected for the digital analysis, as shown in Figure 12.

Figure 12. Example model.

A typical shading plate, as shown in Figure 13, is selected for analysis and comparison. The orientation of the window is facing the north-east direction. The SC2 value calculated by using the table from the Singapore Green Mark Standard is 0.6345.

However, the SC2 value calculated by the proposed digital analysis method is 0.7308, which is larger than the calculation from the BCA Table, as shown in Figure 14. There is a gap of 0.0962.

The study found that the bulk of the variance was due to the assumption of the shading plate extended along the long axial direction when creating the BCA table.

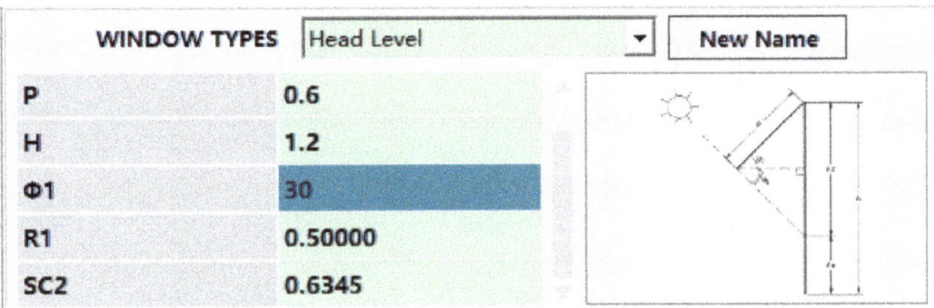

Figure 13. SC2 calculation by using the ETTV Guide table.

Orientation: NORTHEAST   Month: ALL   Area: 1.44   Mesh Precision: 0.1

Time	θ1	θ2	ID	Id	IT	As	G	Q
M : 7 AM	6	+45	94	23	117	0.34	0.7639	94.8066
M : 8 AM	26	+45	293	76	369	0.51	0.6458	265.2194
M : 9 AM	44	+46	336	106	442	0.67	0.5347	285.6592
M : 10 Al—	59	+47	278	126	404	0.96	0.3333	218.6574
M : 11 Al—	72	+48	154	136	290	0.97	0.3264	186.2656
M : 12 N—	83	+52	31	136	167	0.53	0.6319	155.5889
M : 1 PM	—	—	0	133	133			133
M : 2 PM	—	—	0	123	123			123
M : 3 PM	—	—	0	104	104			104
M : 4 PM	—	—	0	85	85			85
M : 5 PM	—	—	0	60	60			60
M : 6 PM	—	—	0	28	28			28

SC(Day): 0.7308   Calculate SC   Add   Delete   Clear

Calculate Time: 0 h 0 m 2 s   Draw   Close

Figure 14. SC2 calculation by using the proposed digital method.

When the length of the shading plate was altered (as shown in Figure 15) and stretched long enough to shade all the solar rays on both sides of the shading plate, the two values converged.

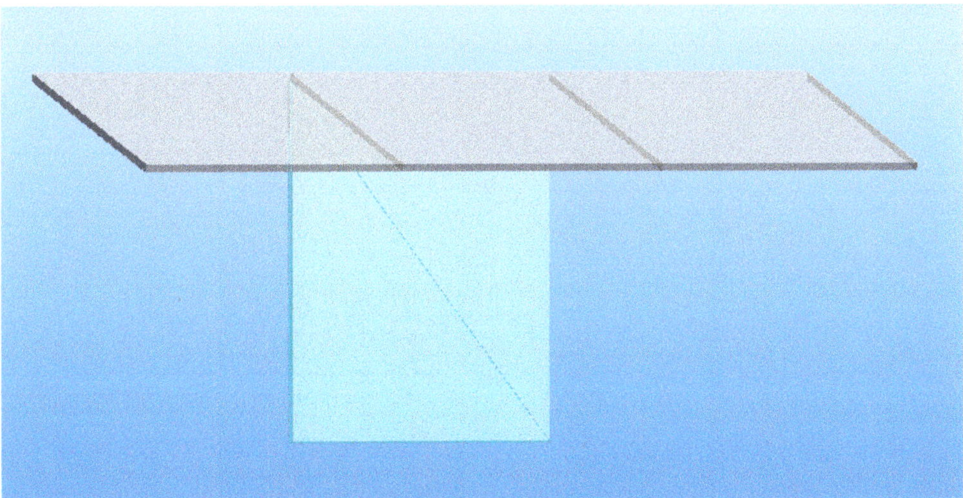

Figure 15. Shading plate with a longer length.

In this case, the calculated result for the SC2 value turned out to be 0.6356, as shown in Figure 16, which is very close to the result of 0.6345 obtained by the BCA table.

Check SC

Orientation: NORTHEAST — Month: ALL — Area: 1.44 — Mesh Precision: 0.05

Time: ALL

θ1:

θ2:

ID:

Id:

Time	θ1	θ2	ID	Id	IT	As	G	Q
D : 7 AM	15	+69	52	20	72	0.535	0.6285	52.682
D : 8 AM	46	+70	111	63	174	1.0025	0.3038	96.7218
D : 9 AM	67	+74	87	83	170	1.4175	0.0156	84.3572
D : 10 AM	81	+81	28	98	126	1.44	0.0	98.0
D : 11 AM	—	—	0	109	109			109
D : 12 N(...	—	—	0	116	116			116
D : 1 PM	—	—	0	116	116			116
D : 2 PM	—	—	0	108	108			108
D : 3 PM	—	—	0	93	93			93
D : 4 PM	—	—	0	73	73			73
D : 5 PM	—	—	0	50	50			50
D : 6 PM	—	—	0	20	20			20

SC(Day): 0.6356 — Calculate SC — Add — Delete — Clear

Calculate Time : 0 h 0 m 13 s — Draw — Close

Figure 16. SC2 calculation by using the proposed method.

The comparison of the SC2 parameter calculated from the Green Mark Guide table and proposed analysis method is provided. The parameters used in our analysis are displayed in Figure 17, which is adapted from [21,33].

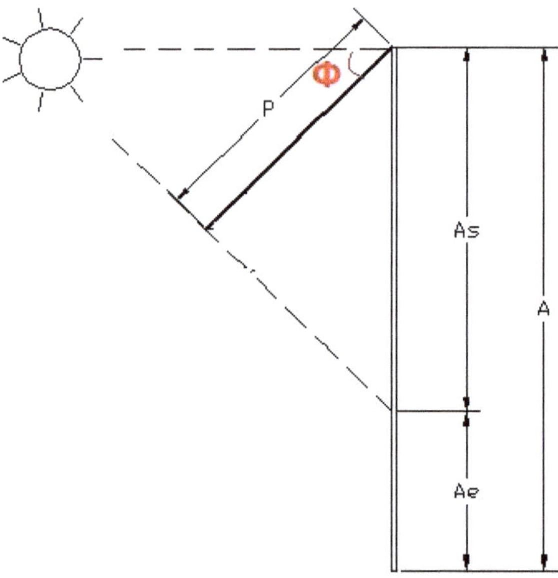

Figure 17. SC2 calculation by using the proposed method.

In the following analysis process, the window width is 1 m, window length A = 1 m, shading plate width p = 0.1 m, and the shading plate thickness is 0.01 mm (we set the thickness as almost zero to reduce the effect of the shading plate thickness).

The analysis results of the SC2 parameter compared with the ETTV Guide table for the case of Φ = 0 and Φ = 20 are provided in Figure 18. From the analysis results, it is concluded that the calculated results by using the Green Mark Guide table will lead the SC2 value to be smaller than the actual value since it assumes that the length of the shading device is infinitely extended.

Orientation	Φ	SC (Auto Calculate)		SC (GM Table)	Difference
		Real Length	Extended Length		
Horizontal Projection					
NORTH & SOUTH	0	0.9426	0.9392	0.938	0.0046
NORTHEAST & NORTHWEST	0	0.9312	0.9264	0.9273	0.0039
EAST & WEST	0	0.9375	0.9358	0.9363	0.0012
SOUTHEAST & SOUTHWEST	0	0.9304	0.9266	0.9253	0.0051
NORTH & SOUTH	20	0.9344	0.9306	0.93	0.0044
NORTHEAST & NORTHWEST	20	0.9195	0.9146	0.9137	0.0058
EAST & WEST	20	0.9206	0.9188	0.9195	0.0011
SOUTHEAST & SOUTHWEST	20	0.9166	0.9128	0.9107	0.0061
Vertical Projection					
NORTH & SOUTH	0	0.9576	0.9558	0.9526	0.005
NORTHEAST & NORTHWEST	0	0.9583	0.9528	0.9517	0.0066
EAST & WEST	0	0.9821	0.9803	0.9805	0.0016
SOUTHEAST & SOUTHWEST	0	0.9579	0.9539	0.9528	0.0051
NORTH & SOUTH	20	0.9589	0.9558	0.9549	0.004
NORTHEAST & NORTHWEST	20	0.9403	0.9406	0.9383	0.0074
EAST & WEST	20	0.9719	0.9697	0.9764	0.0025
SOUTHEAST & SOUTHWEST	20	0.8711	0.968	0.9396	0.0215

Figure 18. SC2 comparison calculation by using the proposed digital method and the traditional method.

As an example, a digital analysis technology is used to calculate a mesh-shaped shading device, as shown in Figure 19. The proposed numerical method should be adopted for this kind of shading shape.

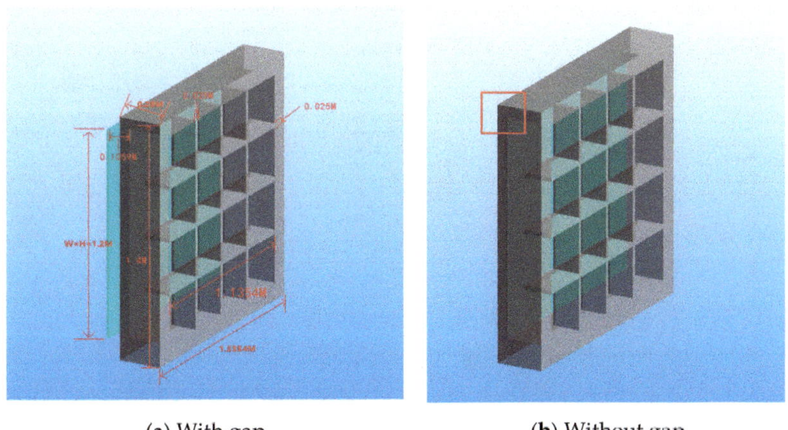

(a) With gap   (b) Without gap

Figure 19. Mesh-shaped shading device.

The orientation of the window facing is in the east direction. The calculated SC2 value with the gap of 10.59 mm between the window and shading device is 0.5935. If the gap between the window and shading device is removed, the calculated SC2 value becomes 0.6002 (as shown in Figure 20), which means more space in the window will be shaded when the gap is removed.

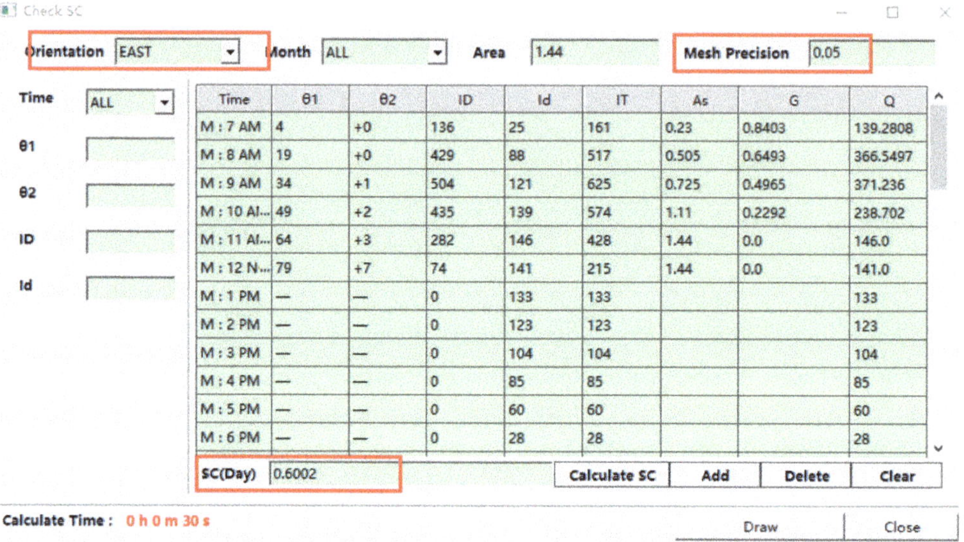

Time	θ1	θ2	ID	Id	IT	As	G	Q
M : 7 AM	4	+0	136	25	161	0.23	0.8403	139.2808
M : 8 AM	19	+0	429	88	517	0.505	0.6493	366.5497
M : 9 AM	34	+1	504	121	625	0.725	0.4965	371.236
M : 10 Al...	49	+2	435	139	574	1.11	0.2292	238.702
M : 11 Al...	64	+3	282	146	428	1.44	0.0	146.0
M : 12 N...	79	+7	74	141	215	1.44	0.0	141.0
M : 1 PM	—	—	0	133	133			133
M : 2 PM	—	—	0	123	123			123
M : 3 PM	—	—	0	104	104			104
M : 4 PM	—	—	0	85	85			85
M : 5 PM	—	—	0	60	60			60
M : 6 PM	—	—	0	28	28			28

Figure 20. SC2 calculation by using the proposed digital method.

5. Validation

The following three validation studies (as shown in Figure 21)with analysis and results are provided to test the accuracy of the envelope thermal performance calculations. The results were obtained using a BIM model and calculation using the proposed analysis method. In the proposed analysis method, the shading areas of sun shading devices are obtained by using proposed digital analysis technology, which is embedded in the Green Mark Collaboration Environment.

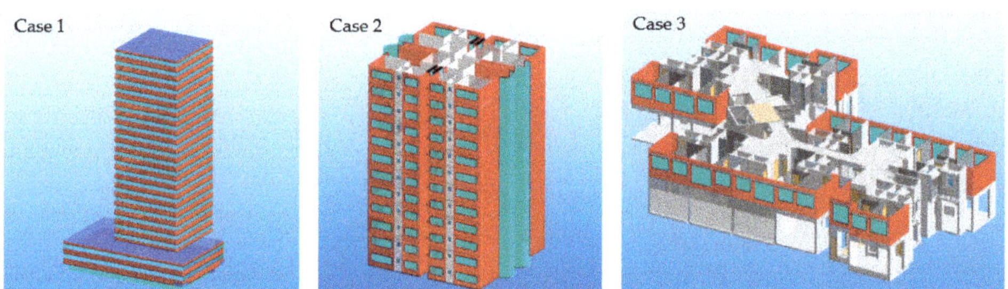

Figure 21. Models for validation.

5.1. Case 1: Twenty-Five-Story Office Building

An example from the ETTV Guide for a 25-storey office building is calculated. The analysis results are compared in Table 3.

Table 3. Case 1 comparison.

Method	N	E	S	W
ETTV Guide	41.6	52.1	44.7	55.6
Proposed	41.6	52.1	44.69	55.58
Differ	0	0	0.01	0.02

5.2. Case 2: Twelve-Storey Residential Building

An example of the Residential Envelope Transmittance Value for a 12-storey residential building is calculated. The analysis results are compared in Table 4.

Table 4. Case 2 comparison.

Method	N	N-E	E	S-E	S	S-W	W	N-W
ETTV Guide	18.048	34.075	17.143	34.368	18.69	36.419	17.619	35.54
Proposed	18.45	33.97	17.17	34.26	18.76	36.31	17.66	35.43
Differ	0.402	0.105	0.027	0.108	0.07	0.109	0.041	0.11

5.3. Case 3: TwoStorey Residential Building

An example of a Residential Envelope Transmittance Value for a two-storey residential building has been calculated. The RETV of this building has been calculated via spreadsheet tools. The results obtained by both Excel tools and the proposed method are compared in Table 5.

Table 5. Case 3 comparison.

Method	N	E	S	W
Excel	22.24	10.35	22.85	10.48
Proposed	23.18	10.37	23.80	10.48
Differ	0.94	0.02	0.95	0.0

Based on the three cases above, the tolerance between the proposed method and manual calculation is about 0% to 4.05%.

6. Research Novelty

This research provides a thermal envelope performance engine based on a BIM cloud platform-Green Mark Collaboration Environment. Thermal data of the façade can be automatically extracted from the BIM model or manually assigned if the thermal data is not available in the BIM model. Solar data and calculation formulas have been embedded in the proposed platform to facilitate calculating the building envelope thermal performance in a more productive and convenient manner. The contribution of this research is elaborated as follows:

- The thermal envelope performance analysis and calculation engine not only includes the function of computing through a traditional way but also provides the function of direct analysis and a calculation engine to address the complex situation (arbitrarily shaped shading devices), which is not covered in the ETTV guideline [21], as explained in Section 4.3, which provides a simplified building envelope thermal performance calculation procedure through interlinking formulas/equations with the building elements. The thermal information in the model can be easily tracked by stakeholders.
- The consolidation of information from stakeholders could be eliminated because the information is progressively collected by the proposed Green Mark Collaboration Environment during the development of the green building project. Errors from the manual procedure could be avoided.

- The proposed Green Mark Collaboration Environment marks the abandonment of the outdated and tedious extraction and collection of the information needed for building envelope thermal performance by providing a new and rapid approach for harnessing readily available information in the BIM model.
- Dynamic information and fast decision-making: Stakeholders would be able to envision the green impact of design changes through involvement via Green Mark Collaboration Environment during the design stage. This would enable stakeholders to make quick decisions with regard to the project.

7. Future Research

It is worth mentioning that ETTV only takes into consideration the heat gain through external walls and windows of a building. Basically, it is a metric to measure the solar gain through the building envelope and does not directly link to energy efficiency. ETTV is more like a passive strategy to minimize the solar gain, whereas the actual energy efficiency in buildings is impacted by the performance of the HVAC system, operating strategies (e.g., cooling/heating set point) and energy-related occupant behaviors, which have large variations in reality [34].

This study only focuses on the proposed Green Mark Collaboration Environment, the proposed performance-based method for arbitrarily shaped shading devices and the validation of its accuracy. The relationship between ETTV and energy efficiency will be the future research direction that requires attention from scholars. Moreover, the research will also progressively develop other modules, such as the Concrete Usage Index (CUI) Calculator, Lighting Calculator and other modules to automate the Green Mark process and embed all the modules into the proposed Green Mark Collaboration Environment.

8. Conclusions

This study compared various BIM-based building envelope thermal performance software. Based on an initial comparative analysis, the study identified that there is an onerous amount of effort required by the users while exporting the building performance BIM model to the BPA software. Furthermore, the findings of this research also indicate that the current methods of calculating ETTV through spreadsheets are tedious and error-prone, which is quite cumbersome if there are any future changes raised by project stakeholders. In addition, the current methods of calculating ETTV is dependent upon the ETTV guidelines, which do not cater to complex and irregular shading devices. To address these various issues, this research proposed the development of a BIM-based platform to digitalize the ETTV calculation. This platform is an improvement over current methods, which require the import and export of a BIM model. At the same time, the platform also facilitates parametric capabilities that promotes change management, which means that any changes in the BIM model by the project stakeholders will directly update the ETTV.

Furthermore, the platform developed through this research is a cloud-based platform, which acts as a common data environment (CDE) for the Green Mark scheme. This allows various project stakeholders to access the Green Mark data simultaneously in a federated manner so that the relevant information can be shared, validated and allows project stakeholders to keep track of the GM data generated throughout the project lifecycle. The platform also allows the relevant documentation to be hyperlinked with the BIM model, which can finally be produced in the form of a report for regulatory approval purposes. The research validated that the use of this platform will enable productivity improvement as far as the ETTV calculation is concerned.

Author Contributions: Conceptualization, Z.L. and V.J., L.G.; methodology, Z.L.; software, Z.L.; validation, Z.L., Q.W. and V.J., L.G.; writing—original draft preparation, Z.L.; writing—review and editing, Z.L., Q.W. and V.J., L.G.; supervision, L.P., Q.W. and V.J., L.G. All authors have read and agreed to the published version of the manuscript.

Funding: The research project was funded by the Building and Construction Authority (BCA) of Singapore under an Innovation Grant (iGrant) with reference number *BCA RID 94.15.2.39*.

Acknowledgments: The development was conducted at ITI Hub Pte. Ltd. (ITI) and United Project Consultant Pte. Ltd. (UPC). Special thanks go to Sun Keming (ITI), Er. Tang Kian Cheong (UPC), and Felix Batad (UPC). The authors would also like to express their gratitude towards Tan Kee Wee, Programme Director, Digitalization Department, Built Environment Research and Innovation Institute (BERII), Building and Construction Authority (BCA).

Conflicts of Interest: The authors declare no conflict of interest.

Disclaimer: The opinions and recommendations expressed in this paper are the authors' personal opinions and do not necessarily represent the official position of any organization. This paper does not endorse any software in any capacity.

References and Note

1. LIU, Z.; LU, Y.; Peh, L. Review and Scientometric Analysis of Global Building Information Modeling (BIM) Research in the Architecture, Engineering and Construction (AEC) Industry. *Buildings* **2019**, *9*, 210. [CrossRef]
2. Kaneta, T.; Furusaka, S.; Tamura, A.; Deng, N. Overview of BIM Implementation in Singapore and Japan. *J. Civ. Eng. Archit.* **2016**, *10*, 1305–1312. [CrossRef]
3. Building and Construction Authority of Singapore. The BIM Issue. *Build Smart Magazine*, 2011; p. 4.
4. Azhar, S.; Brown, J. Bim for Sustainability Analyses. *Int. J. Constr. Educ. Res.* **2009**, *5*, 276–292. [CrossRef]
5. Schlueter, A.; Thesseling, F. Building Information Model Based Energy/Exergy Performance Assessment in Early Design Stages. *Autom. Constr.* **2009**, *18*, 153–163. [CrossRef]
6. Eastman, C. An Outline of the Building Description System. *Res. Rep. No. 50.* **1974**, *23*, 5.
7. US National Institute of Building Sciences. *National Building Information Modeling Standard, Version 1-Part 1: Overview, Principles, and Methodologies, Glossary*; US National Institute of Building Sciences: Washington, DC, USA, 2007.
8. Lu, Y.; Wu, Z.; Chang, R.; Li, Y. Building Information Modeling (BIM) for Green Buildings: A Critical Review and Future Directions. *Autom. Constr.* **2017**, *83*, 134–148. [CrossRef]
9. USGBC. LEED V4.1. Available online: https://new.usgbc.org/leed-v41 (accessed on 16 November 2019).
10. BRE. *BREEAM International New Construction 2016 Technical Manual*; BRE Global Ltd.: Watford, UK, 2016; p. 1385.
11. BEAM Society Limited. *BEAM Plus For New Buildings* **2012**, *2*. Available online: https://www.beamsociety.org.hk/files/BEAM_Plus_For_New_Buildings_Version_1_2.pdf (accessed on 19 November 2019).
12. GBCA. Green Star. Available online: https://new.gbca.org.au/green-star/ (accessed on 11 November 2019).
13. GreenBuildingIndex. GBI Assessment Criteria for Non-Residential New Construction (NRNC). Available online: https://new.greenbuildingindex.org/Files/Resources/GBI%20Tools/GBI%20NREB%20Data%20Centre%20Tool%20V1.0.pdf (accessed on 16 November 2019).
14. Building and Construction Authority of Singapore. *Green Mark for Non-Residential Buildings NRB: 2015 Including Hawker Centres, Healthcare Facilities, Laboratory Buildings and Schools*; Building and Construction Authority of Singapore: Singapore, 2015.
15. Building and Construction Authority. BCA Green Mark Assessment Criteria, Online Application and Verification Requirements. Available online: https://www.bca.gov.sg/GreenMark/green_mark_criteria.html (accessed on 19 November 2019).
16. Chen, X.; Yang, H.; Lu, L. A Comprehensive Review on Passive Design Approaches in Green Building Rating Tools. *Renew. Sustain. Energy Rev.* **2015**, *50*, 1425–1436. [CrossRef]
17. Breesch, H.; Janssens, A. Performance Evaluation of Passive Cooling in Office Buildings Based on Uncertainty and Sensitivity Analysis. *Sol. Energy* **2010**, *84*, 1453–1467. [CrossRef]
18. Badescu, V.; Cathcart, R.B. Environmental Thermodynamic Limitations on Global Human Population. *Int. J. Glob. Energy Issues* **2006**, *25*, 129–140. [CrossRef]
19. Natephra, W.; Motamedi, A.; Yabuki, N.; Fukuda, T. Integrating 4D Thermal Information with BIM for Building Envelope Thermal Performance Analysis and Thermal Comfort Evaluation in Naturally Ventilated Environments. *Build. Environ.* **2017**, *124*, 194–208. [CrossRef]
20. Natephra, W.; Yabuki, N.; Fukuda, T. Optimizing the Evaluation of Building Envelope Design for Thermal Performance Using a BIM-Based Overall Thermal Transfer Value Calculation. *Build. Environ.* **2018**, *136*, 128–145. [CrossRef]
21. Building and Construction Authority of Singapore. *Guidelines on Envelope Thermal Transfer Value for Buildings*; Building and Construction Authority of Singapore: Singapore, 2004.

22. Seghier, T.E.; Lim, Y.W.; Ahmad, M.H.; Samuel, W.O. Building Envelope Thermal Performance Assessment Using Visual Programming and BIM, Based on ETTV Requirement of Green Mark and GreenRE. *Int. J. Built Environ. Sustain.* **2017**, *4*, 227–235. [CrossRef]

23. Chua, K.J.; Chou, S.K. An ETTV-Based Approach to Improving the Energy Performance of Commercial Buildings. *Energy Build.* **2010**, *42*, 491–499. [CrossRef]

24. Singhaputtangkul, N.; Low, S.P.; Teo, A.L. Integrating Sustainability and Buildability Requirements in Building Envelopes. *Facilities* **2011**, *29*, 255–267. [CrossRef]

25. Liu, Z.; Chen, K.; Peh, L.; Tan, K.W. A Feasibility Study of Building Information Modeling for Green Mark New Non-Residential Building (NRB): 2015 Analysis. *Energy Procedia* **2017**, *143*, 80–87. [CrossRef]

26. Ansah, M.K.; Chen, X.; Yang, H.; Lu, L.; Lam, P.T.I. A Review and Outlook for Integrated BIM Application in Green Building Assessment. *Sustain. Cities Soc.* **2019**, *48*, 101576. [CrossRef]

27. Parisa, E.; Moakher, E.; Pimplikar, S.S. Building Information Modeling (BIM) and Sustainability—Using Design Technology in Energy Efficient Modeling. *IOSR J. Mech. Civ. Eng.* **2012**, *1*, 1684–2278.

28. Hazel Joanne. *Sustainable Architectural*; 2009; p. 16.

29. Integrated Environmental Solutions Limited. VE-Navigator for GM Pro A Professional Solution for a Comprehensive Green Mark®®Assessment. Available online: https://www.iesve.com/software/virtual-environment/modules/nav-gmpro (accessed on 19 November 2019).

30. Yildirim, H.T.; Senturk, N. Analysis of Imp3 Expression in Prostate Adenocarcinomas. *Turkish, J. Pathol.* **2012**, *28*, 128. [CrossRef]

31. Chen, S.; Jin, R.; Alam, M. Investigation of Interoperability between Building Information Modeling (BIM) and Building Energy Simulation (BES). *Int. Rev. Appl. Sci. Eng.* **2018**, *9*, 137–144. [CrossRef]

32. Inhabitgroup. ETTV Façade Designer Released to Singapore Market. Available online: https://inhabitgroup.com/ettv-facade-designer-released-to-singapore-market-2/ (accessed on 10 November 2019).

33. Building and Construction Authority of Singapore. Code on Envelope Thermal Performance for Buildings O. Available online: https://www.bca.gov.sg/PerformanceBased/others/RETV.pdf (accessed on 1 October 2019).

34. Gan, V.J.L.; Lo, I.M.C.; Ma, J.; Tse, K.T.; Cheng, J.C.P.; Chan, C.M. Simulation Optimisation towards Energy Efficient Green Buildings. *J. Clean. Prod.* **2020**. [CrossRef]

MDPI
St. Alban-Anlage 66
4052 Basel
Switzerland
Tel. +41 61 683 77 34
Fax +41 61 302 89 18
www.mdpi.com

Energies Editorial Office
E-mail: energies@mdpi.com
www.mdpi.com/journal/energies